FSM-based Digital Design using Verilog HDL

FSM-based Digital Design using Verilog HDL

Peter Minns
Ian Elliott
Northumbria University, UK

John Wiley & Sons, Ltd

Telephone (+44) 1243 779777

Email (for orders and customer service enquiries): cs-books@wiley.co.uk
Visit our Home Page on www.wileyeurope.com or www.wiley.com

Other Wiley Editorial Offices

John Wiley & Sons Inc., 111 River Street, Hoboken, NJ 07030, USA

Jossey-Bass, 989 Market Street, San Francisco, CA 94103-1741, USA

Wiley-VCH Verlag GmbH, Boschstr. 12, D-69469 Weinheim, Germany

John Wiley & Sons Australia Ltd, 42 McDougall Street, Milton, Queensland 4064, Australia

John Wiley & Sons (Asia) Pte Ltd, 2 Clementi Loop #02-01, Jin Xing Distripark, Singapore 129809

John Wiley & Sons Canada Ltd, 6045 Freemont Blvd, Mississauga, ONT, L5R 4J3

Wiley also publishes its books in a variety of electronic formats. Some content that appears in print may not be available in electronic books.

Wiley publishes in a variety of print and electronic formats and by print-on-demand. Some material included with standard print versions of this book may not be included in e-books or in print-on-demand. If this book refers to media such as a CD or DVD that is not included in the version you purchased, you may download this material at http://booksupport.wiley.com. For more information about Wiley products, visit www.wiley.com

British Library Cataloguing in Publication Data

A catalogue record for this book is available from the British Library

ISBN 978-0470-06070-4

Typeset in 10/12 pt Times by Thomson Digital, Noida, India

Contents

Preface

This book covers the design and use of finite state-machines (FSMs) in digital systems. It includes stand-alone applications and systems that use microprocessors, microcontrollers, and memory controlled directly from the FSM, as well as other common situations found in practical digital systems. The emphasis is on obtaining a good understanding of FSMs, how they can be used, and where to use them.

The popular and widely used Verilog hardware description language (HDL) is introduced and applied to the description and verification of many of the designs in the book. In addition to logic gate and Boolean equation-level styles of Verilog description, there is also a chapter covering the use of HDL at the so-called behavioural level, whereby a design is described using the high-level features provided by Verilog HDL.

There is also a chapter using the One Hot technique, commonly used to implement FSMs in field programmable gate arrays with examples on the development of dynamic memory access (DMA) controllers and data sequence detectors. Asynchronous (event-driven) FSMs not requiring a clock signal are covered in a chapter using a technique that allows rapid development of reliable systems. A chapter on the use of Petri-net-based controllers is included, allowing parallel-based digital FSMs to be developed.

In the development of digital systems, microcontrollers have been used for many years to control digital inputs and outputs, as well as process analogue information. Now, using the techniques in this book, FSM-based designs can be implemented using a deterministic model, the state diagram, as a design aid. Once developed, the state diagram can be used to implement the final system using either Boolean equations obtained directly from the state diagram, or a behavioural Verilog HDL description, again developed directly from the state diagram. External devices, such as memory, address counters and comparators, can be implemented either from the Boolean equations that define their operation or via behavioural-level descriptions in Verilog HDL.

The book is targeted at undergraduate final-year students of Electrical, Electronic, and Communications Electronic Engineering, as well as postgraduate students and practising Electronic Design Engineers who want to know how to develop FSM-based systems quickly. The book will assume an understanding of basic logic design and Boolean algebra, as would be expected of a final-year undergraduate. The book sequence follows.

The first three chapters are in the form of a linear frame programmed learning format to help the reader learn the essential concepts of synchronous FSM design.

This set of notes has been used with undergraduate final-year students at our university for some years now and has been well received by the students. These chapters cover the idea of basic FSM design and synthesis. Once this is covered, the book reverts to a more familiar text. However, the first three chapters, being linear, can be read in the same style as the more familiar text if the reader desires.

A breakdown of the chapters in the book now follows.

Chapter 1 contains an introduction to FSMs, the Mealy and Moore models of an FSM, differences between synchronous (clock-driven) FSMs and asynchronous (event-driven) FSMs, the state diagram and how it can be used to indicate sequential behaviour and the inputs and outputs of a system. This follows with a number of examples of FSMs to illustrate the way in which they can be developed to meet particular specifications.

Chapter 2 covers the use of external hardware and how this hardware can be controlled by the FSM. The examples include how to create wait states using external timers, how to control analogue-to-digital converters, and memory devices. This opens up the possibilities of FSM-based systems that are not normally covered in other books.

Chapter 3 is a continuation of the programmed learning text, looking at synthesization of state diagrams using *T* flip-flops and *D* flip-flops, as well as initialization techniques.

The remaining chapters of the book will be in a more conventional format.

Chapter 4 covers synchronous (clock-driven) FSM examples, some with simulation. This chapter gives some practical examples commonly found in real applications, such as a digital waveform synthesizer and asynchronous serial transmit and receive blocks.

Chapter 5 is an introduction to the use of 'One Hotting' in synchronous FSM design. Amongst the examples covered is a DMA controller and serial bit stream code detection.

Chapter 6 is an introduction to Verilog HDL and how to use it at the gate level and the Boolean equation level, together with how to combine different modules to form a complete system.

Chapter 7 introduces the basic lexical elements of the Verilog HDL. Emphasis is placed on those aspects of the language that support the description of synthesizable combinational and sequential logic.

Chapter 8 takes a more detailed look at the Verilog HDL, with emphasis on behavioural modelling of FSM designs. It covers using an HDL to implement synchronous FSMs at the behavioural level - with examples.

Chapter 9 is an introduction to asynchronous (event-driven) design of FSMs from initial concepts through to the design of asynchronous FSMs to given specifications. This will also include a brief discussion of race problems with asynchronous designs and how to overcome them.

Chapter 10 is an introduction to synchronous Petri nets, and how they can be used to implement both sequential and parallel FSMs. Petri nets allow the design of parallel FSMs with synchronized control. This chapter shows how a Petri net can be designed and synthesized as an electronic circuit using *D*-type flip-flops.

Each chapter contains examples with solutions, many of which have been used by the authors in real practical systems.

There is a CD-ROM included with the book containing a digital simulation program to aid the reader in learning and verifying the many examples given in the book. The program is based on Verilog HDL. This tool has been used to simulate most of the examples in the book.

Also on the CD-ROM are folders containing many of the book's examples, complete with test-bench descriptions to allow the simulations to be run directly on a PC-based computer.

Peter Minns BSc(H) PhD CEng MIET
Ian Elliott BSc(H) MPhil CEng MIET
Newcastle Upon Tyne

Acknowledgements

We would like to thank all those who have helped in the proof reading of this book. In particular, we thank Safwat Mansi for his proof reading of our ideas and his helpful suggestions. In addition, Kathleen Minns is thanked for her help in the proof reading of the entire manuscript.

We would like to thank our editors, Emily Bone, Laura Bell, Kate Griffiths and Nicky Skinner for their help over the time that we have worked on this book. Thanks also to Caitlin Flint for her help with the marketing of the book.

Special thanks go to Donna Mitchell at SynaptiCAD for her help with Appendix C on the use of the VeriLoger Extreme Simulation Program on the CD-ROM with this book. Also, Gary Covington for his help in creating the CD-ROM.

Finally, thanks to our wives, Helen and Kathleen, for putting up with our frequent disappearances during the preparation of the book.

Any errors are, of course, entirely the responsibility of the authors.

1

Introduction to Finite-State Machines and State Diagrams for the Design of Electronic Circuits and Systems

1.1 INTRODUCTION

This chapter, and Chapters 2 and 3, is written in the form of a linear frame, programmed learning text. The reason for this is to help the reader to learn the basic skills required to design clocked finite-state machines (FSMs) so that they can develop their own designs based on traditional T flip-flops and D flip-flops. Later, other techniques will be introduced, such as One Hot, asynchronous FSMs, and Petri nets; these will be developed along the same lines as the work covered in this chapter, but not using the linear frame, programmed learning format.

The text is organized into frames, each frame following on consecutively from the previous one, but at times the reader may be redirected to other frames, depending upon the response to the questions asked. It is possible, however, to read the programmed learning chapters as a normal book.

There are *tasks* set throughout the frames to test your understanding of the material.

To make it easier to identify input and output signals, inputs will be in lowercase and outputs in uppercase.

Please read the Chapters 1–3 first and attempt all the questions before moving on to the later chapters. The reason for this approach is that the methods used in the book are novel, powerful, and when used correctly can lead to a rapid approach to the design of digital systems that use FSMs.

Chapters 1–5, 9 and 10 make use of techniques to develop FSM-based systems at the equation and gate level, where the designer has complete control of the design.

Chapters 6–8 can be read as a self-contained study of the Verilog hardware description language (HDL).

FSM based Digital Design using Verilog HDL Peter Minns and Ian Elliott

1.2 LEARNING MATERIAL

Frame 1.1 What is an FSM?

An FSM is a digital sequential circuit that can follow a number of predefined states under the control of one or more inputs. Each state is a stable entity that the machine can occupy. It can move from this state to another state under the control of an outside-world input.

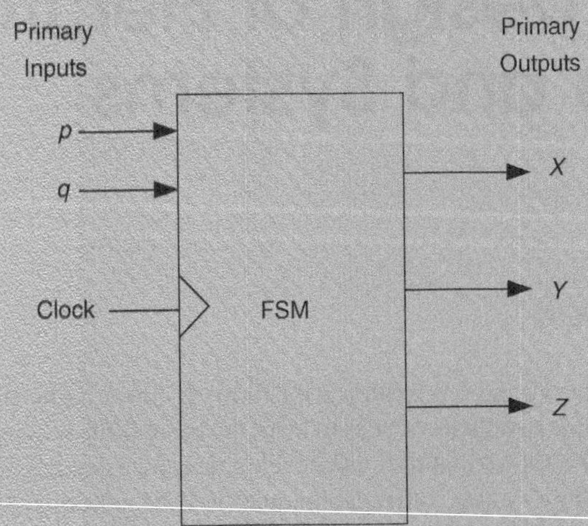

Figure 1.1 Block diagram of an FSM-based application.

Figure 1.1 shows an FSM with three outside-world inputs p, q, and the clock, and three outside-world outputs X, Y, and Z are shown. Note that some FSMs have a clock input and are called synchronous FSMs, i.e. those that do not belong to a type of FSM called asynchronous FSMs. However, most of this text will deal with the more usual synchronous FSMs, which *do* have a clock input. Asynchronous FSMs will be dealt with later in the book.

As noted above, inputs use lower case and output upper case names.

A synchronous FSM can move between states only if a clock pulse occurs.

Task Draw a block diagram for an FSM with five inputs x, y, z, t, and a clock, and with two outputs P and Q.

Go to Frame 1.2 after attempting this question.

Frame 1.2

The FSM with five inputs x, y, z, t, and a clock, and with two outputs P and Q is shown in Figure 1.2.

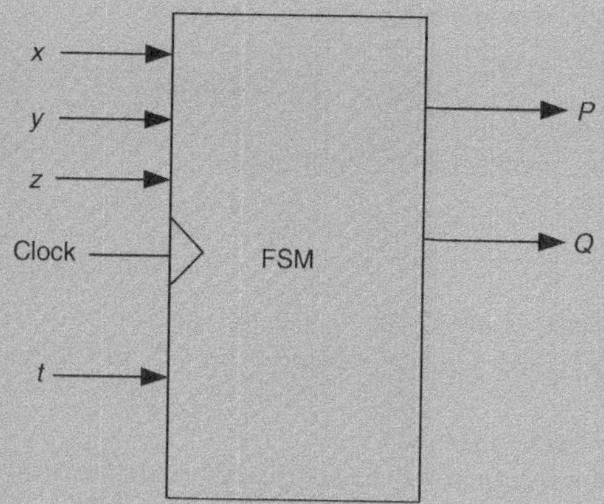

Figure 1.2 Block diagram with inputs, outputs, and a clock input.

The reader may wish to go back and reread Frame 1.1 if the answer was incorrect.

Each state of the FSM needs to be identifiable. This is achieved by using a number of internal (to the FSM block) flip-flops. An FSM with four states would require two flip-flops, since two flip-flops can store $2^2 = 4$ state numbers. Each state has a unique state number, and states are usually assigned numbers as s0 (state 0), s1, s2, and s3 (for the four-state example).

The rule here is

$$\text{Number of states} = 2^{\text{Number of flip-flops}},$$

for which

$$\text{Number of flip flops} = \frac{\log_{10}(\text{Number of states})}{\log_{10}(2)}.$$

So an FSM with 13 states would require 2^4 flip-flops (i.e. 16 states, of which 13 are used in the FSM); that is:

$$\text{Number of flip flops} = \frac{\log_{10}(13)}{\log_{10}(2)} = 3.7.$$

This must be rounded up to the nearest integer, i.e. 4.

Tasks 1. How many flip-flops would be required for an FSM using 34 states?
2. What would the state numbers be for this FSM?

After answering these questions, go to Frame 1.3.

Frame 1.3

The answers to the questions are as follows:

1. How many flip-flops would be required for an FSM using 34 states?

$$2^6 = 64$$

would accommodate 34 states. In general:

$$2^4 = 16 \text{ states}, \quad 2^5 = 32 \text{ states}, \quad 2^6 = 64 \text{ states}, \quad 2^7 = 128 \text{ states, etc.}$$

2. What would the state numbers be for this FSM?

These would be

s0, s1, s2, s3, s4, s5, s6, s7, s8, s9, s10, s11, s12, s13, s14, s15, s16, s17, s18, s19, s20, s21, s22, s23, s24, s25, s26, s27, s28, s29, s30, s31, s32, s33.

The unused states would be s34–s63.

Note, in this book, lower case 's' will be used to represent states to avoid confusion of state s0 with the word 'so' or 'So'.

As well as containing flip-flops to define the individual states of the FSM uniquely, there is also combinational logic that defines the outside-world outputs. In addition, the outside-world inputs connect to combinational logic to supply the flip-flops inputs.

Go to Frame 1.4.

Frame 1.4

Figure 1.3 illustrates the internal architecture for a Mealy FSM.

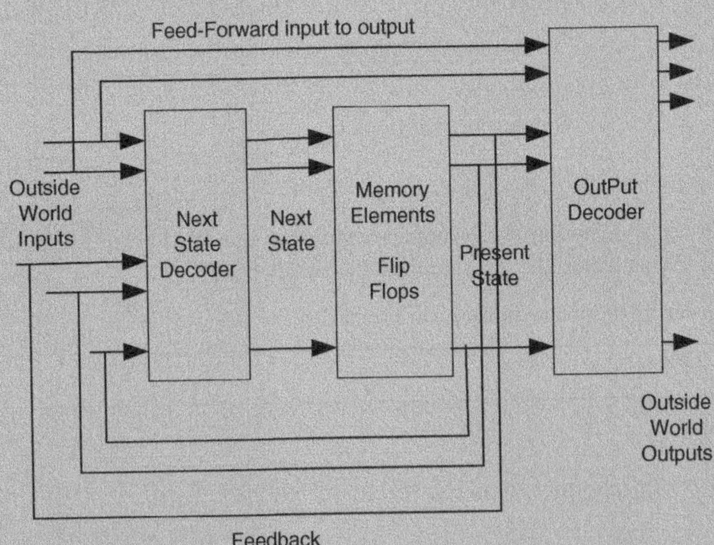

Figure 1.3 Block diagram of a Mealy state machine structure.

This diagram shows that the FSM has a number of inputs that connect to the Next State Decoder (combinational) logic. The Q outputs of the memory element Flip-Flops connect to the Output Decoder logic, which in turn connects to the Outside World Outputs.

The Flip-Flops outputs are used as Next State inputs to the Next State Decoder, and it is these that determine the next state that the FSM will move to. Once the FSM has moved to this Next State, its Flip-Flops acquire a new Present State, as dictated by the Next State Decoder.

Note that some of the Outside World Inputs connect directly to the Output Decoder logic. This is the main feature of the Mealy-type FSM.

Go to Frame 1.5.

Frame 1.5

Another architectural form for an FSM is the Moore FSM.

The Moore FSM (Figure 1.4) differs from the Mealy FSM in that it does not have the feed-forward paths.

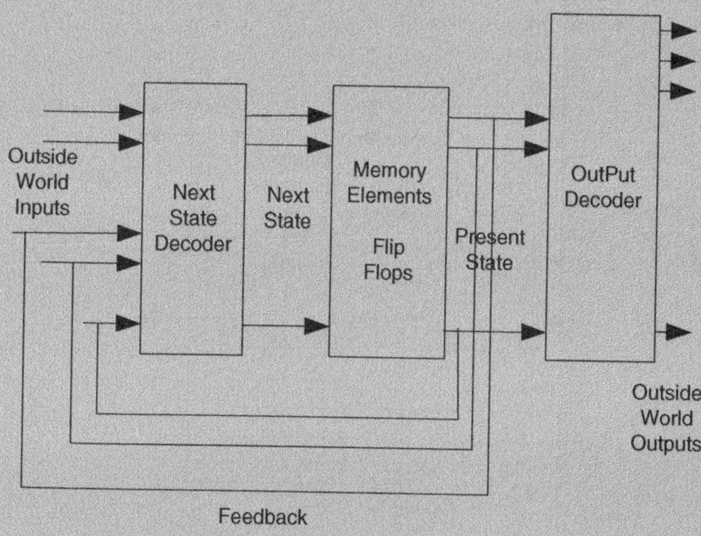

Figure 1.4 Block diagram of a Moore state machine structure.

This type of FSM is very common. Note that the Outside World Outputs are a function of the Flip-Flops outputs only (unlike the Mealy FSM architecture, where the Outside World Outputs are a function of Flip-Flops outputs *and* some Outside World Inputs).

Both the Moore and Mealy FSM designs will be investigated in this book.

Go to Frame 1.6.

Frame 1.6

Complete the following:

A Moore FSM differs to that of a Mealy FSM in that it has

____.

This means that the Moore FSM outputs depend on

____

____

whereas the Mealy FSM outputs can depend upon ______________________.

Go back and read Frame 1.4 and Frame 1.5 for the solutions.

____.

Frame 1.7

Look at the Moore FSM architecture again, but with removal of all of the Outside World Inputs, apart from the clock. Also remove the Output Decoding logic. What is left should be a very familiar architecture. This is shown in Figure 1.5.

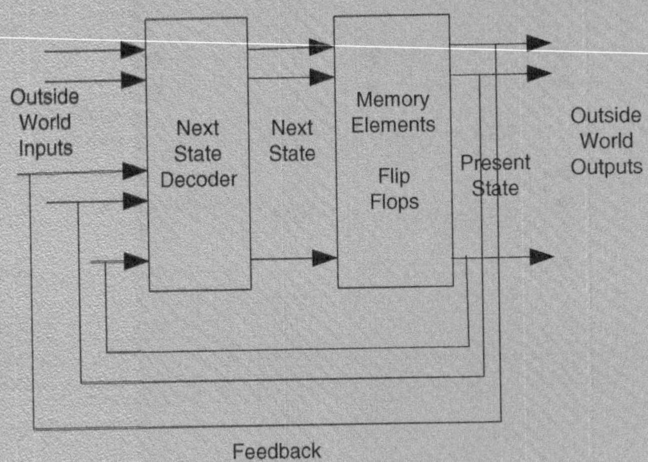

Figure 1.5 Block diagram of a Class C state-machine structure.

This architecture is in fact the synchronous counter that is used in many counter applications. Note that an Up/Down counter would have the additional outside-world input 'Up/ Down', which would be used to control the direction of counting.

The Flip-Flops outputs in this architecture are used to connect directly to the outside-world. Note that, in a synchronous (clock-driven) FSM, one of the inputs would be the clock.

Go to Frame 1.8.

Frame 1.8

Historically, two types of state diagram have evolved: one for the design of Mealy FSMs and one for the design of Moore-type FSMs. The two are known as *Mealy state* diagrams and *Moore state* diagrams respectively.

These days, a more general type of state diagram can be used to design both the Mealy and Moore types of FSM. This is the type of state diagram that will be used throughout the remainder of this book.

A state diagram shows each state of the FSM and the transitions to and from that state to other states. The states are usually drawn as circles (but some people like to use a square box) and the transition between states is shown as an arrowed line connecting the states (Figure 1.6).

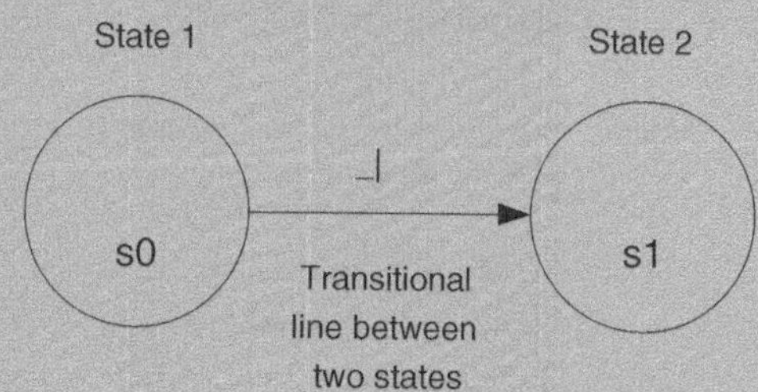

Figure 1.6 Transition between states.

In addition to the transitional line between states there is an input signal name (Figure 1.7).

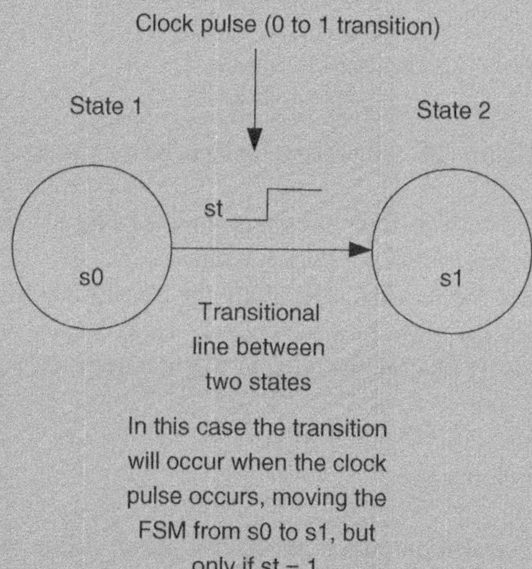

Figure 1.7 Outside-world input to cause transition between states.

In the above diagram, the transition between state s0 and s1 will occur only if the Outside World Input st = 1 *and* a 0-to -1 transition occurs on the clock input.

Task What changes would be needed to the state diagram of Figure 1.9 to make the transition between s0 and s1 occur when input st = 0?

After attempting this question, go to Frame 1.9.

Frame 1.9

The answer is shown in Figure 1.8.

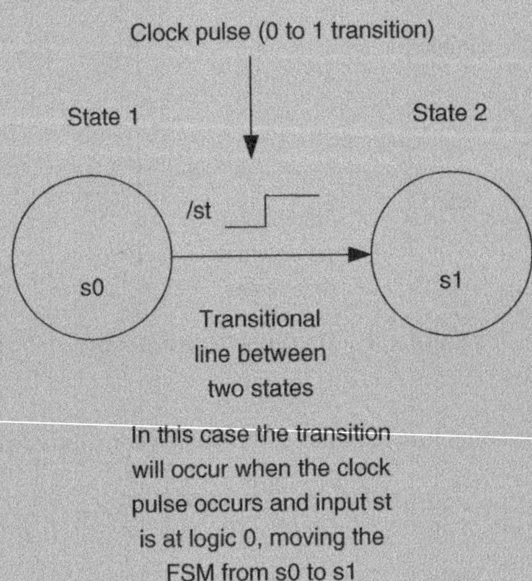

Figure 1.8 Outside-world input between states.

Here, st has been replaced with /st, indicating that st must be logic 0 before a transition to s1 can take place), i.e. /st means 'NOT st'; hence, when st = 0, /st = 1.

Note that outside-world inputs always lie along the transitional lines.

The state diagram must also show how the outside-world outputs are affected by the state diagram. This is achieved by placing the outside-world outputs either

- inside the state circle/square (Figure 1.9), or
- alongside the state circle/square.

In this diagram, outside-world outputs P and Q are shown inside the state circles. In this particular case, P is logic 1 in state s0, and changes to logic 0 when the FSM moves to state s1. Output Q does not change in the above transaction, remaining at logic 0 in both states.

Inputs like st are primary inputs; outputs like P and Q are primary outputs.

Task Draw a block diagram showing inputs and outputs for the state diagram of Figure 1.9.

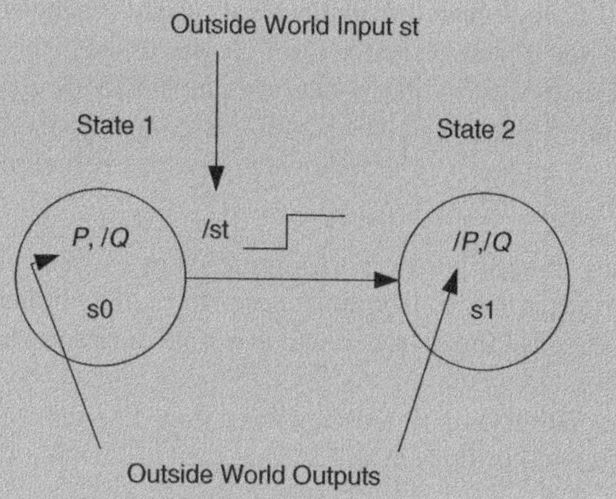

Figure 1.9 Showing placing of outside-world outputs.

Now go to Frame 1.10.

Frame 1.10

The block diagram will look like Figure 1.10.

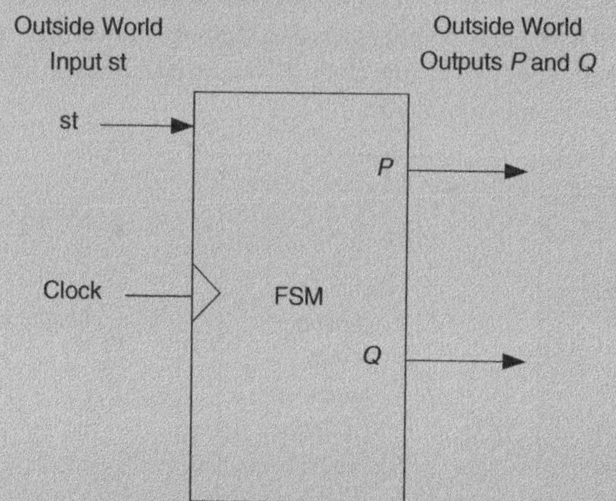

Figure 1.10 The block diagram for state diagram of Figure 1.9.

This is easily obtained from the state diagram since inputs are located along transitional lines and outputs inside (or along side) the state circle.

Recall that in Frame 1.2 each state had to have a unique state number and that a number of flip-flops were needed to perform this task. These flip-flops are part of the internal design of the FSM and are used to produce an internal count sequence (they are essentially acting like a synchronous counter, but one that is controlled by the outside-world inputs). The internal count sequence produced by the flip-flops is used to control the outside-world decoder so that outputs can be turned on and off as the FSM moves between states.

In Frames 1.4 and 1.5 the architecture for the Mealy and Moore FSMs were shown. In both cases, the memory elements shown are the flip-flops discussed in the previous paragraph.

At this stage it is perhaps worth while looking at a simple FSM design in detail to see what it looks like. This will bring together all the ideas discussed so far, as well as introducing a few new ones. However, try answering the following questions before moving on to test your understanding so far:

Tasks
1. A Mealy FSM differs from a Moore FSM in . . .. (See Frames 1.4 and 1.5.)
2. The circles in a state diagram are used to. . .. (See Frames 1.8 and 1.9.)
3. Outside World Inputs are shown in a state diagram where? (See Frames 1.8 and 1.9.)
4. Outside World Outputs are shown in a state diagram where? (See Frame 1.9.)
5. The internal flip-flops in an FSM are used to . . .. (See Frame 1.10.)

Go to Frame 1.11

Frame 1.11 Example of an FSM: a single-pulse generator circuit with memory

The idea here is to develop a circuit based on an FSM that will produce a single output pulse at its primary output *P* whenever its primary input *s* is taken to logic 1. In addition, a primary output *L* is to be set to logic 1 whenever input *s* is taken to logic 1, and cleared to logic 0 when the input *s* is released to logic 0. Output *L* acts as a memory indicator to indicate that a pulse has just been generated. The FSM is to be clock driven, so it also has an input clock. The block diagram of this circuit is shown in Figure 1.11.

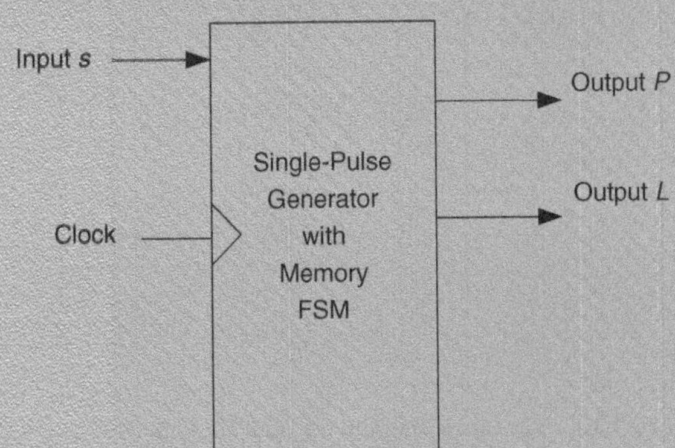

Figure 1.11 Block diagram of single-pulse with memory FSM.

A suitable state diagram is shown in Figure 1.12.

In this state diagram the sling (loop going to and from s0) indicates that while input *s* is logic 0 (/*s*) the FSM will remain in state s0 regardless of how many clock pulses are applied to the FSM.

Only when input *s* goes to logic 1 (*s*) will the FSM move from state s0 to s1, and then only when a clock pulse arrives. Once in state s1, the FSM will set its outputs *P* and *L* to logic 1, and on the next clock pulse the FSM will move from state s1 to state s2.

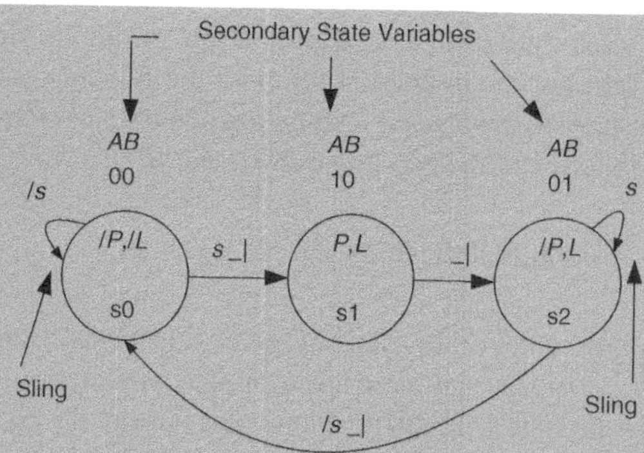

Figure 1.12 State diagram for single-pulse with memory FSM.

The reason why the FSM will stay in state s1 for only one clock pulse is because, in state s1, the transition from this state to state s2 occurs on a clock pulse only. Once the FSM arrives in state s2 it will remain there whilst input $s = 1$. As soon as the input s goes to logic 0 ($/s$) the FSM will move back to state s0 on the next clock pulse.

Since the FSM remains in state s1 for only a single clock pulse, and since $P = 1$ only in state s1, the FSM will produce a single output pulse. Note that the memory indicator L will remain at logic 1 until s is released, so providing the user with an indication that a pulse has been generated.

Note in the FSM state diagram (Figure 1.12) that each state has a unique state identity s0, s1, and s2.

Note also that each state has been allocated a unique combination of flip-flop states:

- state s0 uses the flip-flop combination $A = 0$, $B = 0$, i.e. both flip-flops reset;
- state s1 uses the flip-flop combination $A = 1$, $B = 0$, i.e. flip-flop A is set;
- state s2 uses the flip-flop combination $A = 0$, $B = 1$, i.e. flip-flop A is reset, flip-flop B is set.

The A and B flip-flops values are known as the secondary state variables.

The flip-flop outputs are seen to define each state. The A and B outputs of the two flip-flops could be used to determine the state of the FSM from the state of the A and B flip-flops. The code sequence shown in Figure 1.12 follow a none unit distance coding, since more than one flip-flop changes state in some transitions.

Go to Frame 1.12.

Frame 1.12 The output signal states

It would also be possible to tell in which state the output P was to be logic 1, i.e. in state s1, where the flip-flop output logic levels are $A = 1$ and $B = 0$.

Therefore, the output $P = A \cdot /B$ (where the middot is the logical AND operation). Note that the flip-flops are used to provide a unique identity for each state.

Similarly, output L is logic 1 in states s1 and s2 and, therefore, $L = \text{s1} + \text{s2}$.

$$L = \text{s1} + \text{s2} = A \cdot /B + /A \cdot B.$$

Also, see that since each state can be defined in terms of the flip-flop output states, the outside-world outputs can also be defined in terms of the flip-flop output states since the outside-world's output states themselves are a function of the states (P is logic one in state s1, and state s1 is defined in terms of the flip-flop outputs $A \cdot /B$).

$$L \text{ is defined by } A \cdot /B + /A \cdot B.$$

The allocation of unique values of flip-flop outputs is rather an arbitrary process. In theory, any values can be used so long as each state has a unique combination. This means that one cannot have more than one state with the flip-flop values of say $A \cdot /B$.

In practice, it is common to assign flip-flop values so that the transition between each state involves only one flip-flop changing state. This is known as following a *unit distance pattern*. This has not been done in the example above because there are two flip-flop changes between states s1 and s2.

The single-pulse generator with memory state diagram could be made to follow a unit distance pattern by adding an extra state. This extra state could be inserted between states s2 and s0, having the same outputs for P and L as state s2.

Go to Frame 1.13.

Frame 1.13

The completed state diagram with unit distance patterns for flip-flops is shown in Figure 1.13.

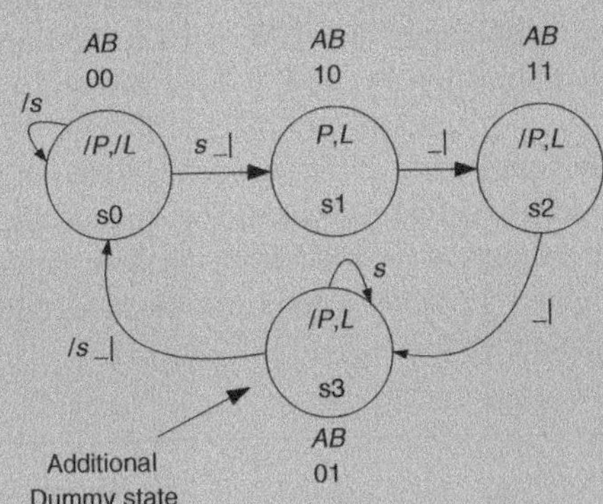

Figure 1.13 State diagram for single-pulse generator with memory.

Note that the added state has the unique name of s3 and the unique flip-flop assignment of $A = 0$ and $B = 1$. It also has the outputs $P = 0$, as it would be in state s0 (the state it is going to go to when $s = 0$). Also, L is retained at logic 1 until the input s is low, since L is the memory indicator and needs to be held high until the operator releases s.

In this design, the addition of the extra state has not added any more flip-flops to the design, since two flip-flops can have a maximum of $2^2 = 4$ states (recall Frames 1.2 and 1.3).

The single pulse generator with memory FSM is to have an additional input added (called *r*) which will, when high (logic 1), cause the FSM to flash the *P* output at the clock rate. Whenever the *r* input is reverted to logic 0, the FSM will resume its single pulse with memory operation.

Tasks 1. Draw the block diagram for the FSM.
2. Draw the state diagram for this modified FSM.

Go to Frame 1.14 to see the result.

Frame 1.14

The block diagram is shown in Figure 1.14.

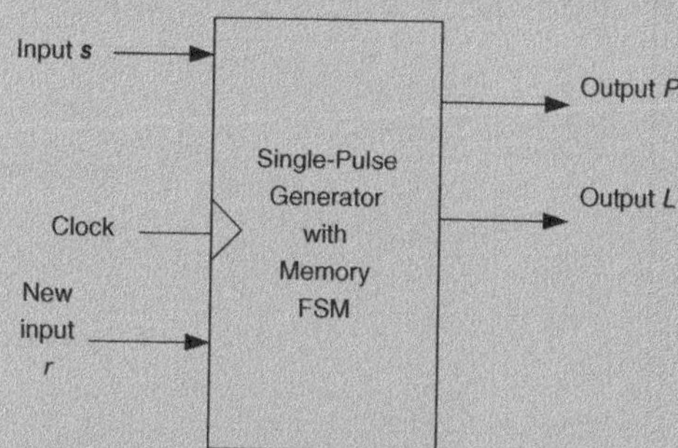

Figure 1.14 Block diagram for the FSM.

The new state diagram is shown in Figure 1.15.

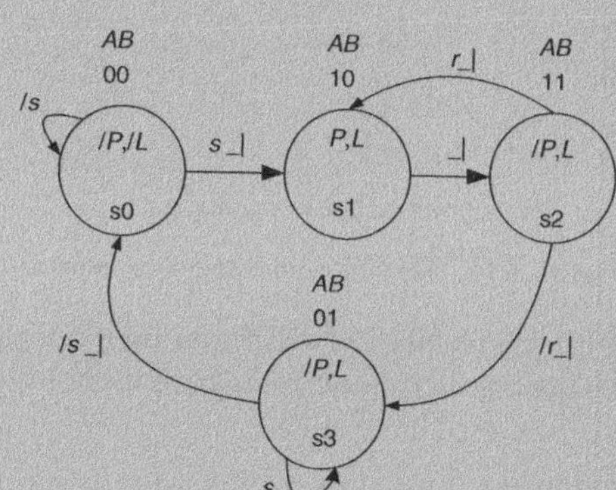

Figure 1.15 Single-pulse generator with multi-pulse feature.

The additional input has been added and a new transition from s2 to s1. Note that, when $r = 1$, the FSM is clocked between states s1 and s2. This will continue until $r = 0$.

In this condition, the *P* output will pulse on and off at the clock rate as long as input *r* is held at logic 1.

An alternative way of expressing output L

In the state diagram of Figure 1.15, $L = s1 + s2 + s3 = A \cdot /B + A \cdot B + /\ A \cdot B = A + /A \cdot B$. Therefore, $L = A + B$. See Appendix A and the auxiliary rule for the method of how this Boolean equation is obtained.

An alternative way of expressing L is in terms of its low state:

$$L = /(s0) = /(/A \cdot /B).$$

This implies that when $A = 0$ *and* $B = 0,\ L = 0$.

Dealing with active-low signals

The state diagram fragment in Figure 1.16 illustrates how an *active-low signal* (in this case CS) that is low in states s4, s5 and s6 is obtained.

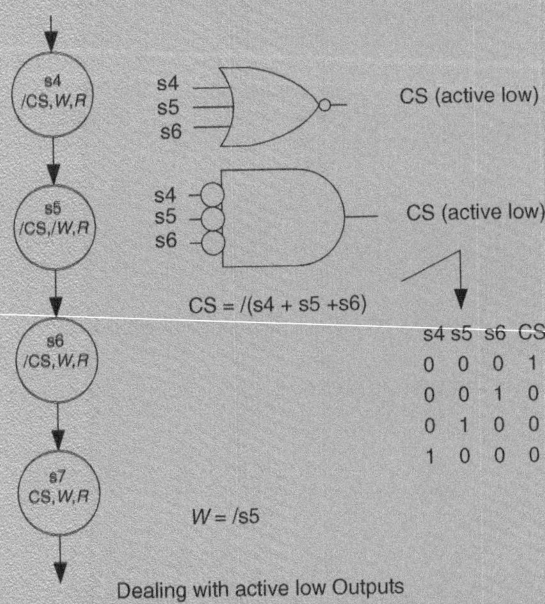

Figure 1.16 Dealing with active-low outputs.

Also, the active-low signal W is obtained as well. From this it can be inferred that, to obtain the active-low output, all states in which the output is low must be negated. This is a common occurrence in FSMs and will be used quite often.

Finally:

- If an output is high in more states than it is low, then the active-low equation might produce a minimal result.
- If the output is low in more states than it is high, then the active-high form of the output equation will produce the more minimal result.

Go to Frame 1.15.

Frame 1.15

The previous frames have considered the flip-flop output patterns. These are often referred to as the *secondary state variables* (Figure 1.17).

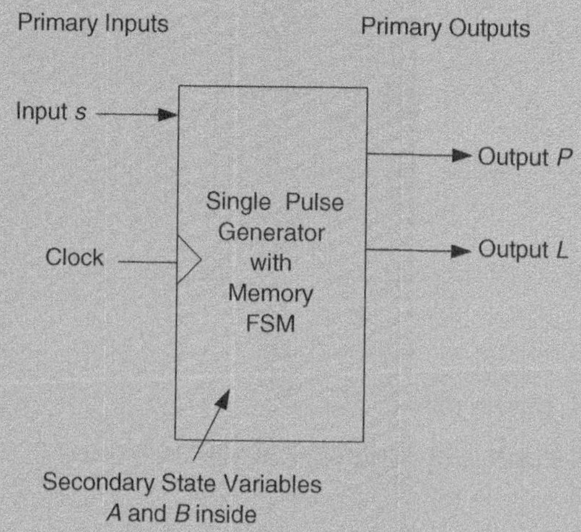

Figure 1.17 Block diagram showing secondary state variables in the FSM.

These are called secondary state variables because they are (from the FSM architecture viewpoint) internal to the FSM. Consider the Outside World inputs and outputs as being *primary*; then, it seems sensible to call the flip-flop outputs *secondary* state variables (state variables because they define the states of the state machine).

The outputs in the FSM are seen to be dependent upon the secondary state variables or flip-flops internal to the FSM. Looking back to Frame 1.5, see that Moore FSM outputs are dependent upon the flip-flop outputs only. The Output Decoding logic in the single-pulse generator with memory example is

$$P = \text{s1} = A \cdot /B$$

(see Frame 1.13) and

$$L = \text{s1} + \text{s2} + \text{s3} = A \cdot /B + A \cdot B + /A \cdot B = A + /A \cdot B = A + B$$

(auxiliary rule again), i.e. it consists of one AND gate and an OR gate. This means that the single-pulse generator with memory design is a Moore FSM.

How could the single-pulse generator design be converted into a Mealy FSM?

One way would be to make the output P depend on the FSM being in state s1 ($A \cdot /B$), but also gate it with the clock *when it is low*. This would make the P output have a pulse width equal to the clock pulse, but only in state s1, and only when the clock is low. This would be providing a feed-forward path from the (clock) input to the P (output).

Task How could the state diagram be modified to do this?

Try modifying the state diagram, then go to Frame 1.16 to check the answer.

Frame 1.16

The modified state diagram is shown in Figure 1.18.

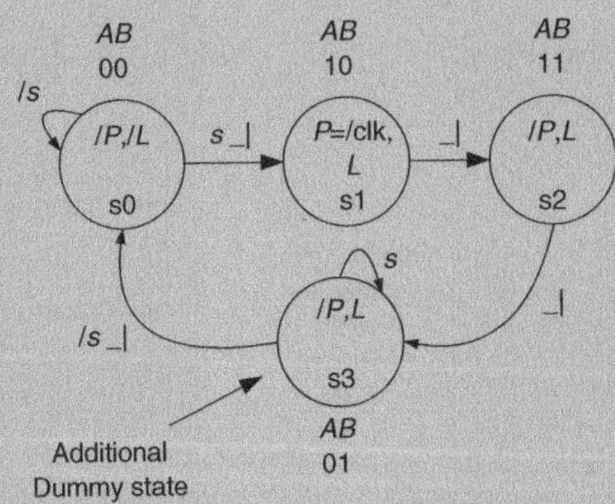

Figure 1.18 State diagram with Mealy output *P*.

Notice that, now, the output *P* is only equal to logic 1 when

- the FSM is in state s1 where flip-flop outputs are $A = 1$ and $B = 0$;
- the clock signal is logic 0.

The FSM enters state s1, where the *P* output will only be equal to logic 1 when the clock is logic 0. The clock will be logic 1 when the FSM enters state s1 (0-to-1 transition); it will then go to logic 0 (whilst still in state s1) and P will go to logic 1. Then, when the clock goes back to logic 1, the FSM will move to state s2 and the flip-flop outputs will no longer be $A \cdot /B$, so the *P* output will go low again. Therefore, the *P* output will only be logic 1 for the time the clock is zero in state s1.

The timing diagram in Figure 1.19 illustrates this more clearly.

The waveforms show both versions of *P* (under the *A* and *B* waveforms in Figure 1.19). As can be seen, the Moore version raises *P* for the whole duration that the FSM is in state s1, whereas the Mealy version raises *P* for the time that the clock is low during state s1.

However, the bottom waveform for the Mealy *P* output illustrates what can happen as a result of a delay in the /clk signal, along with the change of state from s0 to s1(*/A/B* to *A/B*). Here, a glitch has been produced in the *P* signal as a result of the delay between clk and its complement /clk, after the *A* signal change. This is brought about by the clk signal causing *A* to change to logic 1 while the /clk signal is still at logic 1 due to the delay between the clk and /clk signals. This must be avoided.

This example is not unique; different delays can result in other unexpected outputs (glitches) from signal *P*. Essentially, if two signal changes occur, then a glitch can be produced in *P* as a result in the delays between signals (static 1 hazards).

Note that the *P* output signal is delayed in time as a result of the delays in signals *A*, *B*, and the /clk. This delay is not so important as long as it does not overrun the clock period (which in most practical cases it will not).

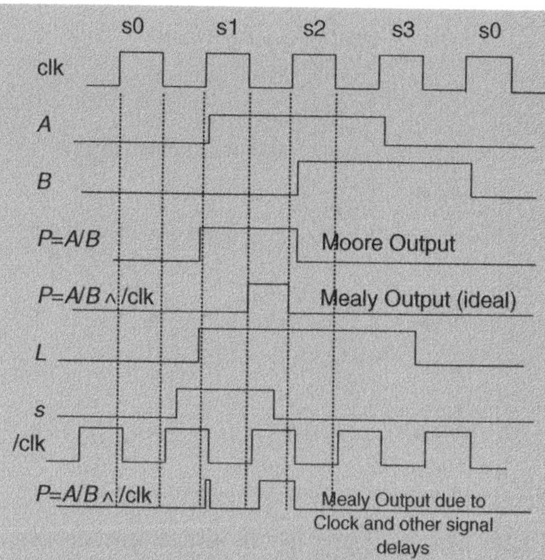

Figure 1.19 Timing diagram showing Moore and Mealy outputs.

It is best not to use the clock signal to create Mealy outputs. Also, as will be discussed in Chapter 3, it is wise, where possible, to use a unit distance coding for A and B variables to avoid two signal changes from occurring together; but more on this later.

Now for another example.

Task Produce a state diagram for an FSM that will generate a 101 pattern in response to m going high. The input m must be returned low before another 101 pattern can be produced.

After attempting this task, go to Frame 1.17.

Frame 1.17

The solution to this problem is to use the basic arrangement of the single-pulse generator state diagram and insert more states to generate the required 101 pattern. This will be developed stage by stage so as to build up the complete design (Figure 1.20).

Start by first waiting for the input s to become logic 1. Therefore, in state s0, wait for $s = 1$. Once the input $s = 1$ *and the clock changes 0 to 1*, the FSM is required to move into the next state s1, where P will be raised to the logic 1 level.

The next state s2 will be used to generate the required logic 0 at the P output. And then the next state s3 will be needed to generate the last $P = 1$.

Note that the FSM must leave state s3 on a clock pulse so that $P = 1$ for the duration of a single clock pulse only.

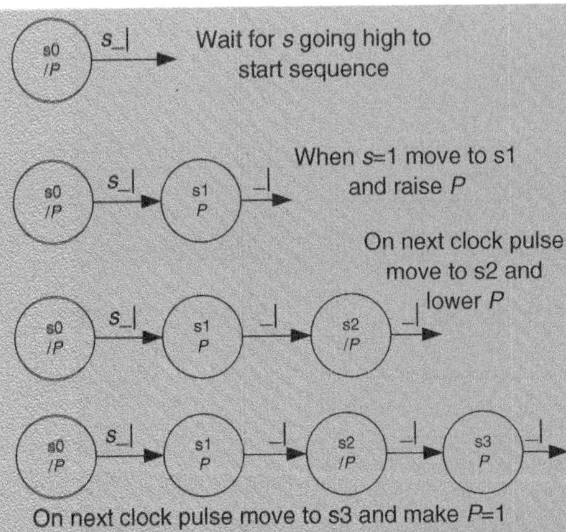

Figure 1.20 Development of the 101 pattern-generator sequence.

The final state required is to monitor for the input $s = 0$ condition. This state should return the FSM back to state s0.

Task Complete the FSM state diagram.

Now go to Frame 1.18.

Frame 1.18

The completed state diagram is shown in Figure 1.21.

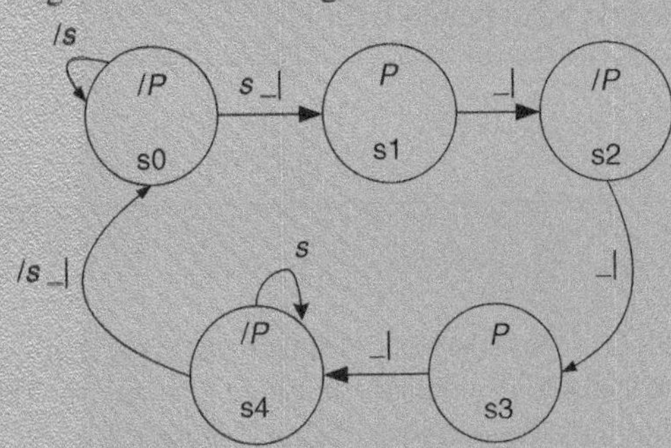

Figure 1.21 Complete state diagram for the 101 pattern-generator.

The Boolean equation for P in this diagram is $P = \mathrm{s1} + \mathrm{s3}$. However, it is possible to make the P output a Mealy output that is only equal to one when in states s1 and s2, and only if an input $y = 1$. Then:

$$P = \mathrm{s1} \cdot y + \mathrm{s3} \cdot y,$$

since P must be high in both states s1 and s3, but only when the input y is high.

A note on slings

A sling has been used for each state with an outside-world input along the transitional line. This is not really necessary, because slings are not used to obtain the circuits to perform the FSM function in modern state diagrams. In fact, they are really only included for cosmetic reasons, to improve the readability of the design. From now on, slings will only be used where they improve the readability of the state diagram.

Task Now try modifying the state diagram to make it produce a 1010 sequence of clock pulses (in the same manner shown in Figure 1.21, but with the P output pulse in state s3 to be conditional on a new input called x. If $x = 0$, the FSM should produce the output sequence 1000 at P. If $x = 1$, then the output sequence at P should be 1010.

After drawing the state diagram, move to Frame 1.19.

Frame 1.19

The modified state diagram is shown in Figure 1.22.

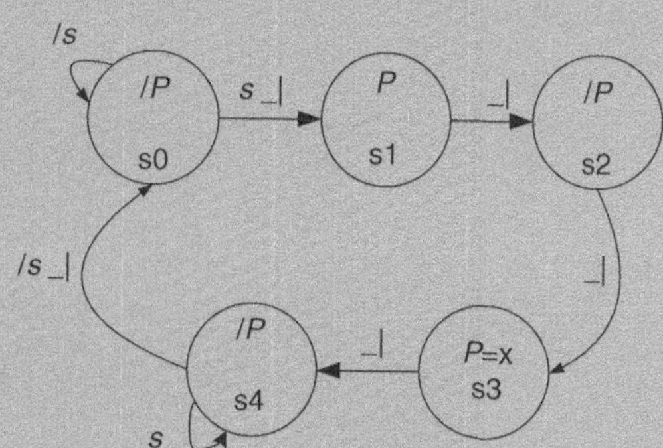

Figure 1.22 Modified state diagram with output P as a Mealy output.

In this state diagram, the input signal x is used as a qualifier in state s3 so that the output P is only logic 1 in this state when the clock is logic 1.

In state s3, the output P will only produce a pulse if the x input happens to be logic 1.

A pulse will always be produced in state s1.

It can be seen that if $x = 0$, then when the input s is raised to logic 1, the FSM will produce the sequence 1000 at output P.

If $x = 1$, then when s is raised to logic 1, the FSM will produce a 1010 sequence at the output P. This FSM is an example of a Mealy FSM, since the output P is a function of both the state and the input x, i.e. the input x is fed forward to the output decoding logic. Therefore, the equation for P is

$$P = s1 + s3 \cdot x.$$

It would be easy to modify the FSM so that the 1000 sequence at *P* is produced if $x = 1$ and the 1010 sequence is produced if $x = 0$.

Tasks
1. Produce the Boolean equation for *P* in state s3 that would satisfy this requirement.
2. Then, assign a unit distance code to the state diagram (refer to Frames 1.12 and 1.13 for why).
3. Finally, produce a timing diagram of the modified FSM.

After this, go to Frame 1.20.

Frame 1.20

The answer to Task 1 in Frame 1.19 is as follows: the Boolean equation for *P* which will produce a *P* 1010 sequence when $x = 0$ is

$$P = \text{s1} + \text{s3} \cdot /x.$$

Note that in this case the qualifier for *P* is NOT *x*, rather than with *x*.

The answer to Task 2 in Frame 1.19, with regard to assigning a unit distance code to the state diagram, is shown in Figure 1.23.

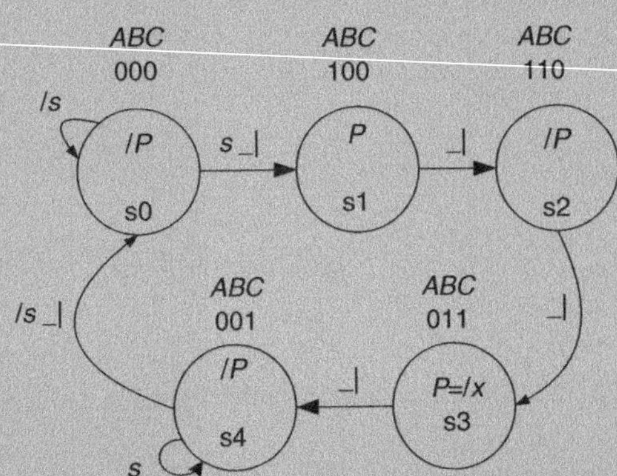

Figure 1.23 State diagram with unit-distance coding of state variables.

The equation for *P* in s3 (it could be written outside the state circle if there is not enough room to show it inside the state circle) is conditional on the *x* input being logic 0. It is very likely that you will have come up with a different set of values for the secondary state assignments to those obtained here. This is perfectly all right, since there is no real preferred set of assignments, apart from trying to obtain a unit distance coding.

Some cheating has taken place here, since the transition between states s2 and s3 is not unit distance (since flip-flops *A* and *C* both change states). A unit distance coding could be obtained if an additional dummy state is added (as was the case in Frame 1.13 for the single-pulse generator with memory FSM).

However, in this example, one must be careful where one places the dummy state. If a dummy state is added between states s1 and s2, for example, then it would alter the *P* output sequence so that instead of producing, say, 1010, the sequence 10010 would be produced.

A safe place to add a dummy state would be between states s3 and s4, or between states s4 and s0, since they are outside the 'critical' *P*-sequence-generating part of the state diagram.

Move to Frame 1.21 for the timing waveform diagram solution.

Frame 1.21

The answer to Task 3 in Frame 1.19 is as follows.

A solution is shown in Figure 1.24 based on the secondary state assignment that was used earlier, so your solution could well be different.

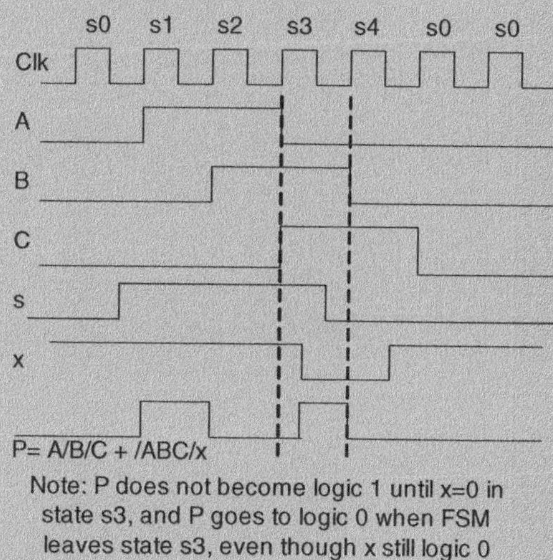

Figure 1.24 Timing diagram showing the effect of input *x* on output *P*.

Note that in this solution the input *x* has been change to logic 0 in the middle of the clock pulse in state s3 just to illustrate the effect that this would have on the output *P*. Note that the output pulse on *P* is not a full clock high period in state s3.

This is a very realistic event, since the outside-world input *x* (indeed, any outside-world input) can occur at any time.

1.3 SUMMARY

At this point, the basics of what an FSM is and how a state diagram can be developed for a particular FSM design have been covered:

- how the outputs of the FSM depend upon the secondary state variables;

- that the secondary state variables can be assigned arbitrarily, but that following a unit distance code is good practice;
- a number of simple designs have shown how a Mealy or Moore FSM can be realized in the way in which the output equations are formed.

However, the state diagram needs to be realized as a circuit made up of logic gates and flip-flops; this part of the development process is very much a mechanized activity, which will be covered in Chapter 3.

Chapter 2 will look at a number of FSM designs that control outside-world devices in an attempt to provide some feel for the design of state diagrams for FSMs. The pace will be quicker, as it will be assumed that the preceding work has been understood.

2

Using State Diagrams to Control External Hardware Subsystems

2.1 INTRODUCTION

In real-world problems there is often a need to use external subsystems, such as hardware timers/counters, analogue-to-digital converters (ADCs), memory devices, and handshake signals to communicate with external devices.

This chapter looks at how a state diagram (and, hence, an FSM) can be used to control such devices. This opens up a much wider range of activities for the FSM and can lead to solutions in hardware that can be implemented in a relatively short time.

In later chapters, the ideas explored in this chapter will be used to develop some interesting real-world systems.

2.2 LEARNING MATERIAL

Frame 2.1

One of the most common requirements in an FSM is the need to wait in a state for some predefined period. For example, a need to turn on an outside world output for a certain period of time, then turn it off again. This could be done by just allocating a number of consecutive states with the required output held high, but this would be very wasteful of states (and the corresponding flip-flops needed to implement the FSM) for all but very short delays. The best way of dealing with this kind of requirement is to use an external timer unit that can be controlled by the FSM.

A typical timer unit might look something like the illustration in Figure 2.1.

The timer unit has two inputs, the clock input clk and the start timer input ts, and a single output TO. From the timing diagram (Figure 2.1) for this timer unit, the timer output TO will go high when the timer start input ts makes a 0-to-1 transition. The output TO will remain high until the time period has elapsed, after which it will go low.

FSM based Digital Design using Verilog HDL Peter Minns and Ian Elliott

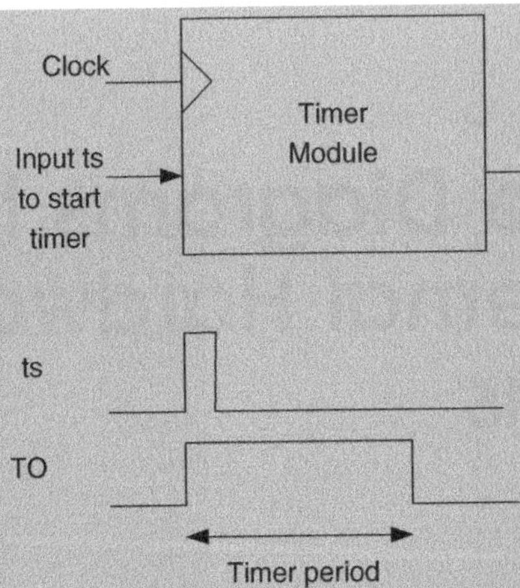

Figure 2.1 Timing module.

In Figure 2.2, TS, an output from the FSM, is used to start the timer prior to the time-out state. Then, on the next clock pulse the FSM moves into the time-out state. In the time-out state, the TS signal is returned low and the timer output signal 'to' is monitored by the FSM, which is looking for this 'to' signal going low, signalling the end of the timer period. At this point, the FSM will leave the time-out state.

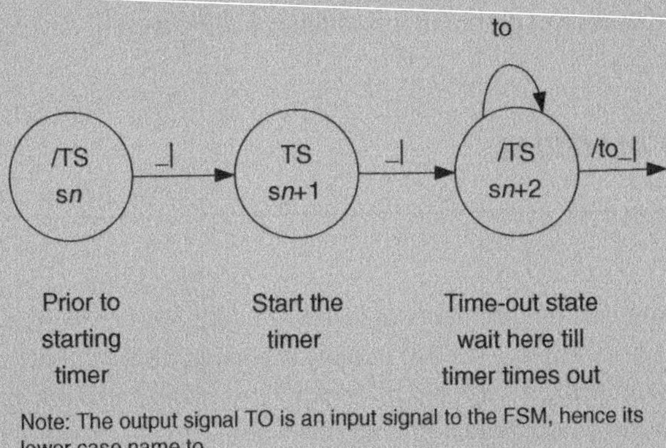

Figure 2.2 State sequence to control the timing module.

Note that the FSM will wait in state *sn* + 2 until the signal 'to' goes low. Therefore, the time spent in state *sn* + 2 will be dictated by the timer unit.

Also note here that the timer output signal 'to' is in lower case now, since it is an input to the FSM, whereas the timer input signal TS is uppercase, since it is an output from the FSM.

Now please turn to Frame 2.2.

Frame 2.2

An example of how the timer unit could be used now follows.

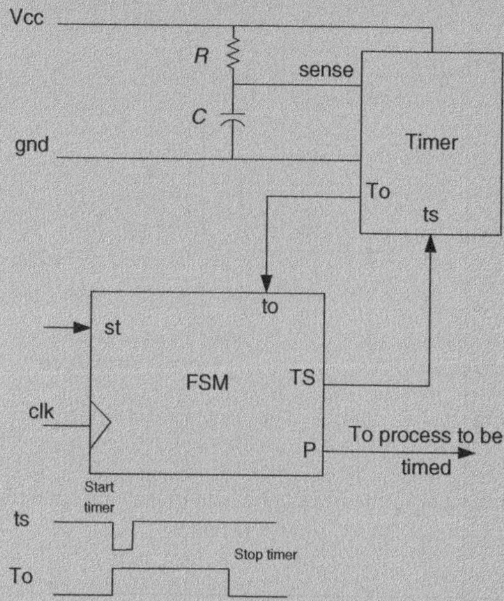

Figure 2.3 Block diagram showing how to use the timing module.

In Figure 2.3, the FSM is controlling a timer unit. This timer unit happens to be a little different to the one seen in Frame 2.1 in that it does not have a clock input. It is in fact a timer based around an RC charging circuit (a practical device would be the 555 timer units that are readily available). This does not alter things in any way as far as the FSM is concerned. Note that the actual time delay would be given by

$$\text{To} = 1.1 \times C \times R.$$

if Timer is a 555 timer chip.

The start input St starts the FSM from state s0. The idea here is to turn on the process P for a period of time (dictated by the RC time constant of the timer unit), then turn it back off again.

Task Produce the state diagram to do this. Then go to Frame 2.3 for a solution.

Frame 2.3

The state diagram to implement the arrangement shown in Figure 2.3 is shown in Figure 2.4.

The process P is turned on in state s1, since the timer is started in this state, before the FSM moves into the time-out state s2. The FSM will remain in state s2 monitoring the timer output signal ‘to’ until the timer has timed out, and then move to state s3 to stop the process P.

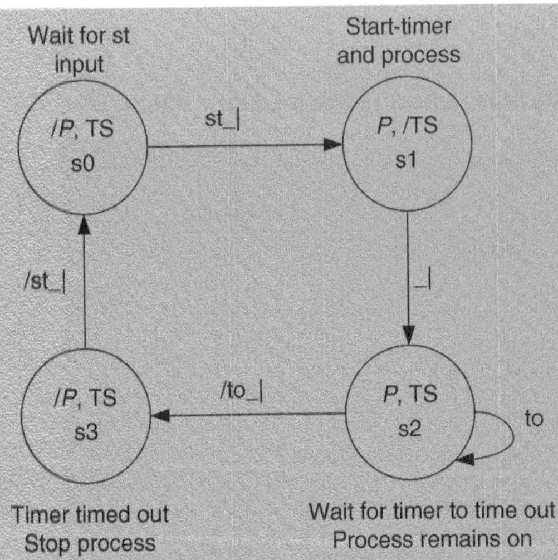

Figure 2.4 State diagram using the Timer module.

Using this arrangement allows the FSM to be held in the wait state for any length of time, depending only on the value of the *RC* time constant.

Go to Frame 2.4.

Frame 2.4

Having seen how to control an outside world device like a timer, the next stage is to see how other outside world devices can be controlled. This is really what FSMs are all about, controlling outside world devices.

Now look at how an FSM can be used to control

- an ADC;
- a memory device.

Controlling an ADC

ADCs are used to convert analogue data into digital form. They are needed to allow digital computers to be able to process data from the real world (which is essentially analogue in nature). Most systems that use an ADC will be controlled from a microprocessor (or microcomputer). However, it is often the case that a system (or part of a system) will be implemented using a customized chip design, a programmable logic device (PLD), or even a field programmable logic array (FPGA). Consider the ADC shown in Figure 2.5.

This ADC has (as is usually the case) a Start Conversion input SC and an End Of Conversion output eoc.

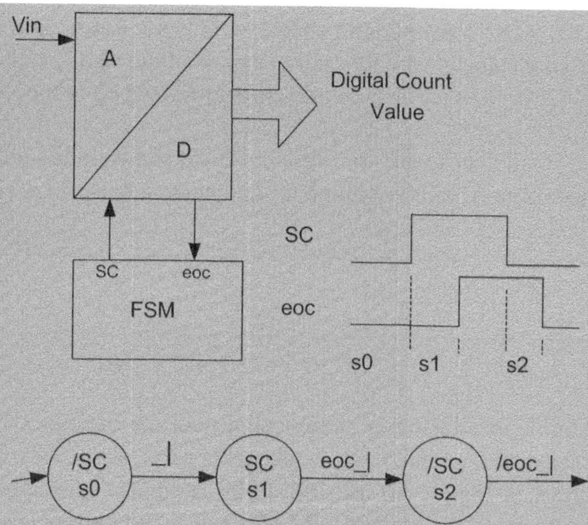

Figure 2.5 Controlling an ADC from a state diagram.

The analogue input and digital outputs are connected to the external circuit and are not part of the FSM, since they form the *data flow* part of the system. The FSM is used to *control the system components* (the ADC in this case).

The segment of state diagram shows how the ADC is controlled by an FSM.

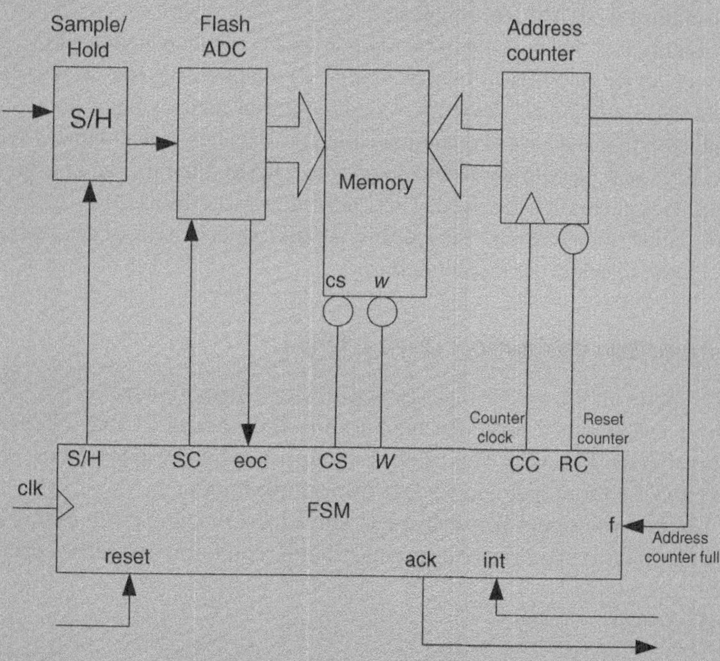

Figure 2.6 Block diagram for a small DAS.

The FSM starts the ADC conversion in state s1 and waits for the eoc[1] signal from the ADC to say that a digital value is available at the outputs of the ADC. At this point the FSM will move into state s2. Here, it will wait for eoc to return low before moving on to the next state.

Now consider the small data acquisition system (DAS) shown in Figure 2.6.

In this system there is an ADC and a number of other *outside world devices.*

Go to Frame 2.5.

Frame 2.5

This particular example is a bit more complicated than the examples looked at so far; however, it can be separated out into more manageable parts, as will be revealed.

The diagram shown in Figure 2.6 uses the FSM to *control* a sample-and-hold (S/H) amplifier, ADC, a random access memory (RAM) device, and a simple binary counter.

All these outside world components allow the FSM to be able to

- sample a.c. analogue data from the outside world;
- store the data in RAM.

These could be under the control of a remote end device (which could be a microcomputer).

Before attempting to produce a state diagram for this FSM, discussion is required on how an FSM can control the RAM and counter.

Consider Figure 2.7, which shows a memory device controlled by an FSM.

The memory device is written to/read from in s*n*+3 (on the rising edge of the read or write signal). Note that the memory device has a collection of address input lines (commonly called the address bus) and a set of data lines called the data bus. If the memory is read, only the data bus lines will be outputs. If the memory is a RAM, then the data bus lines will be bidirectional. This means that the */R* and */W* control signals can be used to condition the data bus lines to be either inputs (when */W* is used) or outputs (when */R* is used). In addition, there is a chip select input to select the memory chip.

Further information on memory device timing

Figure 2.8 shows all the waveforms associated with a memory device.

The address bus selects a particular memory location at time T1 (the address bus lines changing just after the T1 starts). This causes the chip select CE signal to become active (low) at time T2 (allowing for propagation delays through the logic).

At time T3, the write signal *W* is activated (active low) and, as a result of this action, the memory chip data bus is taken out of its normal tri-state condition so it can accept input data.

[1]Note that some ADCs have a busy signal instead of the eoc. These ADCs raise the busy signal when SC is asserted, lowering it when the conversion is complete.

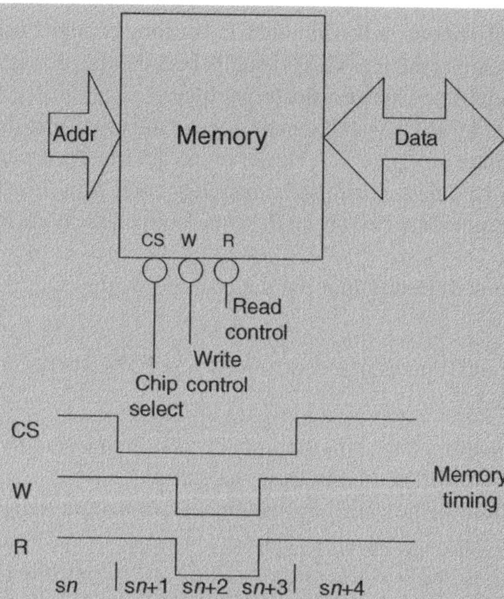

Figure 2.7 Control of a memory device.

At time T4 (after a suitable time has been allowed for the memory to settle) the write *W* signal is taken high, then the chip select signal CE will be taken high to deselect the memory chip. It is during this transition (0 to 1) of the *W* signal that the data are written into the memory chip. Note that in some memory chips the CE and *W* signals appear to go high in T4

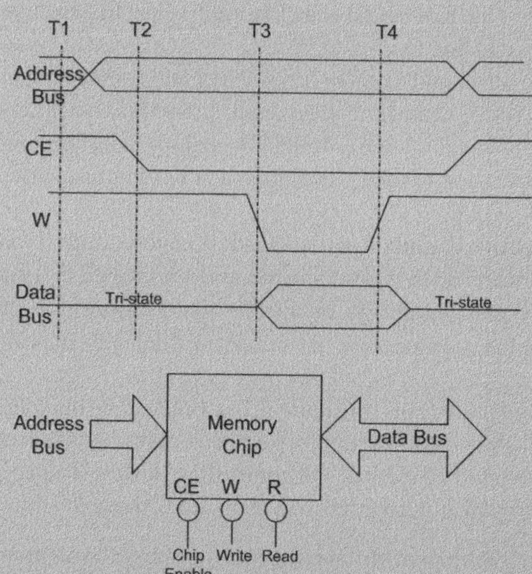

Figure 2.8 Timing of the control of a memory device.

at the same time. The microprocessor will hold CE low long enough to allow the *W* signal to write the data into the memory device. This is usually because the propagation delay is longer in the CE path due to additional address decoding logic.

In a system controlled by an FSM, this can be done in the waveform diagram sequence, as shown earlier in this frame in Figure 2.7. However, an alternative arrangement could be to cause the CE signal to be delayed within the memory chip. This would be possible if the memory was being implemented in an HDL to be contained in an FPGA, perhaps also containing the FSM.

The main thing here is to ensure that the data can be written into the RAM before it is deselected.

Note: the signals CE and *W* need to be controlled by the FSM whenever the memory is to be written to or read from.

Note that if *W* is replaced by *R* then the memory cycle is a read memory cycle in which data stored in the memory chip will be output from the chip.

The read operation follows the same basic sequence as the write signal, and the arguments discussed earlier about delaying the chip select also apply.

Now go to Frame 2.6 to see how the memory chip can be controlled from an FSM.

Frame 2.6

To access the memory device, the chip select line must be asserted (this means that the chip select line must be active, in this case active is logic 0). Then, write data into the RAM device by lowering the write signal line. A little later, raise the write line to logic 1 to write the data into the RAM device.

To read the contents of the RAM, first select the chip select line by making it go low, then a little time later set the read line low.

In most cases, 'chip select and read' or 'chip select and write' control lines can be raised high (to disassert them) at the same time. It is usually at this point in the cycle that the memory device is read or written; but, if there is a doubt about chip select remaining low long enough for the write or read operation to take place, then it is best to raise write or read first before raising the chip select signal.

In practice, the data bus will remain active for a few nanoseconds (typically 10 ns) in order for the data to be written or read by memory in memory controlled by a microprocessor, but in an FSM-controlled system the design engineer should ensure that this occurs either by adding another state to the state machine or by creating a delay on the chip select signal in the memory device.

The segment of timing diagrams of Figure 2.7 in Frame 2.5 illustrates this process.

When reading from and writing to memory devices, the process of reading and writing is implied to be from the point of view of the controlling device. The controlling device in a microprocessor system is the microprocessor. In our case, the controlling device is the FSM.

Task Try producing a segment of state diagram to control the memory device for writing.

Now go to Frame 2.7 to find out whether it is correct.

Frame 2.7

The segment of state diagram to control the memory illustrated in Frames 2.5 and 2.6 is illustrated in Figure 2.9.

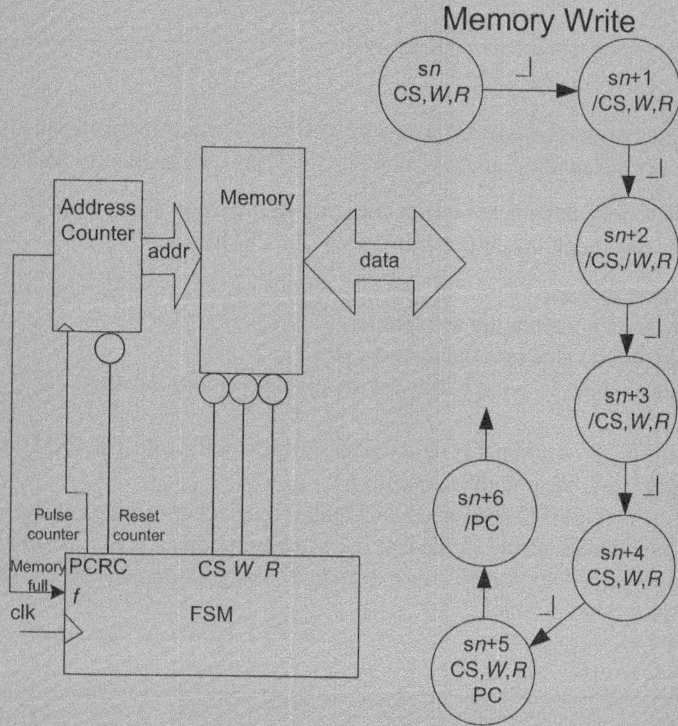

Figure 2.9 Using an FSM to control the writing of data to a memory device.

In state *sn*, all controls are disasserted. In *sn* + 1 the chip select line CS is active; then, in state *sn* + 2, the write control line *W* is active. In state *sn* + 3, *W* is deactivated to write the data into the memory, and it is at this point that the data are written into the memory device. Finally, in state *sn* + 4, the chip select CS is raised to deselect the memory device.

To read or write to a memory device, the data transaction will occur in the memory element currently accessed by the address bus. To access another memory element, another address needs to be selected; this is done by the address counter. This is what the counter in Figure 2.6 (and in Figure 2.9) is being used for. In this case, each memory location is selected in sequence by incrementing the binary counter after accessing each memory location.

Note that in state *sn* + 5 the signal PC is set high. This increments the counter, thereby incrementing the address to point to the next consecutive memory location.

The counter can be reset to zero by sending the signal RC to logic 0. It can be incremented by the FSM with a pulse to the PC signal. In this way, each memory location of the memory can be accessed sequentially. Note that the address counter is incremented after disasserting the memory chip. This is because the address on the memory chip should not be changed while it is selected.

Go to Frame 2.8.

Frame 2.8

Task Having seen how the individual outside world devices are controlled in Figure 2.6, try to produce a state diagram to implement the FSM used to control the system.

The FSM is to wait in state s0 until it receives an interrupt signal from the remote end device over the int signal line. When this occurs, the FSM is to

- obtain a sample of data;
- perform an analogue-to-digital conversion;
- store the converted value into the memory device;
- increment the counter to point to the next available memory location.

The FSM should keep on doing this until the memory device is full. The FSM will know this when the *f* input (from the counter) goes high.

At this point, the FSM is to send an acknowledge signal to the remote end device using the ACK signal line; then, once the signal line int is low (remember, it was asserted high at the beginning of the sequence), it is to return to state s0 ready for another cycle.

When completed, go to Frame 2.9.

Frame 2.9

The complete state diagram is illustrated in Figure 2.10. Your state diagram may not look quite like this one, as there is more than one way of drawing a state diagram and there is no 'one' correct solution. However, the diagram in Figure 2.10 is very concise.

The first thing to note is that the reset line RC used to hold the memory address counter reset is held active (asserted) in state s0. Thereafter, it is held in its disasserted state (i.e. RC = 1).

Not all states show this, but the sequential nature of the state diagram implies it.

When the int input is asserted, the FSM moves into state s1, removing the reset from the address counter and simultaneously asserting the sample-and-hold amplifier S/H = 1.

On the next clock pulse, the FSM moves to state s2, where (with the S/H still set) the start conversion signal of the ADC is asserted SC = 1. At this point, the FSM waits for the end of conversion signal eoc = 1; then, on the next clock pulse, it moves to state s3, where the S/H signal is disasserted (S/H = 0), since the ADC has converted the analogue data into digital form. While in state s3 the chip select in asserted (CS = 0) and the FSM waits in state s3 for eoc to be returned low. When this happens, on the next clock pulse the FSM moves to s4, where the memory device write signal line *W* is asserted low to set up the memory data bus lines as inputs. This allows the ADC digital output value to be input to the memory.

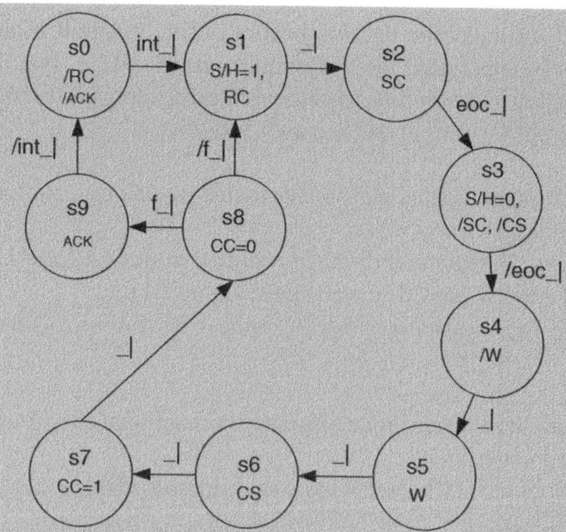

Figure 2.10 State diagram for the DAS.

The next clock pulse will move the FSM into state s5, where *W* is disasserted high, thereby writing the ADC value into the memory chip.

The FSM now moves into state s6 on the next clock pulse to deselect the memory (CS = 1).

Go to Frame 2.10.

Frame 2.10

At this stage in the FSM cycle, the FSM is in state s6.

The state machine will now move to state s7, where CC will be asserted high. On the next clock pulse, the state machine moves to s8, where CC will go low. The 0-to-1 transition on this signal line, caused by the FSM moving from state s6 to s7, will cause the address counter to increment. Note that, in state s6, the CS and *W* signals are now both high (disasserted).

In state s8 the FSM can move, either to state s1, if the signal *f* is disasserted low (hence repeating the sequence s1 to s8), or, if signal *f* is asserted, the FSM can move from s8 into s9.

The signal *f* is used to indicate whether the address counter has reached the end of the memory. If it has not reached the end of the memory ($f = 0$) then another cycle is started, otherwise the FSM moves into state s9 where the ACK signal will be asserted high to let the external device know that the FSM has completed its tasks. The FSM will wait for the signal int going low.

By waiting for int to go low, the FSM will be ready for the next low to high transition on int to start the next cycle of operations. Note that the external device will need to lower the int signal to complete the handshake. On seeing int go low, the FSM lowers the ACK signal to complete the handshake.

The forgoing example is quite a complex one and shows how an FSM can be used to control a complex sequence to control a number of outside world devices.

As seen from this example, the development of a state diagram is largely an intuitive process. However, by applying the techniques discussed in this book the reader can become experienced in developing their own state diagrams to control external devices.

Some of the ideas put forward in this text are as follows:

- ANDing clock and other input signals to an outside world output to form Mealy outputs (Frame 1.16, Figure 1.18);
- using dummy states to obtain unit distance coding (Frames 1.12 and 1.13);
- using an external timer to provide a wait state (Frame 2.1);
- using the FSM to control outside world devices like ADC (Frame 2.4) and memory devices (Frame 2.5).

The steps necessary to get from a state diagram to a functional circuit are very mechanical and will be discussed in Chapter 3.

However, before looking at Chapter 3, there are a number of other techniques that need to be considered.

Move on to Frame 2.11.

Frame 2.11

Consider the block diagram of an FSM in Figure 2.11.

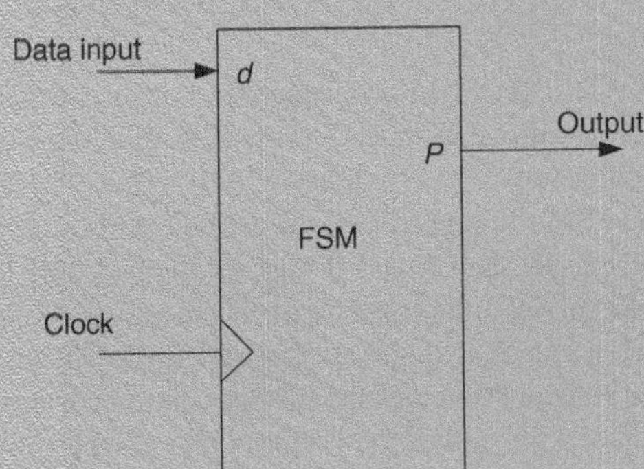

Figure 2.11 Block diagram for the data-input FSM.

In this particular FSM, a single clock pulse is required at the output P whenever the d input is asserted high twice.

Task Try producing a state diagram for this one, then turn to Frame 2.12 to see a solution.

Frame 2.12

The block diagram of Frame 2.11 is repeated in Figure 2.12.

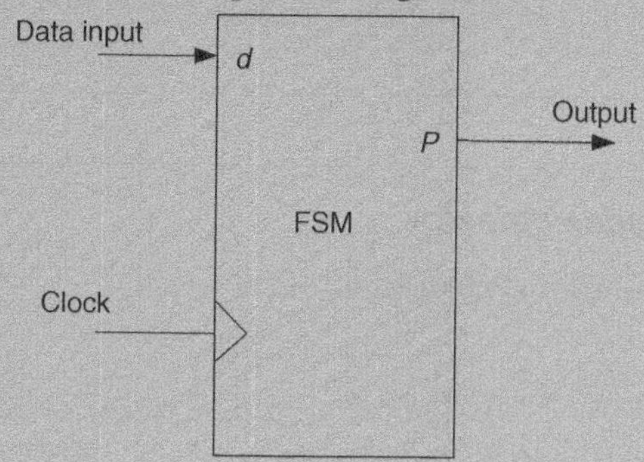

Figure 2.12 Block diagram for the data-input FSM.

The point to note here is to monitor the input for change. This implies the need to monitor *d* for *two* assertions.

The monitoring of *d* is very important in this example, since the FSM must determine when *d* has been asserted *twice*. To do this, monitor *d* going high, then monitor *d* going low (at this point, *d* has gone high then low once). Continue to monitor *d* going high again, followed by monitoring *d* going low (at this point, *d* has gone high then low twice).

The state diagram is shown in Figure 2.13.

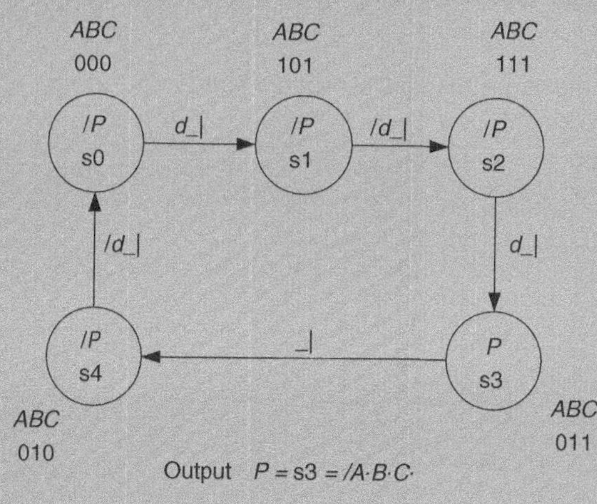

Figure 2.13 State diagram to detect two 1-to-0 transitions of *d* input.

In this state diagram, the FSM monitors the *d* input going high, then low (s0 and s1), then monitors *d* going high again (s2 and s3). In state s3, the FSM knows that *d* has been asserted twice, so the output *P* is allowed to become asserted high. The FSM moves out of state s3 on the next clock pulse and waits in state s4 for the *d* input to go low before moving back into state s0. So, inputs with multiple assertions *must* be monitored by the FSM.

Note that in Figure 2.13 the state assignments between s0 and s1 and between s4 and s0 are not unit distance. You might like to try to obtain a unit distance coding for the state diagram.

A possible solution is as follows. A dummy state needs to be added to the state diagram, a possible place would be between s3 and s4 (call it s5). Then, the following unit distance state assignment could be applied: s0 = 000, s1 = 100, s2 = 110, s3 = 111, s5(dummy) = 011, s4 = 001.

Go to Frame 2.13.

Frame 2.13 A sequence detector

Consider the example shown in Figure 2.14.

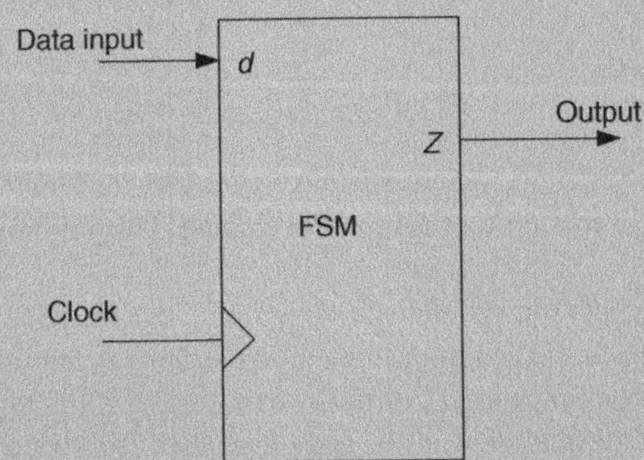

Figure 2.14 Block diagram for the 110 sequence detector.

This detector has the timing diagram shown in Figure 2.15.

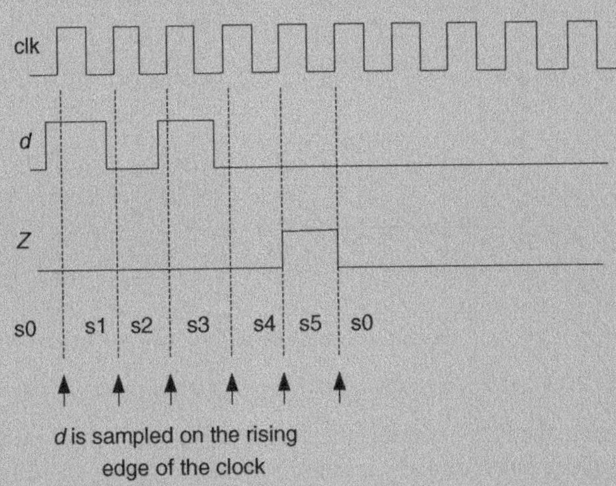

Figure 2.15 Possible timing diagram for the 110 sequence detector.

Note, *d* is sampled on the 0-to-1 transition of the clock (shown by the arrows in n Figure 2.15).

The FSM changes state on the 0-to-1 clock transition also. The timing diagram illustrates how the FSM is to do this.

In the timing diagram of Figure 2.15, *d* follows a 110 sequence. In practice, of course, one needs to produce an FSM that can identify the 110 sequence from all other possible sequences. Only the 110 sequence, however, should produce a *Z* output pulse.

Task Assume that the *d* input is a succession of pulses, i.e. $1 \rightarrow 0 \rightarrow 1 \rightarrow 0$ is two *D* pulses. Try producing a state diagram for such an FSM.

Hint Produce the state diagram for detecting the required 110 sequence first, then *add* to this state diagram the necessary transitions to cover all other sequences.

Go to Frame 2.14 to see the solution.

Frame 2.14

The state diagram for the 110 sequence detector is shown in Figure 2.16.

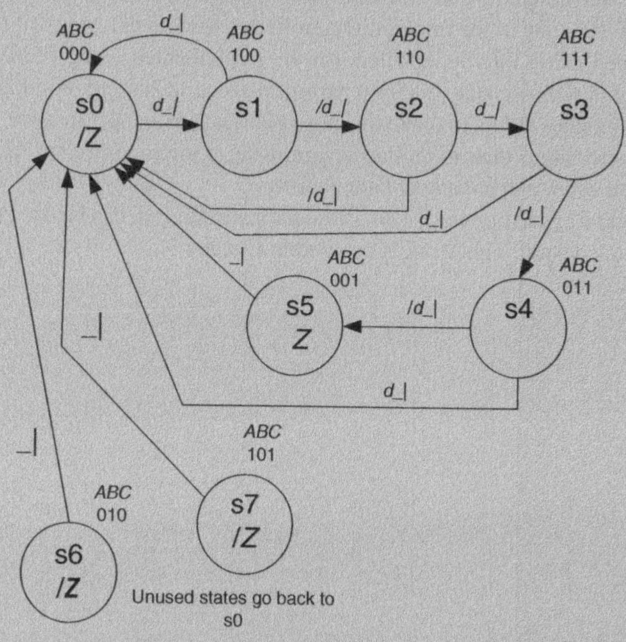

Figure 2.16 State diagram for the 110 sequence detector.

The sequence s0, s1, s2, s3, s4 and s5 detects the 110 sequence (note the assumption that the *d* input is a succession of pulses).

The loop back terms catering for all other sequences are to return the FSM back to state s0 in order to keep on detecting the 110 sequence. The timing diagram of Figure 2.15 should help to explain the different transitions.

Note that the state diagram above uses six states. The secondary state variables allow eight states; therefore, there are two states that are not used. What would happen if the FSM were to fall into one of these two unused states? The answer to this question is that the FSM would not be able to get out of the unused state and the FSM would 'hang'.

To avoid this calamity, it is common to direct all unused states back to state s0 so that the FSM can recover from this misadventure. This is shown in Figure 2.16, where states s6 and s7 are directed back to state s0 on the next clock pulse.

Note that when using *D*-type flip-flops to implement the state machine, getting into an unused state will automatically cause the FSM to reset to state s0; therefore, it is not necessary to connect unused states back to s0 in this case. More on this later.

2.3 SUMMARY

This chapter has dealt with the way in which FSMs can be used to control external hardware in a digital system. Later chapters will illustrate how these and other external devices can be controlled by an FSM. One of the implications from this work is that many of the applications normally developed using microcontrollers can be implemented using FSMs and hardware logic. The block diagram and state diagram approach seen in Chapters 1 and 2 can be used, in conjunction with modern HDLs to make this possible. The advantage, in some cases, will be a design that uses less logic than a similar design using a microcontroller. You will see this possibility later on when you have read later chapters.

For now, the next stage in our work is to see how a state diagram can be used to create a logic circuit to realize the design. This work is covered in Chapter 3.

3
Synthesizing Hardware from a State Diagram

3.1 INTRODUCTION TO FINITE-STATE MACHINE SYNTHESIS

At this point, the main requirements to design an FSM have been covered. However, the ideas discussed need to be practised and applied to a range of problems. This will follow in later chapters of the book and provide ways of solving particular problems.

In the development of a practical FSM there is a need to be able to convert the state diagram description into a real circuit that can be programmed into a PLD, FPGA, or other application-specific integrated circuit (ASIC). As it turns out, this stage is very deterministic and mechanized.

FSM synthesis can be performed at a number of levels.

Develop an FSM using flip-flops, which can be:

- *D*-type flip-flops;
- *T*-type flip-flops;
- *JK*-type flip-flops.

Use a high-level HDL such as VHDL. This can be used to enter the state diagram directly into the computer. The HDL can then be used to produce a design based upon any of the above flip-flop types using one of a number of technologies.

It is also possible to take the state diagram and convert it into a C program and, hence, produce a solution suitable for implementation using a micro-controller.

By using the direct synthesis approach, or an HDL, the final design can be implemented using:

- discrete transistor–transistor logic (TTL) or complementary metal oxide–semiconductor (CMOS) components (direct synthesis);
- PLDs;
- FPGAs;
- ASICs;
- a very large-scale integration chip.

FSM based Digital Design using Verilog HDL Peter Minns and Ian Elliott

Most technologies support *D*- and *T*-type flip-flops, and in practice these devices are used a lot in industrial designs; therefore, this book will look at the way in which *T* flip-flops and *D* flip-flops can be used in a design. Note: *JK* flip-flops can also be used, but these are not covered in this book.

This chapter will look at the implementation of an FSM using *T*-type flip-flops and then move on to look at designs using *D*-type flip-flops.

Why *T*- and *D*-type flip-flops?

These are the most common types of flip-flop used today. The reason is that the *T* type can be easily implemented from a *D* type, and the *D* type requires only six gates (compared with the *JK* type, which requires about 10 gates). This means that *D*-type flip-flops occupy less chip area than *JK* types. Another reason is that the *D*-type flip-flop is more stable than the *JK* flip-flop.

D-type flip-flops are naturally able to reset, in that if the *D* input is at logic 0, then the flip-flop will naturally reset on the next clock input pulse (see later on in this chapter). This can be of great benefit in the design of FSMs.

Go to Frame 3.1 to find out how to use the *T* flip-flop.

3.2 LEARNING MATERIAL

Frame 3.1 The T-type flip-flop

A *T*-type flip-flop can be implemented with a standard *D*-type flip-flop, as illustrated in Figure 3.1.

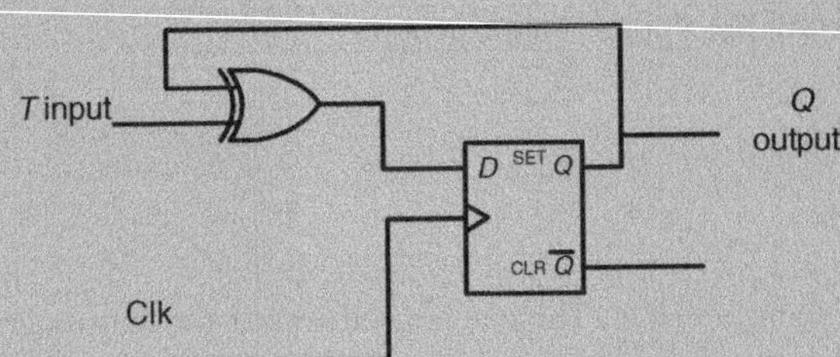

T	Qn	Qn=1	
0	0	0	No change
1	0	1	Toggle
0	1	0	Toggle
1	1	1	No change

Figure 3.1 Diagram and characteristics of a *T* flip-flop.

As can be seen from the diagram, the T flip-flop is implemented by using a *D* flip-flop with an exclusive OR gate. The table under the flip-flop shows its characteristics.

In this table, Q*n* is the *present state* output of the flip-flop (prior to a clock pulse), whilst Q*n* + 1 is the *next state* output of the flip-flop (after the clock pulse). The table shows that the flip-flop will change state on each clock pulse provided that the *t* input is *high*. But if the *T* input is *low*, then the flip-flop will *not* change state.

Therefore, use the *t* input to control the flip-flop, since, whenever the flip-flop is to change state, simply set the *t* input high; otherwise it is held low.

Go to Frame 3.2.

Frame 3.2 The T flip-flop example

Consider the simple single-pulse generator with memory example of Frame 1.13 reproduced in Figure 3.2.

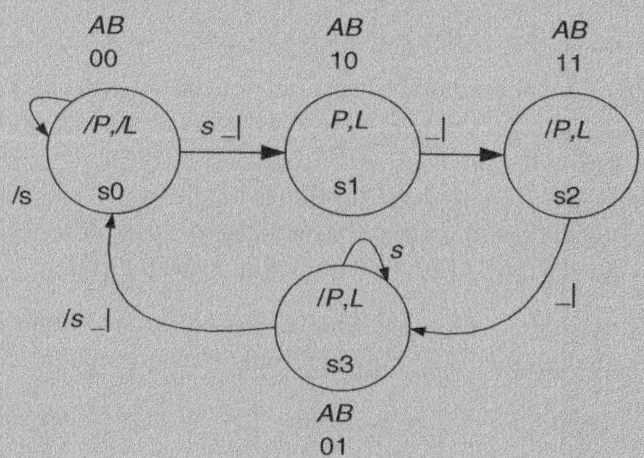

Figure 3.2 State diagram for the single-pulse generator with memory.

Follow the state transitions for secondary state variable *A* and write down the state term wherever there is a 0-to-1 or a 1-to-0 transition in *A*. In the above state diagram there is a 0-to-1 transition in *A* between states s0 and s1, and a 1-to-0 transition between states s2 and s3. Therefore, write down

$$A \cdot T = \text{s0} \cdot s + \text{s2}.$$

This equation defines the logic that will be connected to the *T* input of flip-flop *A*.

Whenever the FSM is in state s0, *and* input *s* is high, the *T* input of flip-flop *A* will be high. Whenever the *T* input is high, the flip-flop will toggle. Since in state s0 both flip-flops are reset, then in state s0, when *s* goes to logic 1, the next clock pulse will cause the flip-flop *A* to go from 0 to 1.

In state s1, the *T* input to flip-flop *A* will be at logic 0 since there is no term to hold this input high in these states. Therefore, in state s1 the flip-flop *A* will not toggle with the clock pulse. When the FSM reaches state s2, the *T* input will go high again and the next clock pulse will cause the flip-flop to toggle back to its reset state as intended. Note that in state s3 the *T* input to flip-flop *A* will be low again, so the flip-flop will *not* toggle on the next clock pulse in state s3.

Note that the equation for $A \cdot T$ uses the present state condition to set the *t* line high. This is necessary in order to make sure that the flip-flop will toggle on the next clock pulse. The logic being produced here, therefore, is that of the *next state decoder* of the FSM (see Frame 1.4).

Task Try producing the equation for the input logic for the T input on flip-flop B.

Then go to Frame 3.3 to see the solution.

Frame 3.3

The equation for the T input of flip-flop B is

$$B \cdot T = s1 + s3 \cdot /s.$$

Since in state s1 the $B \cdot T$ input needs to be logic 1, so that on the next clock pulse the flip-flop will change from a reset state to a set state. Note that there is no outside world input condition between states s1 and s2.

The second term $s3 \cdot /s$ will cause the B flip-flop to toggle from its set state to its reset state in state s3 when the outside world input $s = 0$ on the next clock pulse.

In summary, look for the 0-to-1 or 1-to-0 transition in each flip-flop.

Task Now try the example in Figure 3.3 and also produce the output equations for this design,but with output L being high in s3 only if a new input R is logic 1.

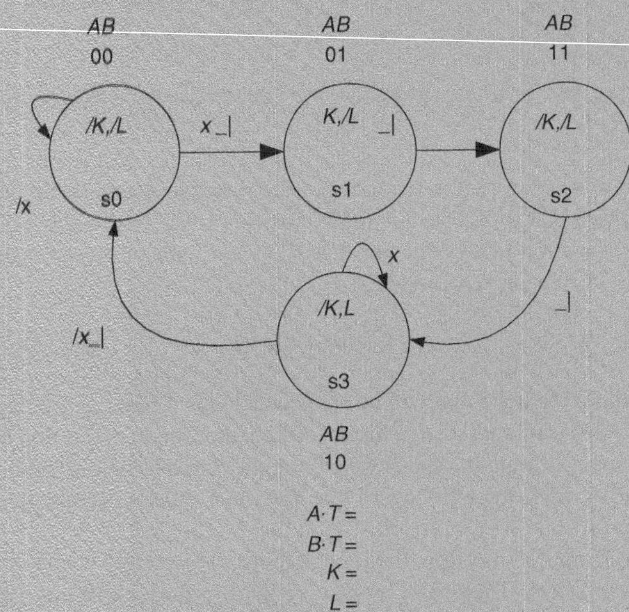

Figure 3.3 State diagram example for implementation using T flip-flops.

Now turn to Frame 3.4.

Frame 3.4

The modified state diagram is shown in Figure 3.4.

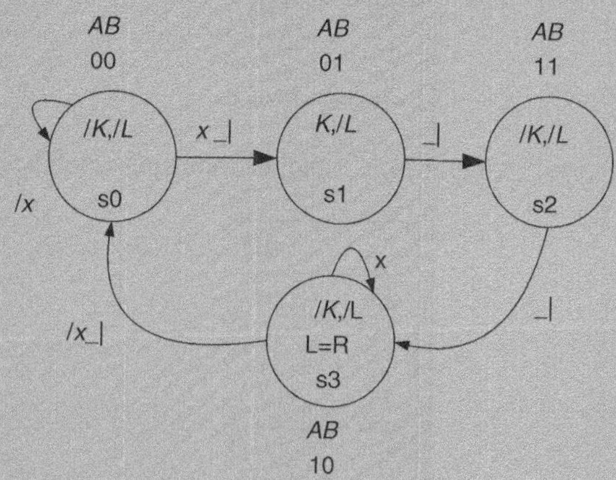

Figure 3.4 State diagram example for implementation using *T* flip-flops.

The equations for $A \cdot T$ and $B \cdot T$ are

$$A \cdot T = \text{s1} + \text{s3} \cdot /x$$
$$B \cdot T = \text{s0} \cdot x + \text{s2}.$$

The outside world outputs are

$$K = \text{s1} = /A \cdot B$$
$$L = \text{s3} \cdot R = A \cdot /B \cdot R.$$

The equation for L is a Mealy output in which the value of L can only be logic 1 in state s3, but only if input R is also logic 1. Refer to Frames 3.1–3.3 for the method if necessary.

Please now turn to Frame 3.5.

Frame 3.5

Task Attempt the following examples. Produce the flip-flop equations and output equations for each of the state diagrams indicated. If you are not too sure, reread Frames 3.1–3.4 before starting to do the problems.

State diagram in Frame 1.19, Figure 1.22, using the following secondary state variables:

	ABC
s0	000
s1	100
s2	110
s3	011
s4	001

State diagram in Frame 2.3, Figure 2.4, using the following secondary state variables:

	AB
s0	00
s1	10
s2	11
s3	01

State diagram in Frame 2.12, Figure 2.13, using the following secondary state variables:

	ABC
s0	000
s1	100
s2	110
s3	111
s4	011
s5	001

State diagram in Frame 2.9, Figure 2.10, using the following secondary state variables:

	ABCD
s0	0000
s1	1000
s2	1100
s3	1110
s4	1111
s5	0111
s6	0011
s7	1011
s8	1001
s9	0001

See Frame 3.6 for the solution to these problems.

Frame 3.6

The answers to the problems in Frame 3.5 are as follows.

For the state diagram of Frame 1.19, Figure 1.22:

	ABC	Answer
s0	000	$A \cdot T = s0 \cdot s + s2 = /A/B/C \cdot s + AB/C$
s1	100	$B \cdot T = s1 + s3 = A/B/C + /ABC$
s2	110	$C \cdot T = s2 + s4 \cdot /s = AB/C + /A/BC \cdot /s$
s3	011	
s4	001	$P = s1 + s3 \cdot x = A/B/C + /ABC \cdot x$

For the state diagram of Frame 2.3, Figure 2.4:

	AB	Answer
s0	00	$A \cdot T = s0 \cdot st + s2 \cdot /to = /A/B \cdot st + AB \cdot /to$
s1	10	$B \cdot T = s1 + s3 \cdot /st = A/B + /AB \cdot /st$
s2	11	
s3	01	$P = s1 + s2 = A$, TS (active-low) $= /s1 = /(A/B)$

For the state diagram of Frame 2.12, Figure 2.13:

	ABC	Answer
s0	000	$A \cdot T = s0 \cdot d + s3 = /A/B/C \cdot d + ABC$
s1	100	$B \cdot T = s1 \cdot /d + s4 \cdot /d = A/B/C \cdot /d + /ABC \cdot /d$
s2	110	$C \cdot T = s2 \cdot d + s4 \cdot /d = AB/C \cdot d + /ABC \cdot /d$
s3	111	
s4	011	$P = s3 = ABC$

For the state diagram of Frame 2.9, Figure 2.10:

	ABCD	Answer
		$A \cdot T = s0 \cdot int + s4 + s6 + s8 \cdot f$
s0	0000	$= /A/B/C/D \cdot int + ABCD + /A/BCD + A/B/CD \cdot f$
s1	1000	
s2	1100	
s3	1110	$B \cdot T = s1 + s5 = A/B/C/D + /ABCD$
s4	1111	
s5	0111	$C \cdot T = s2 \cdot eoc + s7 = AB/C/D \cdot eoc + A/BCD$
s6	0011	$D \cdot T = s3 \cdot /eoc + s8 \cdot /f + s9/ \cdot int$
s7	1011	$= ABC/D \cdot /eoc + A/B/CD \cdot /f + /A/B/CD \cdot /int$
s8	1001	
s9	0001	
		RC $= /s0 = /(/A/B/C/D)$ active-low output
		S/H $= s1 + s2 = A/C/D$

SC = s2 = AB/C/D
CS = /(s3 + s4 + s5) = /(ABC + BCD) active-low output
W = /s4 = /(ABCD) active-low output
CC = s7 = A/BCD

If you are having difficulty in seeing how the active-low output equations are obtained, skip forward to Frame3.25 (and Frame 3.26) for an explanation, then return to this frame.

Task Finally, try producing the equations for the state diagram of Frame 2.14, Figure 2.16; it has already been assigned secondary state variables. The answer is in Frame 3.7.

Frame 3.7

The answers for the state diagram of Frame 2.14, Figure 2.16 are

$$
\begin{aligned}
A \cdot T &= s0 \cdot d + s1 \cdot d + s2 \cdot /d + s3 + s7 \\
&= /A/B/C \cdot d + A/B/C \cdot d + AB/C \cdot /d + ABC + A/BC \\
&= /B/C \cdot d + AB \cdot /d + AC \\
B \cdot T &= s1 \cdot /d + s2 \cdot /d + s3 \cdot d + s4 + s6 \\
&= A/C \cdot /d + BC \cdot d + /AB \\
C \cdot T &= s2 \cdot d + s3 \cdot d + s4 \cdot d + s5 + s7 \\
&= AB \cdot d + BC \cdot d + /BC \\
Z &= s5 \\
&= /A/BC.
\end{aligned}
$$

The complete cycle of designing an FSM and synthesizing it using *T*-type flip-flops has been completed.

T-type flip-flops, as has already been seen in Frame 3.1, can be implemented from a basic *D*-type flip-flop, using an exclusive OR gate. Some PLDs support both the *D*- and *T*-type flip-flops, so FSM designs can be implemented using these PLDs.

Some PLD devices can be programmed to be either *D*-type or *T*-type flip-flops. However, most PLD devices support *D*-type flip-flops, particularly the cheaper PLD devices, such as the 22v10. Therefore, it is worthwhile considering how *D*-type flip-flops can be used to synthesize FSMs.

As it turns out, using *D*-type flip-flops to synthesize FSMs requires a little thought, so some time will be spent looking at the techniques required in order to make use of *D*-type flip-flops in the design of FSMs. This will be time well spent, since it opens up a large number of potential devices that can be used to design FSMs.

Turn to Frame 3.8.

Frame 3.8 Synthesizing FSMs using D-type flip-flops: the D-type flip-flop equations

Consider the basic *D* flip-flop device shown in Figure 3.5.

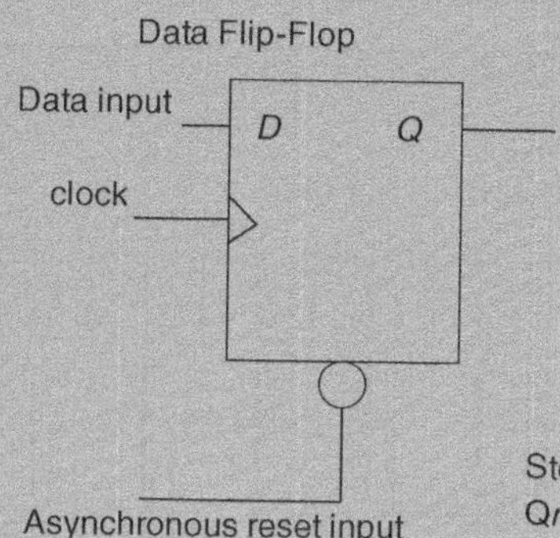

Steering Table

Qn	Qn+1	D
0	0	0
0	1	1
1	0	0
1	1	1

Figure 3.5 Diagram and characteristics of a *D* flip-flop.

The *D*-type flip-flop has a single data input *D* (apart from the clock input).

- The data line must be asserted high before the clock pulse, for the *Q* output to be clocked high by the 0-to-1 transition of the clock.
- For the *Q* output to remain high, the *D* input must be held high so that subsequent clock pulses will clock the 1 on the *D* input into the *Q* output.

These two bullet points are very important and should be remembered when using *D*-type flip-flops.

Consider the waveforms shown in Figure 3.6 applied to a *D* flip-flop.

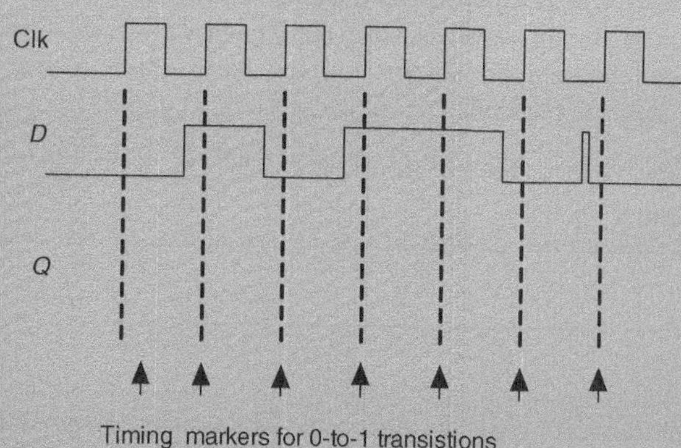

Figure 3.6 Incomplete timing diagram for a *D* flip-flop.

Task Complete the timing diagram for the output Q.

Hint: study the content of this frame and the steering table.

Go to Frame 3.9 after completing the diagram.

Frame 3.9

The completed timing diagram is illustrated in Figure 3.7.

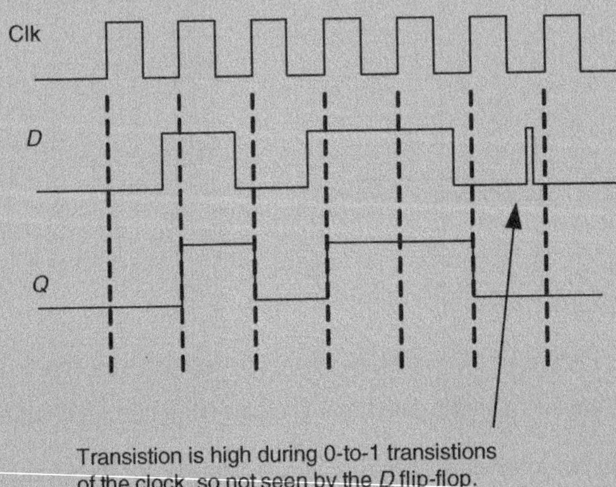

Figure 3.7 Complete timing diagram for a D flip-flop.

The trick here is to look at the value of the D line whenever the clock input makes a transition from 0 to 1; whatever the D line logic level is, the Q output level becomes.

This is because the D-type flip-flop sets its Q output to whatever the D input logic level is at the time of the clock 0-to-1 transition.

In the timing waveform, the D input is held high over two clock periods. This means that the Q output will also be held high over the same two clock periods.

Note the point on the timing waveform when the D input makes a transition to logic 1 for a brief period (between clock pulses). The flip-flop is unable to see this transition of the D input, so the flip-flop is unable to respond.

Note: the flip-flop can only update its Q output at the 0-to-1 transition of the clock input.

Go to Frame 3.10.

Frame 3.10

Having covered the basics of the D flip-flop, consider the state diagram in Figure 3.8.

It is, of course, the single-pulse generator with memory FSM seen in Frame 1.13. This will be synthesized using D-type flip-flops.

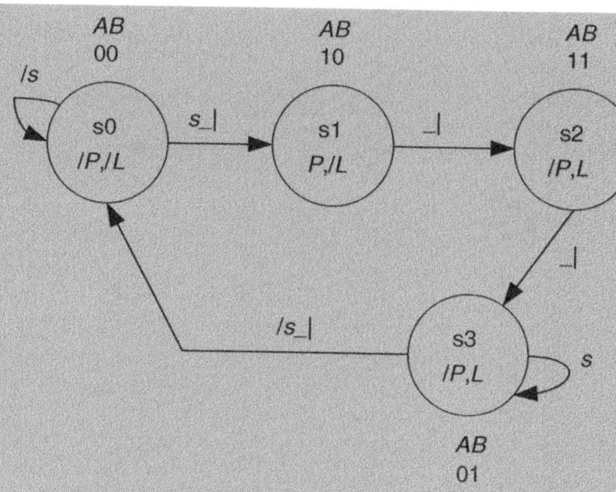

Figure 3.8 State diagram for implementing using *D* flip-flops.

The equation for flip-flop *A* is

$$A \cdot D = \text{s0} \cdot s + \text{s1} = /A/B \cdot s + A \cdot /B = /B \cdot s + A \cdot /B \text{ (using Aux rule)}.$$

Note, the *D* line of flip-flop *A* needs to be set in state s0 and held set over state s1.

Now consider the equation for flip-flop *B*:

$$B \cdot D = \text{s1} + \text{s2} + \text{s3} \cdot s = A \cdot /B + A \cdot B + /A \cdot B \cdot s = A + /A \cdot B \cdot s = A + B \cdot s.$$

The first term sets the *D* line high in state s1, whilst the second term holds the *D* line high over state s2. But what is happening in the third term?

In state s3 the *D* line needs to be held high if the input s is *not* logic 1, since when $s = 0$ the FSM should return to state s0 (by resetting flip-flop *B*). Therefore, whilst $s = 1$, the third term in the equation for $B \cdot D$ will be high. When $s = 0$, this term will become logic 0 and the *B* flip-flop will reset, causing the FSM to move to state s0.

Negate the input term (*s* in this case) with s3 to hold the *D* input of the flip-flop high.

Rule 1 Whenever there is a 1-to-0 transition with an input term present along a transitional line of the state diagram, then AND the state with the *negated* input.

Turn to Frame 3.11.

Frame 3.11

Now consider the state diagram shown in Figure 3.9.

This is just a modification of the single-pulse generator FSM which allows the FSM to produce multiple pulses if input $k = 1$ and $m = 1$, and multiple pulses every four clock cycles if $k = 1$ *and* $m = 0$.

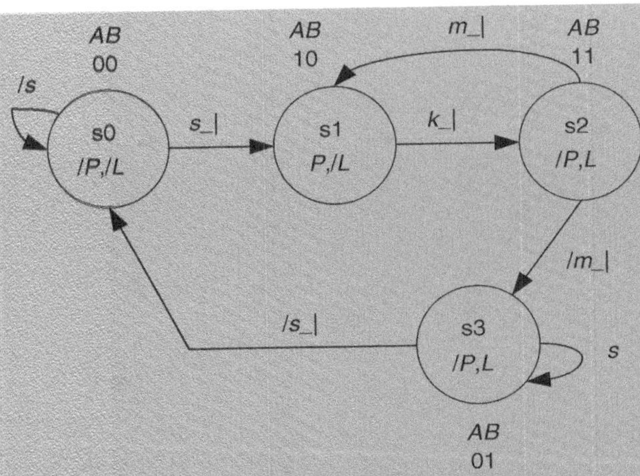

Figure 3.9 State diagram with two-way branch.

The equation for the *A* flip-flop is

$$A \cdot D = \text{s0} \cdot s + \text{s1} + \text{s2} \cdot m.$$

The first term is to set the $A \cdot D$ input high for the next clock pulse to set the flip-flop and cause the FSM to move into state s1.

The second term is to hold the flip-flop set between states s1 and s2. Note that the input term along the transitional line between state s1 and s2 (*k*) is not present in the second term. This is because it is not needed. The flip-flop is to remain set regardless of the state of the input *k*.

Rule 2 A term denoting a transition between two states where the flip-flop remains set does not need to include the input term.

The third term in the equation for $A \cdot D$ is a bit more complicated. This term is a holding term for the two-way branch state s2. In state s2, the transitional path between states s2 and s3 is a 1-to-0 transition. Therefore, apply Rule 1 (defined in Frame 3.10). The term is therefore $\text{s2} \cdot m$. The other path, between states s2 and s1, is a 1-to-1 transition. *Note*: the term denoted by the s2 to s1 transition is not present in the equation for $A \cdot D$. This is because it is not required. The third rule is as follows.

Rule 3 A two-way branch in which one path is a 1-to-0 transition and the other a 1-to-1 transition will always produce a term involving the state and the 1-to-0 transition with the input along the 1-to-0 transitional line negated. *The 1-to-1 transitional path will be ignored.*

Go to Frame 3.12.

Frame 3.12

To see why these three rules apply, look at the state diagram of Frame 3.11 again, which is reproduced in Figure 3.10 for convenience.

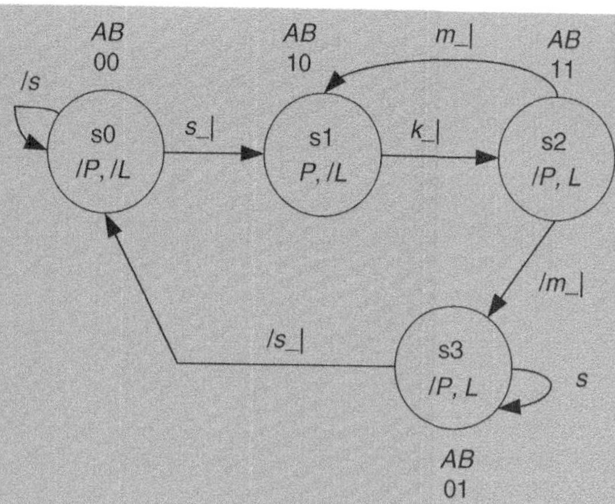

Figure 3.10 State diagram with two-way branch.

Rule 1 Whenever there is a 1-to-0 transition with an input term present along a transitional line of the state diagram, the state is ANDed with the *negated* input.

A 1-to-0 transition with an input along the transitional line connecting the two states needs to be ANDed with the *negated* input condition along the transitional line in order to hold the flip-flop set until the input condition along the transitional line becomes true.

A 1-to-0 transition *without* an input along the transitional line connecting the two states *does not* need to be included in the equation, since the FSM will always be able to follow the transition and the flip-flop will always be able to reset.

Rule 2 A term denoting a transition between two states where the flip-flop remains set does not need to include the input term.

In the above state diagram the transition between state s1 and s2 for flip-flop A is $\text{s1} \cdot k + \text{s1} \cdot /k$, i.e. it does not matter what state the input k is since A is to remain 1 regardless of whether it is in state s1 or s2. Therefore, $\text{s1} \cdot k + \text{s1} \cdot /k = \text{s1}$ by Boolean logical adjacency rule.

Rule 3 A two-way branch in which one path is a 1-to-0 transition and the other a 1-to-1 transition will always produce a term involving the state and the 1-to-0 transition with the input along the 1-to-0 transitional line negated. The 1-to-1 transitional path will be ignored.

In the above state diagram the two-way branch involves a 1-to-0 transition and a 1-to-1 transition. In state s2, flip-flop A will remain set if $m = 1$ and must reset if $m = 0$.

Go to Frame 3.13.

Frame 3.13

The diagram in Figure 3.11 illustrates all possible conditions for a two-way branch in a state diagram.

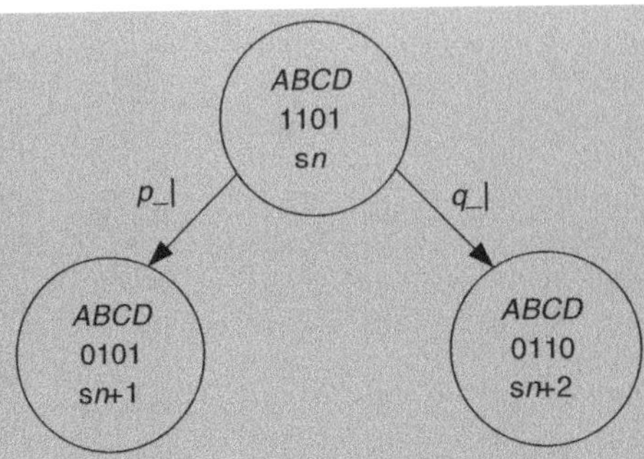

Da = sn·(/p·/q)
Db = sn
Dc = sn·q
Dd = sn·/q

Figure 3.11 State diagram segment showing different conditions in a two-way branch.

In particular, note the condition for flip-flop *A* in a two-way branch with both transitions being 1 to 0. The term $DA = sn \cdot (/p \cdot /q)$ implies that the negation of the input term along each transitional line is being used. Only if both $p = 0$ *and* $q = 0$ will the FSM remain in state *sn*.

Study the diagram carefully and then go to Frame 3.14.

Frame 3.14

Look at the state diagram segments shown in Figure 3.12.

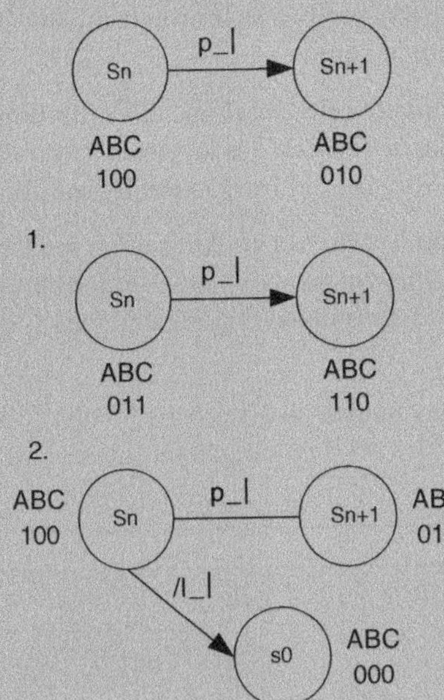

A·d = Sn·/p

B·d = Sn·p

C·d = no term required

A·d =

B·d =

C·d =

A·d =

B·d =

C·d =

Figure 3.12 Some two-way branch examples for the reader.

Task Complete the two sets of D flip-flop equations.

When these have been completed, go to Frame 3.15.

Frame 3.15

The answers to the two problems in Frame 3.14 are as follows:

1. $A \cdot D = sn \cdot p$
   $B \cdot D = sn$
   $C \cdot D = sn \cdot /p.$
2. $A \cdot D = sn \cdot /p \cdot l$, since both $p = 0$ and $l = 1$ are needed to stay in sn
   $B \cdot D = sn \cdot p$, since there is a 0-to-1 transition between sn and $sn + 1$
   $C \cdot D$, no term required.

Refer to Frames 3.8–3.14 for the method if required. Now try the following problem.

Task The FSM illustrated in Figure 3.13, which is to be synthesized with D-type flip-flops, has two states with two-way branches. Produce the equations for the two D flip-flops, as well as the output equation for X.

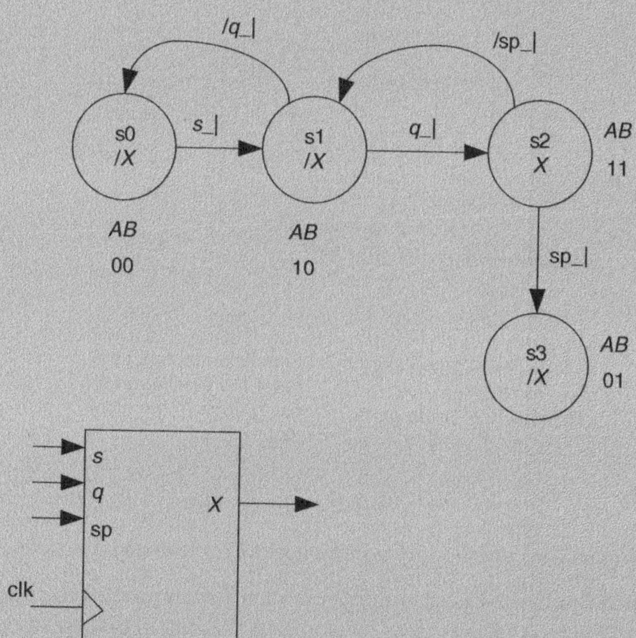

Figure 3.13 An example with multiple two-way branches.

Go to Frame 3.16 after completing this example.

Frame 3.16

The solution to the problem in Frame 3.15 is

$$A \cdot D = \mathrm{s0} \cdot s + \mathrm{s1} \cdot q + \mathrm{s2} \cdot /\mathrm{sp}.$$

Since s0 · *s* is a set term, s1 · *q* is the 1-to-0 transition between s1 and s0, and s2 · /sp is the 1-to-0 transition between s2 and s3:

$$B \cdot D = \mathrm{s1} \cdot q + \mathrm{s2} \cdot \mathrm{sp} + \mathrm{s3}.$$

Since s1 · *q* is the set term, s2 · sp is the 1-to-0 transition holding term between s2 and s1, and s3 is the holding term in state s3. Note that there is no way of leaving state s3. The output equation is $X = \mathrm{s2}$, which makes the FSM a Moore FSM because the output is a function of the secondary state variables.

To provide a way out of s3, and to provide an initialization mechanism, it is wise to provide a reset input to *all* FSMs. In any case, one should *always* provide a means of initializing the FSM to a known state.

Resetting the flip-flops

If the flip-flops have asynchronous reset inputs (see Figure 3.14), then this is easily accomplished by a common connection to all reset inputs.

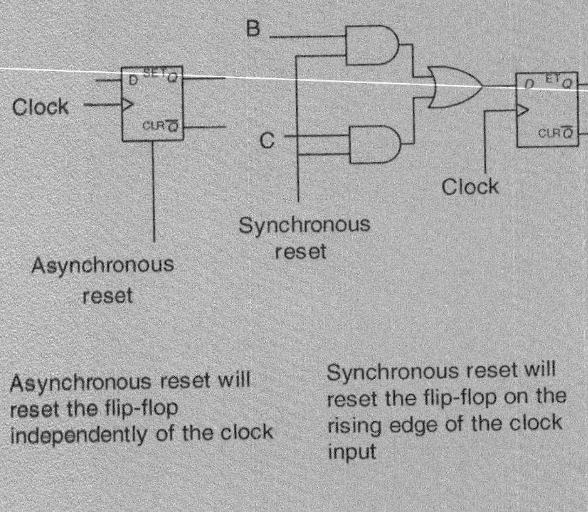

Figure 3.14 Circuit diagrams showing asynchronous and synchronous resetting of a *D* flip-flop.

If the flip-flops do not have an asynchronous reset input (or any reset input), then a synchronous reset can be provided by ANDing a reset input to each *D* input. In the case of the synchronous input, the reset line (which is active-low) is normally held high; this enables the logic for each flip-flop *D* input. Lowering the reset line disables the AND gates and results in the *D* inputs also going low. The next clock pulse, therefore, will cause the flip-flops to reset. Note that the flip-flops will reset on the rising edge of the clock pulse for positive-edge-triggered flip-flops.

Go to Frame 3.17.

Frame 3.17

Task Try producing the *D* flip-flop equations and the output equations for each of the following state diagrams. If you are not too sure, then reread Frames 3.8 to 3.16 again before starting to do the problems.

State diagram in Frame 1.19, Figure 1.22, using the following secondary state variables:

	ABC
s0	000
s1	100
s2	110
s3	011
s4	001

State diagram in Frame 2.3, Figure 2.4, using the following secondary state variables:

	AB
s0	00
s1	10
s2	11
s3	01

State diagram in Frame 2.12, Figure 2.13, using the following secondary state variables:

	ABC
s0	000
s1	100
s2	110
s3	111
s4	011

State diagram in Frame 2.9, Figure 2.10, using the following secondary state variables:

	ABCD
s0	0000
s1	1000
s2	1100
s3	1110
s4	1111

s5	0111
s6	0011
s7	1011
s8	1001
s9	0001

See Frame 3.18 for the solution to these problems.

Frame 3.18

The solutions to the problems in Frame 3.17 are as follows.

State diagram in Frame 1.19, Figure 1.22:

	ABC	Answer
s0	000	$A \cdot D = s0 \cdot s + s1 = /A/B/C \cdot s + A/B/C = /B/C \cdot s + A/B/C$
s1	100	$B \cdot D = s1 + s2 = A/C$
s2	110	$C \cdot D = s2 + s3 + s4 \cdot s = AB/C + /ABC + /A/BC \cdot s = AB/C + /ABC + /ACs$
s3	011	
s4	001	$P = s1 + s3 \cdot x = A/B/C + /ABC \cdot x$ with x input

State diagram in Frame 2.3, Figure 2.4:

	AB	Answer
s0	00	$A \cdot D = s0 \cdot st + s1 + s2 \cdot to = /B \cdot st + A/B + A \cdot to$
s1	10	$B \cdot D = s1 + s2 + s3 \cdot st = A + B \cdot st$
s2	11	
s3	01	$P = s1 + s2 = A$
		$TS = /s1 = /(A/B)$ active-low output

State diagram in Frame 2.12, Figure 2.13:

	ABC	Answer
s0	000	$A \cdot D = s0 \cdot d + s1 + s2$
		$= /A/B/C \cdot d + A/B/C + AB/C$
s1	100	$= /B/C \cdot d + A/B/C + AB/C$
		$= /B/C \cdot d + A \cdot /C$
s2	110	$B \cdot D = s1 \cdot /d + s2 + s3 + s4 \cdot d$
s3	111	$= A/B/C \cdot /d + AB/C + ABC + /ABC \cdot d$
		$= A/C \cdot /d + AB + BC \cdot d$

s4	011	$C \cdot D = s2 \cdot d + s3 + s4 \cdot d$
		$= AB/C \cdot d + ABC + /ABC \cdot d$
		$= AB \cdot d + ABC + BC \cdot d$
		$P = s3 = A \cdot B \cdot C$

State diagram in Frame 2.9, Figure 2.10:

	ABCD	Answer
s0	0000	$A \cdot D = s0 \cdot \text{int} + s1 + s2 + s3 + s6 + s7 + s8 \cdot /f$
s1	1000	$= /B/C/D \cdot \text{int} + A/C/D + AB/D + /BCD + A/BD \cdot /f$
s2	1100	
s3	1110	$B \cdot D = s1 + s2 + s3 + s4$
s4	1111	$= A \cdot /B/C/D + A/C/D \cdot E + ABC$
		$= A/C/D + ABCs$
s5	0111	$C \cdot D = s2 \cdot \text{eoc} + s3 + s4 + s5 + s6$
		$= AB/D \cdot \text{eoc} + ABC + BCD + /ACD$
s6	0011	
s7	1011	$D \cdot D = s3 \cdot /\text{eoc} + s4 + s5 + s6 + s7 + s8 \cdot f + s9 \cdot \text{int}$
s8	1001	$= ABC \cdot /\text{eoc} + CD + A/BD \cdot f + /A/BD \cdot \text{int}$
s9	0001	
		$RC = /s0 = /(/A/B/C/D)$
		$S/H = s1 + s2 = A/C/D$
		$SC = s2 = AB/C/D$
		$CS = /(s3 + s4 + s5) = /(ABC + BCD)$
		$W = /s4 = /(ABCD)$
		$CC = s7 = A/BCD$

Note: active-low outputs are shown here with right-hand side negated.

Task Once these have been completed, try taking the single-pulse generator example of Frame 3.10, Figure 3.8, and produce the D-type flip-flop equations and output equations. Finally, produce a circuit diagram for the FSM using *D*-type flip-flops with asynchronous reset inputs and other logic gates. When complete, go to Frame 3.19.

Frame 3.19

The complete design for the single-pulse generator with memory is given below;

The design equations

$$A \cdot D = s0 \cdot s + s1 = A/B + /B \cdot s$$

$$B \cdot D = s1 + s2 + s3 \cdot s = A + B \cdot s$$

$$P = s1 = A/B \text{ and } L = B.$$

The circuit diagram of Figure 3.15 shows the memory elements (flip-flops), input decoding logic ($A \cdot D$ and $B \cdot D$ logic), and output decoding logic (for the output P).

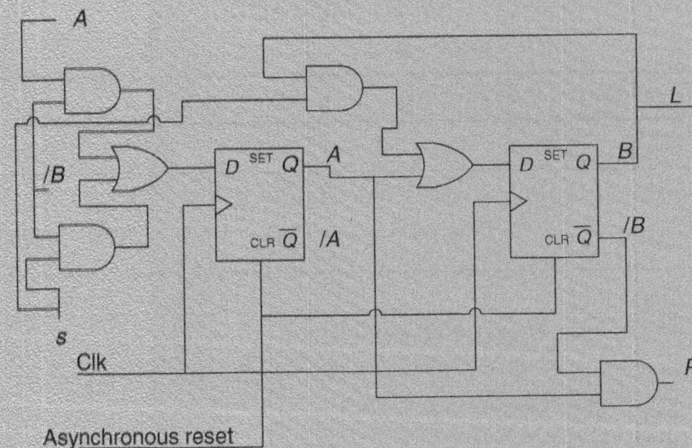

Figure 3.15 Circuit for the single-pulse generator with memory using an asynchronous reset.

If flip-flops with asynchronous reset inputs are not available, then a synchronous reset can be used, ANDed with the $A \cdot D$ and $B \cdot D$ logic, as illustrated in Figure 3.16.

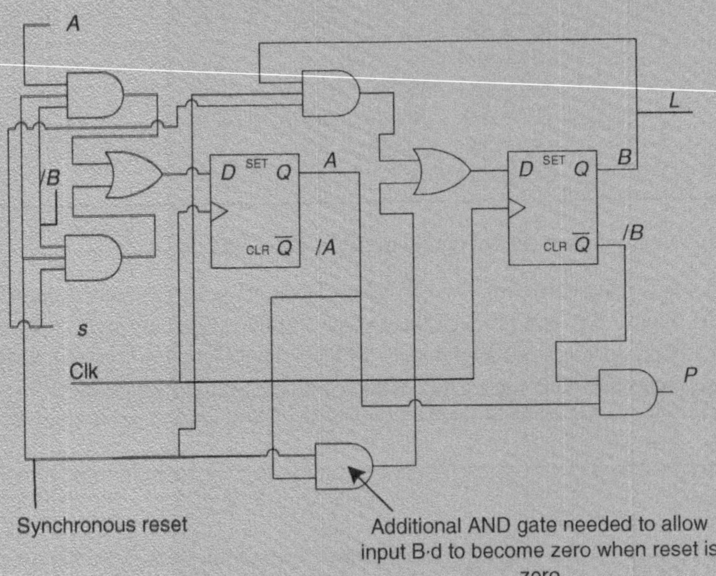

Figure 3.16 Circuit for the single-pulse generator with memory using a synchronous reset.

In this illustration, the reset line is connected to the AND logic of each D input. Note the addition of the extra AND gate to the input logic of $B \cdot D$ so that when reset $=0, B \cdot D=0$ also.

Go to Frame 3.20.

Frame 3.20

So now, all aspects of designing FSMs have been covered: from initial specification, to construction of the state diagram, to synthesizing the circuit used to implement the FSM. A run through the complete design process will now be undertaken. Consider these steps for the single-pulse generator FSM.

The Specification

The block diagram showing inputs and outputs is first constructed (Figure 3.17). This would be supplemented with a written specification describing the required behaviour of the FSM.

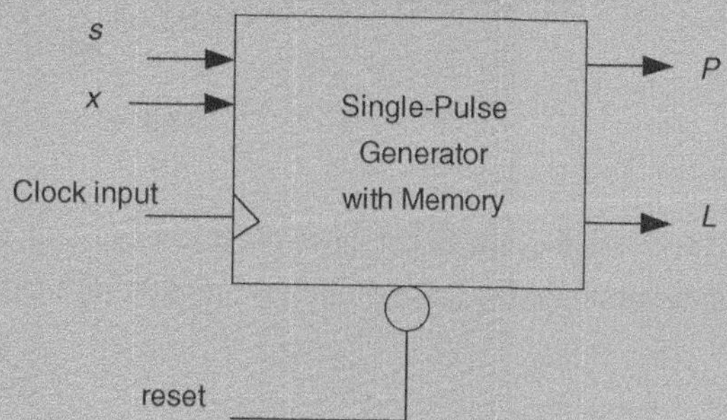

Figure 3.17 Block diagram for the single-pulse generator with memory.

'The FSM is to produce a single pulse at its output P whenever the input s goes high. No other pulse should be produced at the output until s has gone low, then high again. In addition, an output L is to indicate that the P pulse has taken place, to be cancelled when s goes low. The L output can be disabled by asserting input x to logic 1.'

The next step is to produce the state diagram. This is not a trivial step, since it requires the use of a number of techniques developed during this programme of work. This is the skilled part of the development process.

Go to Frame 3.21.

Frame 3.21

The state diagram is shown in Figure 3.18.

Now assign secondary state variables to the state diagram in order to continue with the synthesis of the FSM. Then, develop the equations for the flip-flops next state decoder, and output logic.

The design equations

$$A \cdot D = (\mathrm{s0} \cdot s + \mathrm{s1}) \cdot \text{reset} = (A/B + /B \cdot s) \cdot \text{reset}$$
$$B \cdot D = (\mathrm{s1} + \mathrm{s2} + \mathrm{s3} \cdot s) \cdot \text{reset} = (A + B \cdot s) \cdot \text{reset}$$
$$P = \mathrm{s1} = A/B$$
$$L = \mathrm{s2} \cdot x + \mathrm{s3} \cdot x = B \cdot x.$$

Finally, the circuit is produced from the equations (Figure 3.19). Note that output L is a Mealy output because it used the input x.

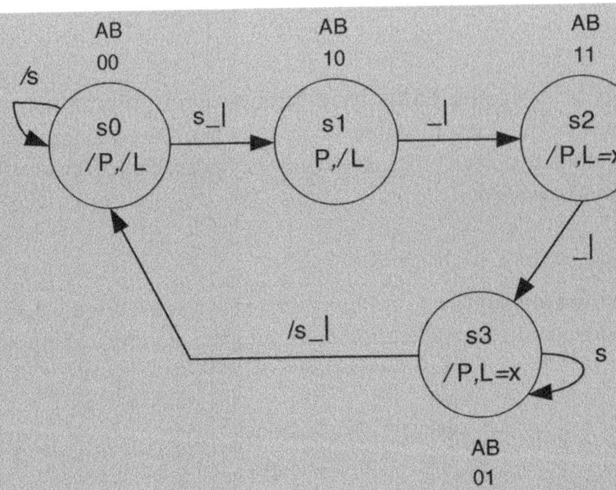

Figure 3.18 State diagram for the single-pulse generator with memory.

The design can then be simulated to ensure that it is functioning according to the original specification.

Simulation

Note here that the output *L* is conditional upon input *x*, so that it can only be logic 1 in states s2 and s3, and then only if input *x* is logic 1 also. This is illustrated in the waveforms in Figure 3.20.

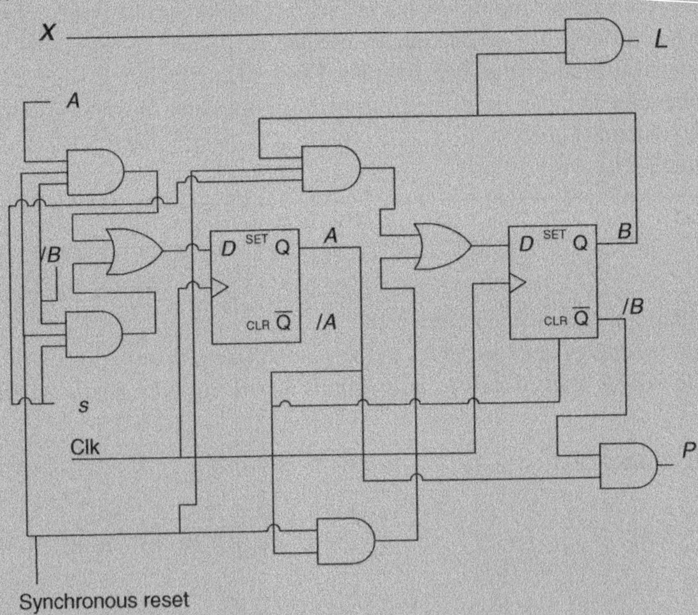

Figure 3.19 Circuit diagram of the single-pulse generator with memory.

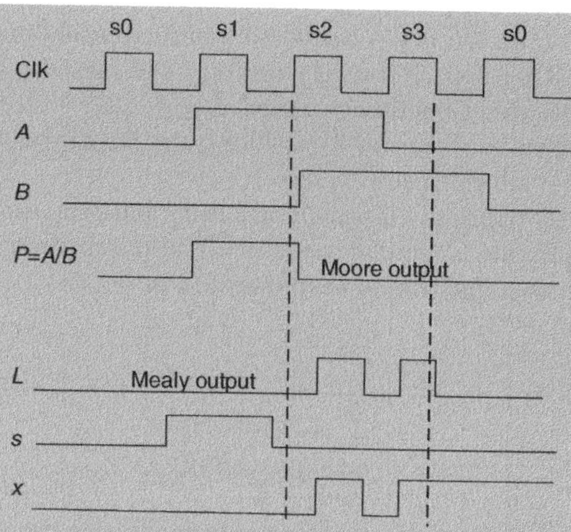

Figure 3.20 Timing diagram for the single-pulse generator with memory.

Go to Frame 3.22.

Frame 3.22

In some cases there is a need to use three-way (or more) branches. This has been avoided up until now, but the rules can be used to resolve all pathways. However, *each path must be mutually exclusive.*

Consider the diagram in Figure 3.21.

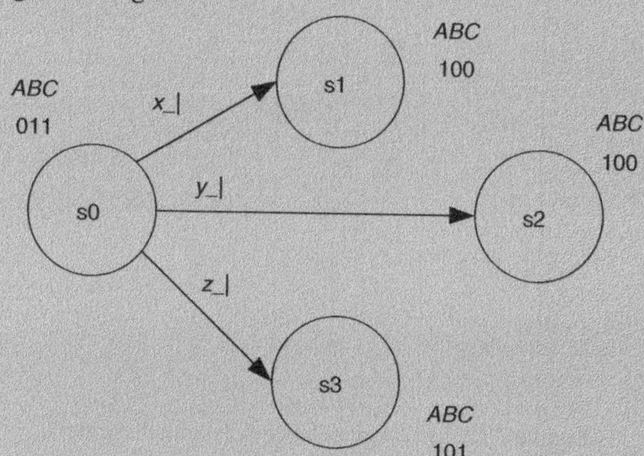

A·d = s0(*x*+*y*+*z*) since any path takes *A* to 1.

B·d = s0·(/x· /y · /z)

C·d = s0·(/x·/y) no need to include *z*

Figure 3.21 State diagram segment with three-way branch.

Here, the input $A \cdot d$ for flip-flop A has 0-to-1 transition in all three paths. To meet the requirements for the D flip-flop, all leaving terms (x, y, and z) need to be logically ORed to provide a transition when any input becomes active.

In the case of $B \cdot d$ there are three 1-to-0 transition paths; this can be dealt with by using the 1-to-0 negation rule for all three paths, as shown.

In the case of $C \cdot d$ there are two 1-to-0 transitions and one 1-to-1 transition. In this case the 1-to-0 negate rule is applied to the two 1-to-0 transition paths, both ANDed because they both have to be true to keep the FSM in s0. The 1-to-1 transition is, as usual, ignored.

Go to Frame 3.23.

Frame 3.23

Task Consider the state diagram fragment in Figure 3.22.
Complete the equations for $A \cdot d$, $B \cdot d$, and $C \cdot d$.

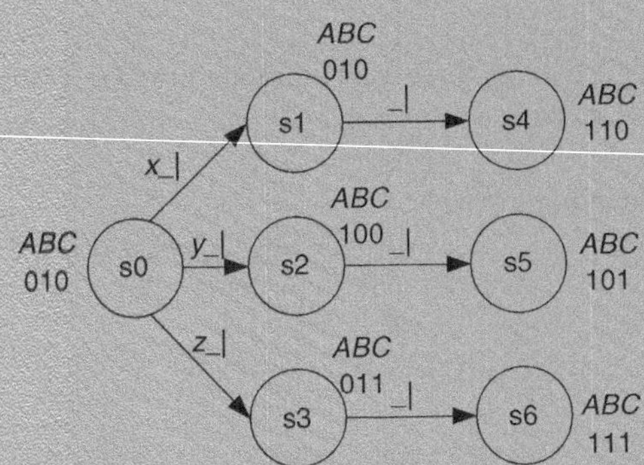

$A \cdot d =$

$B \cdot d =$

$C \cdot d =$

Figure 3.22 Incomplete three-way branch example.

When completed, go to Frame 3.24.

Frame 3.24

The three equations are illustrated in Figure 3.23.

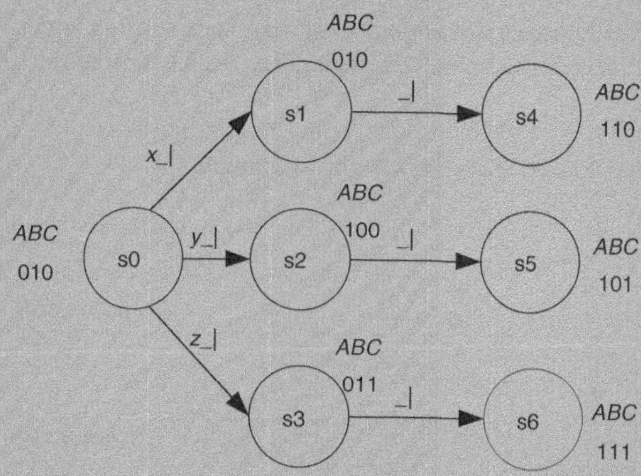

$$A \cdot d = s0 \cdot y + s1 + s2 + s3 + s4 + s5 + s6.$$

$$B \cdot d = s0 \cdot /y + s1 + s3 + s4 + s6.$$

$$C \cdot d = s0 \cdot z + s2 + s3 + s5 + s6.$$

Figure 3.23 Solution to the three-way branch example.

In the equation for $B \cdot D$, the term $s0 \cdot /y$ is keeping the FSM in state s0. In the equation for $C \cdot D$, the term $s0 \cdot z$ will hold the FSM in state s0 until $z = 1$. So the rules for D flip-flops developed earlier still apply.

Go to Frame 3.25.

Frame 3.25 Recap on how to deal with multistate Moore active-low outputs

In some state diagram designs there is a need to write an output equation in its 'active-low' form rather than in its 'active-high' form. This is particularly true when controlling memory devices, where the chip select line from the FSM to the memory device is often active-low. If this signal was dealt with as an active-high signal, then all states where the chip select line was not active would have to be written into the equation for chip select (CS).

The illustration in Figure 3.24 shows a typical example.

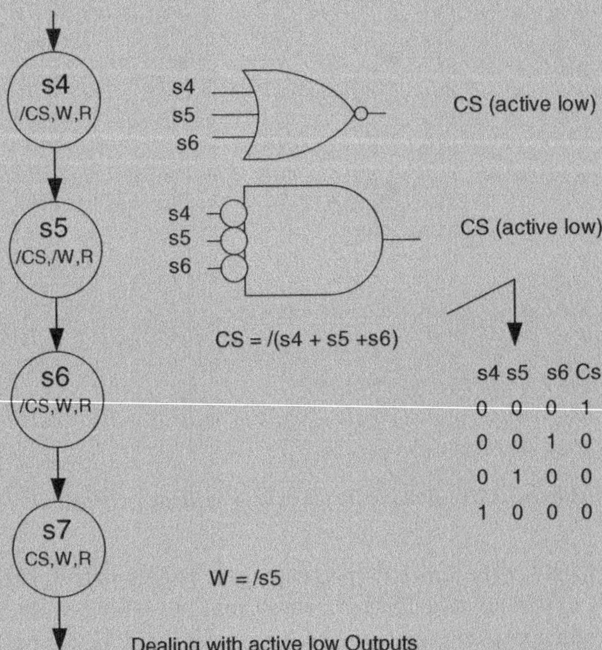

Figure 3.24 Dealing with active-low inputs.

In this example, CS is logic 0 (active) in states s4, s5 and s6, but high again in state s7.

The three states s4, s5 and s6 are all ORed and the whole OR expression inverted (NOR). This can, if preferred, be written either in the NOR form, or, by applying De Morgan's rule in the form of an AND gate with all inputs inverted.

Go to Frame 3.26.

Frame 3.26

Now consider the situation when an output is to be active-low, but only in a particular state, and then only if a particular input is at a certain logic level (Mealy active-low output). How can this be represented in a state diagram? Figure 3.25 illustrates how.

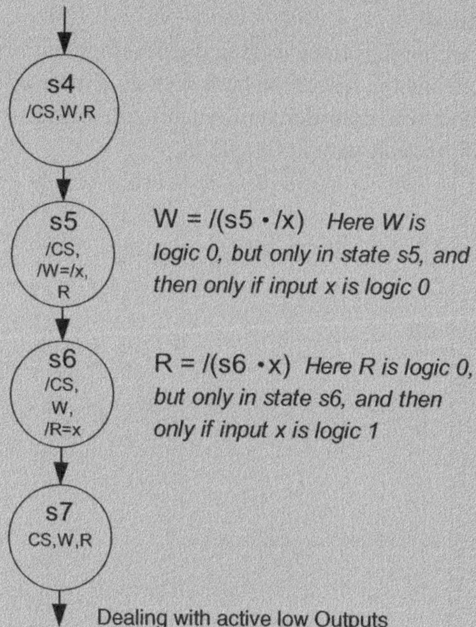

Figure 3.25 Dealing with active-low outputs.

In state s5, the output W is represented by

$$/W = /x.$$

This implies that, in state s5, W is to be logic 0, but only in state s5, and only if input x is logic 0.

When the equation for W is written, it also needs to contain the state s5 as

$$W = /(\mathrm{s5} \cdot /x).$$

Note the whole of the right-hand side of the equation is inverted to provide the active-low output.

In a similar manner, in state s6 the output R is represented as

$$/R = x,$$

indicating that, in state s6, output R is to be logic 0, but only if input x is logic 1. The equation is written as

$$R = /(\mathrm{s6} \cdot x).$$

Here, as with the W signal, the whole right-hand side of the equation is also inverted.

3.3 SUMMARY

This chapter has looked at the method of synthesizing a logic circuit from the state diagram. Methods have been developed to make this process simple and effective for implementation using both *T*-type flip-flops and *D*-type flip-flops. These methods are used in the development of further examples in Chapter 4.

At this point, the main techniques to be used in the development of synchronous design of FSMs have been completed and the rest of the book follows a more traditional format.

There is one more method to be considered in synchronous design, namely that of the 'One Hot' technique, which will be dealt with in Chapter 5.

4

Synchronous Finite-State Machine Designs

This chapter looks at a number of practical designs using the techniques developed in Chapters 1 to 3. It compares the conventional design of FSMs with the design proposed in the book. This illustrates how more effective the latter method is in developing a given design. The traditional method of designing FSMs is common in a lot of textbooks on digital design. It makes use of transition tables and can become cumbersome to use when dealing with designs having a large number of inputs. Even for designs having few inputs, the method used in Chapters 1–3 is quicker and easier to use.

Most designers involved in the development of FSMs make use of unused secondary state assignments to help reduce the flip-flop input and output equations. This practice is investigated with some interesting results.

The chapter covers a number of practical system designs. Some have simulation waveforms showing the FSM design working. The Verilog HDL code used to create the simulations will not be shown, as Verilog HDL code development is not covered until later on in the book. However, the respective Verilog codes are available on the CDROM disk that is included with this book, as are the Verilog tools used to view the simulations.

Eight examples are discussed in this chapter, with each example introducing techniques that help to solve the particular requirements in the design being investigated.

4.1 TRADITIONAL STATE DIAGRAM SYNTHESIS METHOD

Before continuing with the development of FSM systems based on the synthesization method covered in Chapters 1–3, it is worth investigating the more popular traditional method of synthesization used by many system designers. Then see what solutions are obtained by using both methods. It should be possible to obtain the same results, or at least results that are of a similar level of complexity (i.e. number of gates).

Consider the state diagram shown in Figure 4.1. This, being a four-state diagram, will need two *D*-type flip-flops. Using the traditional synthesization method, begin by constructing a state table containing the present state (PS) values and the next state (NS)

FSM based Digital Design using Verilog HDL Peter Minns and Ian Elliott

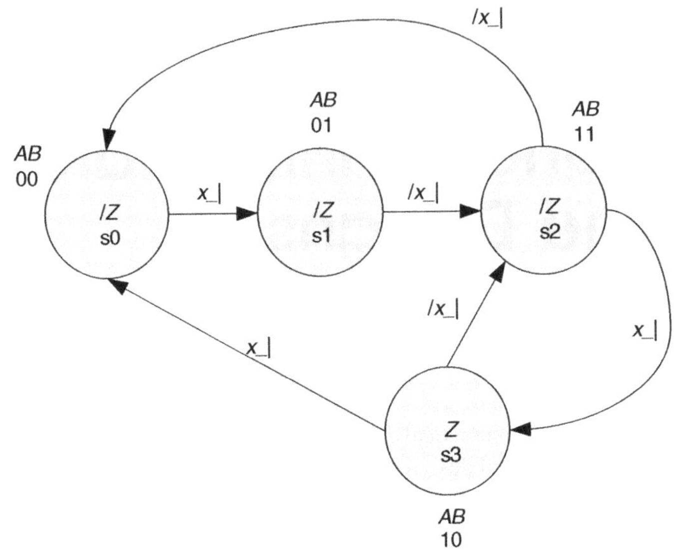

Figure 4.1 A state diagram used in the comparison.

values for *A* and *B*, for all possible values of the input *x*. One then adds to this the next states for the inputs Da and Db, for all possible values of *x*. The result is the state table shown in Table 4.1.

The values for *A* and *B* in Table 4.1 are obtained by inspection of the state diagram in Figure 4.1. For example, in state s0 (PS of $AB = 00$) in col1 the NS of *AB* for $x = 0$ will be 00 in col2; however, if $x = 1$, the NS of $AB = 01$ in col3 (i.e. s1).

The values for the NS Da and Db values will follow the NS values for *AB* because in a *D* flip flop the output of the flip flop (*A*, *B*) follows the Da and Db inputs.

The reader can follow the rest of the rows in Table 4.1 to complete the state table.

Table 4.1 Present state–next state table for the state machine.

	col1	col2	col3	col4	col5
	PS *AB*	NS *AB* $x = 0$	NS *AB* $x = 1$	NS DaDb $x = 0$	NS DaDb $x = 1$
Row1	00	00	01	00	01
Row2	01	11	01	11	01
Row3	11	00	10	00	10
Row4	10	11	00	11	00

The next step is to obtain the Da and Db equations from the state table by writing down the product terms where $Da = 1$ in both columns $x = 0$ and $x = 1$.

Consider, for example, $Da = 1$ when A changes 0 to 1; look for PS $A = 0$ to NS $A = 1$ in row 2, and PS $A = 1$ to NS $A = 1$ in row 3 of columns 1, 3 ($x = 1$):

- when PS $AB = 01$ (row 2) and $x = 0$, flip-flop A should set, and the product term $/AB/x$ is required;
- when PS $AB = 01$ and $x = 1$ (row 2, col3), flip-flop A should be reset, and the term $/ABx$ is *not* required;
- when PS $AB = 10$ (row 4) and $x = 0$, flip-flop A should set, and the term $A/B/x$ is required;
- when PS $AB = 11$ (row 3) and $x = 1$, flip-flop A should be set, and term ABx is required.

Therefore, the D input terms for Da are

$$\mathrm{D{\cdot}a} = /AB{\cdot}/x + A/B{\cdot}/x + AB{\cdot}x,$$

which cannot be reduced. For $\mathrm{D{\cdot}b} = /A/B{\cdot}x + /AB{\cdot}/x + /AB{\cdot}x + A/B{\cdot}/x$ we have

$$\mathrm{D{\cdot}b} = /A{\cdot}x + /AB + A/B{\cdot}/x.$$

The output equation for $Z = \mathrm{s3} = A/B$, since this is a Moore state machine.

Now do the problem using the synthesization method described in Chapters 1–3.

From the state diagram directly:

$$\begin{aligned}
\mathrm{Da} &= \mathrm{s1}{\cdot}/x + \mathrm{s2}{\cdot}\,x + \mathrm{s3}{\cdot}/x \\
&= /AB{\cdot}/x + AB{\cdot}x + A/B{\cdot}/x \\
\mathrm{Db} &= \mathrm{s0}{\cdot}\,x + \mathrm{s1} + \mathrm{s3}{\cdot}/x \\
&= /A/B{\cdot}\,x + /AB + A/B \\
&= /A{\cdot}\,x + /AB + A/B{\cdot}/x.
\end{aligned}$$

This is the same as obtained using the traditional method.

The main advantage of the method used in Chapters 1–3, over the traditional method, is that it does not require the use of the state table. It is also much easier to use when the number of input variables is large (as is the case in large practical FSM designs) since the size of the present state–next state table increases as more inputs are added.

4.2 DEALING WITH UNUSED STATES

When developing state diagrams that use less than the 2^n states for n secondary state variables the question of what to do with the unused states arises. Consider the state diagram of Figure 4.2.

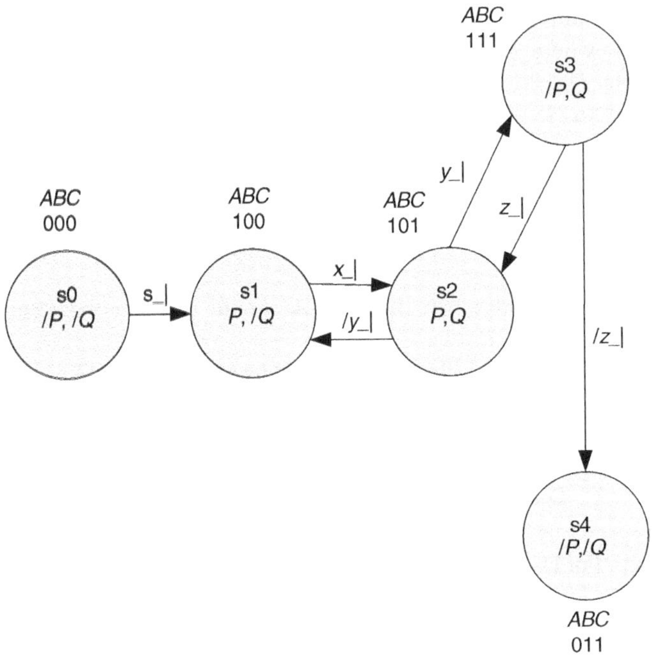

Figure 4.2 A state diagram using less than the 2^3 states.

From the state assignment used in this example there are

Used states	Unused states
s0 = 000	s5 = 010
s1 = 100	s6 = 110
s2 = 101	s7 = 001
s3 = 111	
s4 = 011	

The equations for D flip-flops are:

$$A \cdot d = \mathrm{s0} \cdot s + \mathrm{s1} + \mathrm{s2} + \mathrm{s3} \cdot z$$
$$= \cancel{/A}/B/C \cdot s + A/B\cancel{/C} + A/B\cancel{/C} + A\cancel{/B}C \cdot z.$$

The crossed-out literals are a result of applying logical adjacency and the aux rule (see Appendix A). The result is

$$A \cdot d = /B/C \cdot s + A/B + AC \cdot z$$
$$B \cdot d = \mathrm{s2} \cdot y + \mathrm{s3} \cdot /z + \mathrm{s4}$$
$$= A/BC \cdot y + \cancel{/A}BC \cdot /z + /ABC$$
$$= A/BC \cdot y + BC \cdot /z + /ABC$$
$$C \cdot d = \mathrm{s1} \cdot x + \mathrm{s2} \cdot y + \mathrm{s3} + \mathrm{s4}$$
$$= A/B/C \cdot x + A\cancel{/B}/C \cdot y + \cancel{/A}BC + \cancel{/A}BC.$$

Again, the crossed-out terms are using logical adjacency and the aux rule.

$$C \cdot d = A/B/C \cdot x + AC \cdot y + BC.$$

The output equations:

$$P = \text{s1} + \text{s2} = A/B/C + A/BC$$
$$P = A/B$$
$$Q = \text{s2} + \text{s3} = A/BC + ABC$$
$$Q = A/BC + ABC = AC.$$

If the state machine falls into the unused state s5 (*/AB/C*) then the result will be

$A \cdot d = 0, B \cdot d = 0, \text{and C} \cdot \text{d} = 0$ the state machine falls into s0.

If the state machine falls into unused state s6 (*AB/C*):

$A \cdot d = 0, B \cdot d = 0, \text{and } C \cdot d = 0$ again, the state machine will fall into s0.

If state machine falls into the unused state s7 (*/A/BC*):

$A \cdot d = 0, B \cdot d = 0, \text{and } C \cdot d = 0$ with next state being s0 again.

This shows that the FSM designed with *D*-type flip-flops will be self- resetting.

Note that if *T* flip-flops are used, then the FSM will not be self-resetting since the *T* input either toggles with $T = 1$ or remains in its current state with $T = 0$. The only way to ensue that it does return to s0 is to make transitions available for this, as illustrated in Figure 4.3. Clearly, this requires more product terms in the equations for $A \cdot t$, $B \cdot t$, and $C \cdot t$.

In general, if the state machine has a lot of 1-to-1 transitions and few 1-to-0 and 0-to-1 transitions, then *T* flip-flops may need less terms and, hence, a possible deduction in logic.

If the state machine has few 1-to-1 transitions the *D* flip-flop solution may result in fewer terms. However, the self-resetting features of the *D* flip-flop may provide a greater advantage in the overall design.

The rest of this chapter contains a number of practical examples, making use of the techniques developed in the first three chapters.

4.3 DEVELOPMENT OF A HIGH/LOW ALARM INDICATOR SYSTEM

Figure 4.4 illustrates a block diagram for the proposed system. In Figure 4.4, the FSM is used to control an ADC and monitor the converted analogue signal levels until either the low-level limit or the high-level limit is exceeded. The low- and high-level values are set up on the Lo-word/ Hi-word inputs, which could be dual in-line switches. The comparators are standard 8-bit

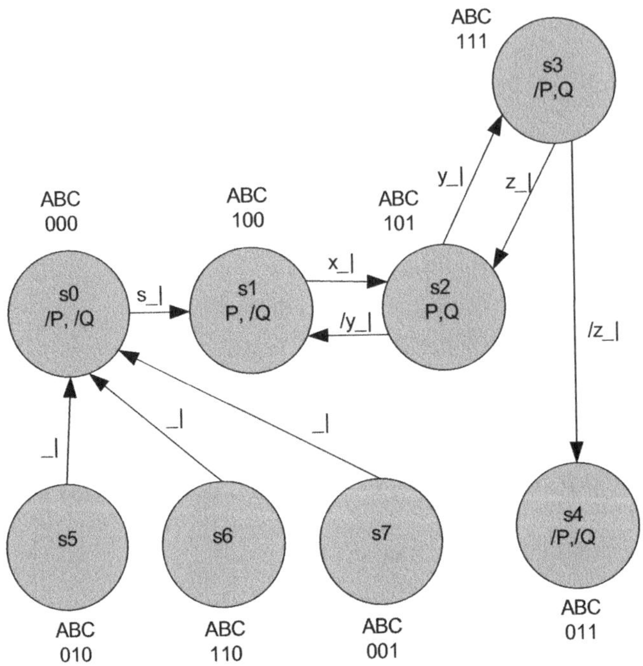

Figure 4.3 The arrangement needed for *T* flip-flops.

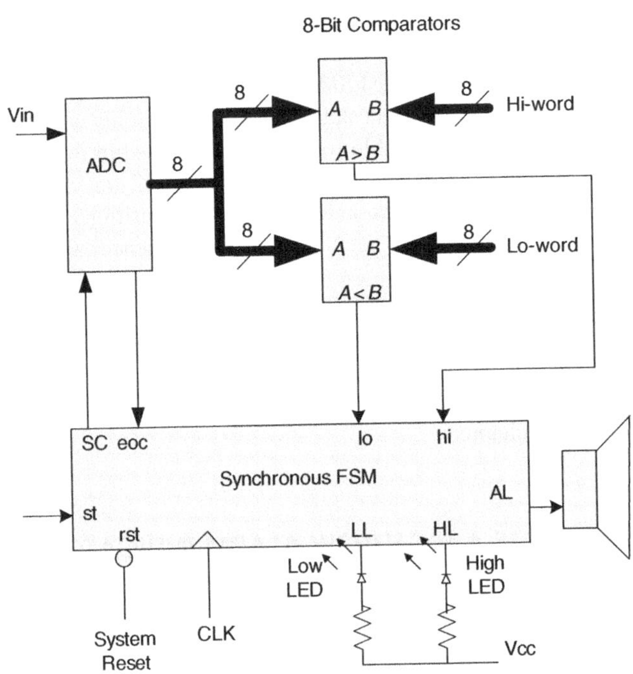

Figure 4.4 Block diagram for the High/Low detector system.

comparator circuits similar to the standard 7485 devices. These could easily be incorporated into a PLD/FPGA along with the FSM.

In this application it is assumed that, when the ADC output *A* exceeds the Hi-word, hi will go to logic 1. An ADC output less than the Lo-word will make lo go to logic 1. The ADC could be a separate device or its digital circuits could be implemented on a PLD/FPGA device and an external R/2R network connected to the chip.

The system is to start when st goes high. It should perform analogue-to-digital conversions at a regular sampling frequency dictated by the system clock and when either the Hi-word or Lo-word are exceeded, turn on the appropriate LED indicator and stop. It can be returned to its initial state by operation of the reset button. Note that in this example the alarm will not sound for an ADC output that is equal either to Hi-word or Lo-word.

From this specification, a state diagram can be developed. The control of the ADC will follow in much the same way that was used in Chapter 2.

The two digital comparators being combinational logic will give an output dependent on the level of the ADC output. When the ADC output is equal to or less than hi-word but greater than Lo-word, then both lo and hi will be low, signifying that the ADC value is between the two limits. When the ADC output is greater than Hi-word, then hi will be logic 1 and is to sound the alarm and turn on the HL indicator. When the ADC output is less than Lo-word, then lo becomes logic 1 and the alarm turns on the LL indicator.

A state diagram has been developed as shown in Figure 4.5. Looking at this state diagram, the system sits in s0 from power on reset and waits for the start input to go high. Then the ADC signal SC is raised to perform an analogue-to-digital conversion. After this the system falls into s2. Here, the outputs from the two comparators are checked, and if either the Hi-word or the Lo-word limit has been exceeded then the state machine will fall into s3. If, however, neither limit has been exceeded, then the state machine will fall back into s1 to perform another analogue-to-digital conversion.

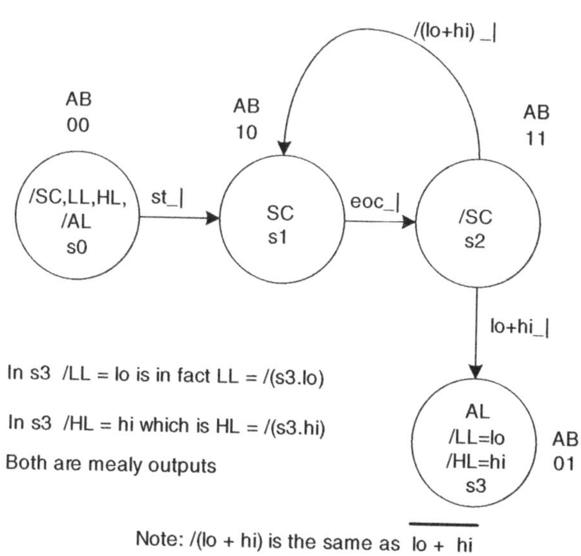

Figure 4.5 A possible state diagram for the problem.

Looking at the two-way branch state s2, it is clear that the inverse of lo + hi is /(lo + hi). As an aside, if one applies De Morgan's rule to /(lo + hi) one gets /lo · /hi, indicating for the transition from s2 to s1 that both lo and hi must be low.

Moving on to look at s3, one can see that the two outputs HL and LL are dictated by the logic state of the comparator outputs lo and hi so that in s3 the HL indicator should be active if hi = 1, whereas the LL indicator should be active if lo = 1.

/HL = hi in s3 indicates that HL must be active low. The output equation for HL will be written as

$$\mathrm{HL} = /(\mathrm{s3} \cdot \mathrm{hi}),$$

which means that HL will be logic 0 when hi = 1, but only when the state machine is in s3. This is defining a Mealy active low output. This is how it was defined in Chapter 3.

In a similar way, LL = $/(\mathrm{s3} \cdot \mathrm{lo})$.

The best way to remember this idea is to think of the /HL = hi equation in the s3 state as representing the equation HL = /(s3 · hi), but then written inside the state circle one does not need to include the s3, as it is implied.

Replacing the state number s3 with its secondary state variable value $AB = 01$, the two Mealy outputs can be written as

$$\mathrm{HL} = /(\mathrm{s3} \cdot \mathrm{hi}) = /(/A \cdot B \cdot \mathrm{hi}) \quad \text{and} \quad \mathrm{LL} = /(\mathrm{s3} \cdot \mathrm{lo}) = /(/A \cdot B \cdot \mathrm{lo}),$$

which results in two three-input NAND gates. Remember, active low signals are inverted (see Chapter 3).

So, from the equation for HL = $/(/A \cdot B \cdot \mathrm{hi})$ it can be seen that, when in state s3, $A = 0$ $(/A = 1)$, $B = 1$, and if hi = 1 then the output of the NAND gate will be zero, which is exactly what is required to light the LED indicator (active low output).

Having gone into some detail to describe the logic behind the Mealy outputs, the next step is to determine the equations for the two flip-flops A and B. Using the method described in Chapter 3 for D flip-flops, these are

$$A \cdot d = \mathrm{s0} \cdot \mathrm{st} + \mathrm{s1} + \mathrm{s2} \cdot /(\mathrm{lo} + \mathrm{hi}) = /A \cdot /B \cdot \mathrm{st} + A \cdot /B + A \cdot B \cdot /\mathrm{hi} \cdot /\mathrm{lo}.$$

The equation for $A \cdot d$ could be simplified using the Auxiliary rule to form

$$A \cdot d = /B \cdot \mathrm{st} + A \cdot /B + A \cdot /\mathrm{lo} \cdot /\mathrm{hi}.$$

Moving on to flip-flop B:

$$B \cdot d = \mathrm{s1} \cdot \mathrm{eoc} + \mathrm{s2} \cdot (\mathrm{lo} + \mathrm{hi}) + \mathrm{s3}. = A \cdot /B \cdot \mathrm{eoc} + A \cdot B \cdot \mathrm{lo} + A \cdot B \cdot \mathrm{hi} + /A \cdot B.$$

Again, using the Auxiliary rule:

$$B \cdot d = A \cdot /B \cdot \mathrm{eoc} + B \cdot \mathrm{lo} + B \cdot \mathrm{hi} + /A \cdot B.$$

The remaining Moore-type outputs are SC = s1 = $A \cdot /B$ *and* AL = s3 = $/AB$.

The next stage would be to develop a Verilog HDL file describing the circuit for the FSM, and comparators. This has been done and is contained on the CDROM in the Chapter 4 folder.

4.3.1 Testing the Finite-State Machine using a Test-Bench Module

In this simulation (Figure 4.6), a test-bench module is added to the Verilog code in order to test the FSM. To do this, test all paths of the state diagram. In the simulation of Figure 4.6 this has been achieved by first following the path s0 → s1 → s2 → s3 with a low limit exceeded and the FSM remains in s3 ($A = 0$, $B = 1$) until a reset (rst = 0) is applied. Then, the sequence is repeated with a Hi limit exceeded, followed by another reset. Finally, the sequence s0 → s1 → s2 → s1 → s2 → s1 → s2 → s1 → s2 → s0 is followed, representing a no limits exceeded until finally another rst = 0 resets the FSM back to s0. Thus, in this way the FSM is tested.

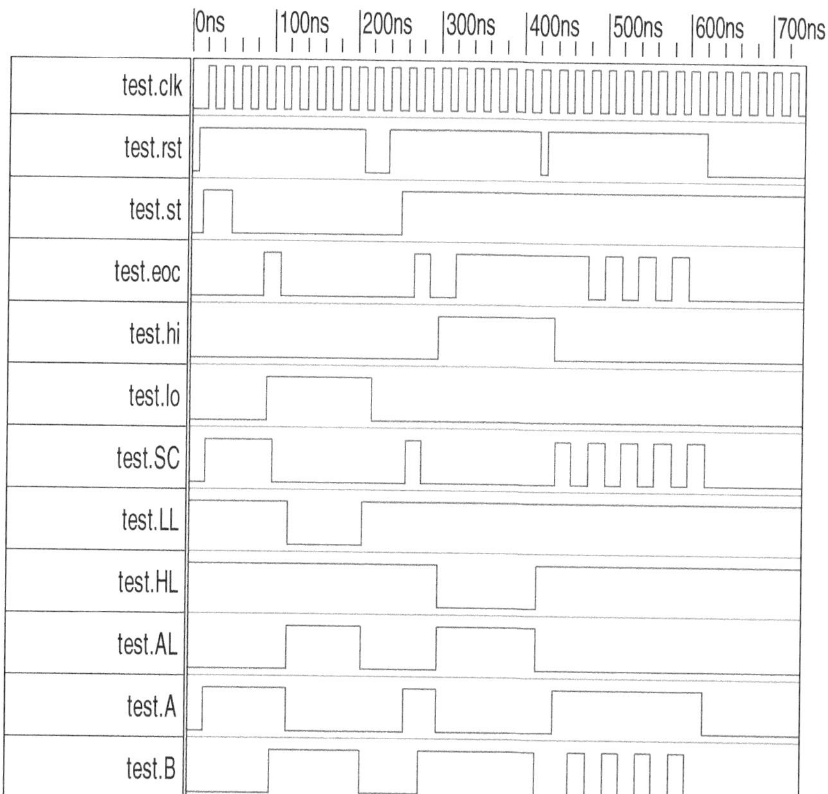

Figure 4.6 Simulation of the FSM controller.

4.4 SIMPLE WAVEFORM GENERATOR

Sometimes there is a need to generate a waveform to order, perhaps to test a product on an assembly line. An oscillator could be used for this purpose, but it can be tedious to build an oscillator to do this if the waveform is not a pure sine wave, square wave, ramp, or triangular. One way of generating a complex waveform would be to use a microcontroller with a digital-to-analogue converter (DAC). The complex waveform could be stored into read only memory (ROM) and accessed via the microcontroller. However, this seems overkill. There are also potential sampling frequency limitations with the microcontroller. An alternative way would be to use a clocked FSM. The sampling rate could then be controlled by the clock rate, which would be limited by that of a PLD or FPGA. The complex waveform is still stored in a ROM but the ROM is controlled by the FSM.

Consider the block diagram of Figure 4.7. In this system, raising the st input starts the waveform generator. Each memory location is accessed in sequence and its content, a digitized sample of the waveform, is sent to the DAC to be converted to an analogue form. When the end of memory is reached, the address counter simply runs over to the zero location and starts again.

Setting the st input low stops the system. The actual sampling rate and, hence, the period of the waveform can be calculated once the state diagram is completed. The output of the DAC will need to be filtered to remove the sampling frequency component – this can be accomplished using a simple first-order low-pass filter section if the sampling frequency is much higher than the highest synthesized waveform frequency. (Usually, it is to satisfy Shannon's sampling theory.)

The state diagram now needs to be developed. A little thought reveals that the block diagram itself provides an indication of the sequence required.

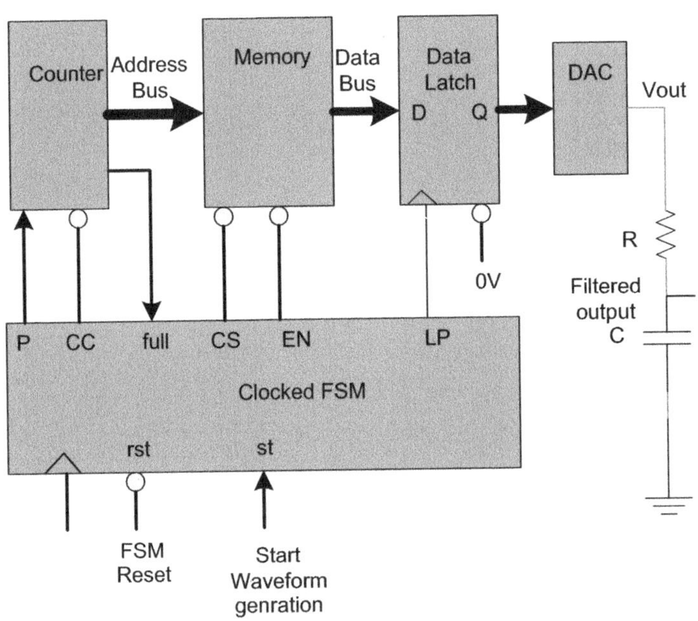

Figure 4.7 Block diagram for simple waveform generator.

1. Initially, the address counter needs to be cleared to provide the necessary zero address for the first location of the memory. The system should remain in state s0 until the start input st is asserted (high).
2. The memory then needs to be enabled, selected, and allowed to settle, after which the data in the memory location will be available at the data latch inputs. Then the data need to be latched into the data latch to be available at the input of the DAC.
3. At this stage, the address counter needs to be incremented so as to point to the next memory location and the sequence in 2 repeated again as long as the start input is still asserted (high).

Note that, in this problem, the end of memory location is not an issue, since the address counter can be allowed to overrun and start from location zero again. This does imply that the waveform information can be fitted into the memory device so that the waveform is produced seamlessly. It would be possible to add further logic to the system to ensure that this was always the case, but this is not done in this example.

The state diagram can now be developed following the sequence of activities described above.

In Figure 4.8, the state diagram is seen to follow the sequential requirements for the system. Note that in s3 the *P* output is a Mealy output. *P* is gated with the clock and can only go high when in s3, and then only when the clock is low. This ensures that the address counter is pulsed (on the rising edge of *P*) after the memory enable EN is disasserted (high). Therefore, the memory data outputs will be tri-state during the change of memory address. The Data Latch ensures that the DAC always has a valid data sample at its input. Note that an alternative arrangement for output *P* would be to provide an additional state between s3 and s1 in which $P = 1$. This would avoid the potential for a glitch at *P* output (as discussed in Chapter 1).

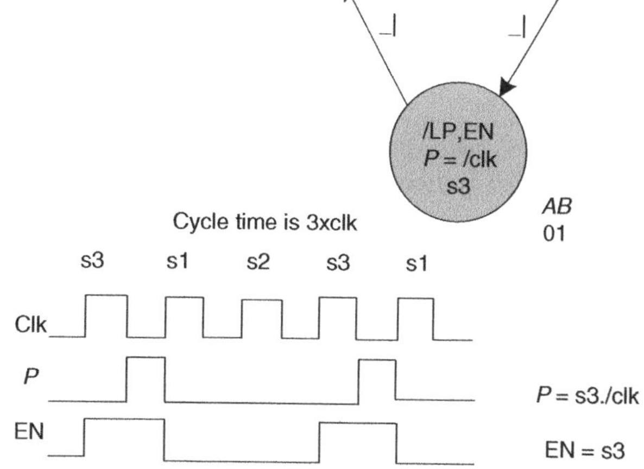

Figure 4.8 The complete state diagram for a simple waveform generator.

The equations can now be developed:

$$
\begin{aligned}
A \cdot d &= \mathrm{s0} \cdot \mathrm{st} + \mathrm{s1} + \mathrm{s3} \\
&= /A \cdot /B \cdot \mathrm{st} + A \cdot /B + /A \cdot B \\
&= /B \cdot \mathrm{st} + A \cdot /B + /A \cdot B \\
B \cdot d &= \mathrm{s1} + \mathrm{s2} \\
&= A \cdot /B + A \cdot B \\
&= A.
\end{aligned}
$$

Outputs are

$\mathrm{CC} = /\mathrm{s0} = /(/A \cdot /B)$ an active low output.

$\mathrm{CS} = \mathrm{s0} = /A \cdot /B$ although an active low signal it is only high in s0.

$\mathrm{LP} = \mathrm{s2} = A \cdot B.$

$\mathrm{EN} = \mathrm{s0} + \mathrm{s3} = /A$ high in these two states.

$P = \mathrm{s3} \cdot /\mathrm{clk} = /A \cdot B \cdot /\mathrm{clk}$ a Mealy output gated with the clock.

In Verilog, these equations can be entered directly, but using the Verilog convention for logic:

AND is & OR is | NOT is ~ exclusive OR is ^.

These equations would be contained in an **assign** block thus:

```
assign
A. d = ~ B& st|A& ~B|~A& B,
B.d = A,
CC = ~ (~ A & ~ B);
CS = ~ A& ~ B,
LP = A&B,
EN = ~ A,
P = ~ A&B&~ clk;
```

Appendix C contains a tutorial on how to produce a Verilog file to simulate a state machine. Also, much more detail is available in Chapters 6 to 8.

4.4.1 Sampling Frequency and Samples per Waveform

From the state diagram of Figure 4.8 it is apparent that the system cycles though three states for every memory access, so the sampling period is three times the clock period.

Therefore, for a sampling frequency of 300×10^3 Hz, a clock of $300 \times 10^3 \times 3 = 900 \times 10^3$ Hz is required. For a critical sampling-rate application, a dummy state could be added to make the sampling frequency four times the clock frequency (for example).

The size of the memory can be whatever is required for the systems use, and will dictate the size of the address counter. If the memory is 1 Kbyte, the address counter needs to be

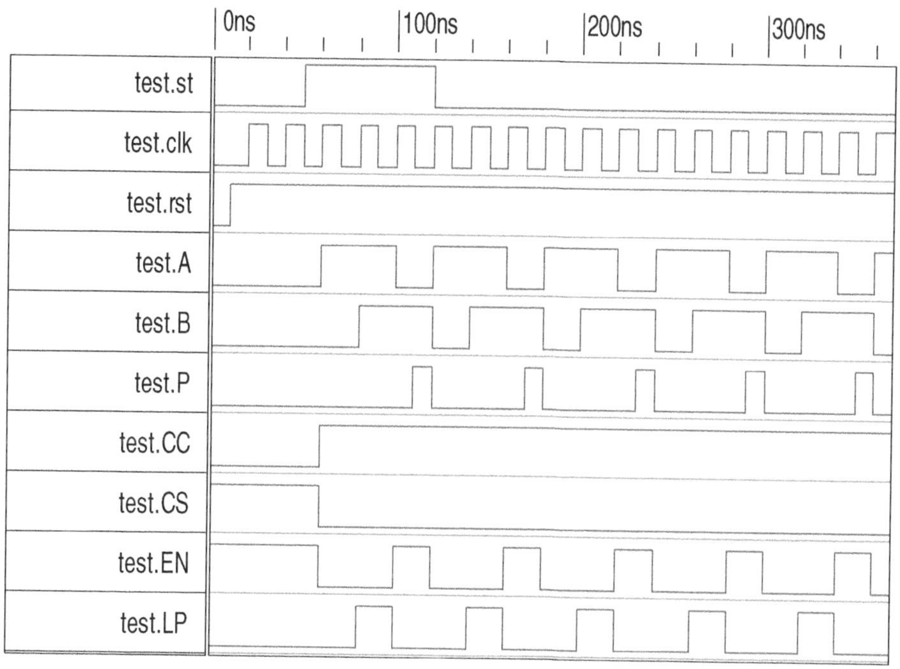

Figure 4.9 Simulation results for the FSM of the waveform synthesizer.

$$\text{Number of flip-flops in address counter} = \ln(1024)/\ln(2) = 10.$$

The simulation of the FSM is illustrated in Figure 4.9.

4.5 THE DICE GAME

In this example the system consists of seven LED indicators, a p input, and a clock. The block diagram of the system is shown in Figure 4.10, with a single push switch p. The clock input could be a simple oscillator circuit using a 555 timer chip running at 100 Hz so as to provide a flicker to add effect.

The LED indicators are arranged as illustrated in Figure 4.11 to look more realistic. In this design it is assumed that low-current LEDs are used with a forward current of 2 mA. This makes the current-limiting resistors 1800 Ω for a 5 V supply. It is also assumed that the FSM outputs are open drain. Figure 4.11 illustrates how the seven LED indicators would look for each number displayed. The situation when all LEDs are off is not shown.

The state machine is simple to develop, as all that is required is to display each number in sequence, but at a speed that the user cannot follow. The state diagram consists of seven states, each one to display a given LED pattern. The transition between each state is conditional on the input p being equal to one for each transition. When the user releases the p button the FSM will stop in a state. Because of the frequency of the clock, the user will not be able to follow the state sequence, thus realizing the chance element of the game. Note that if the clock frequency is too high then all the LED indicators will appear to be on when the p button is pressed. Having a

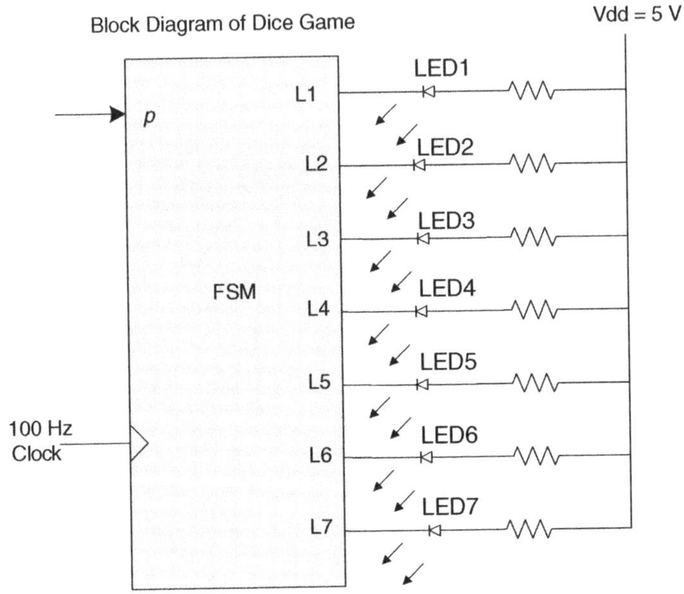

Figure 4.10 Block diagram of the dice game FSM

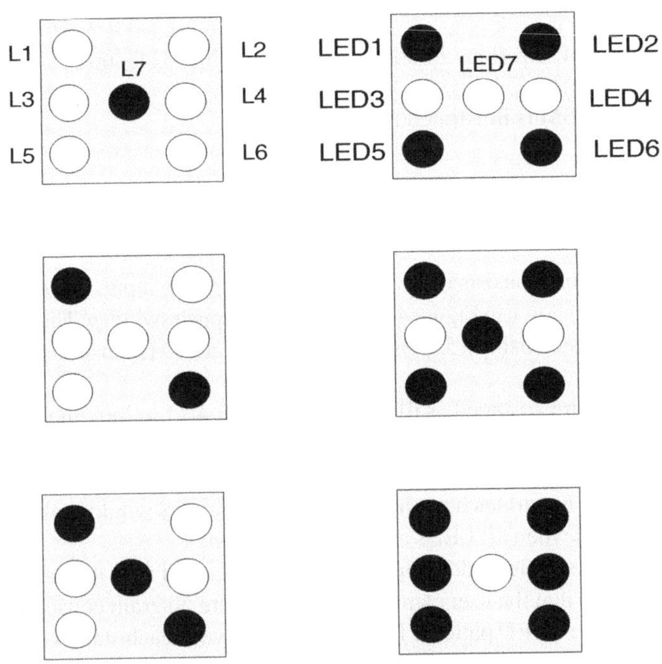

Dice format and possible LED patterns

Figure 4.11 Dice format for numbers.

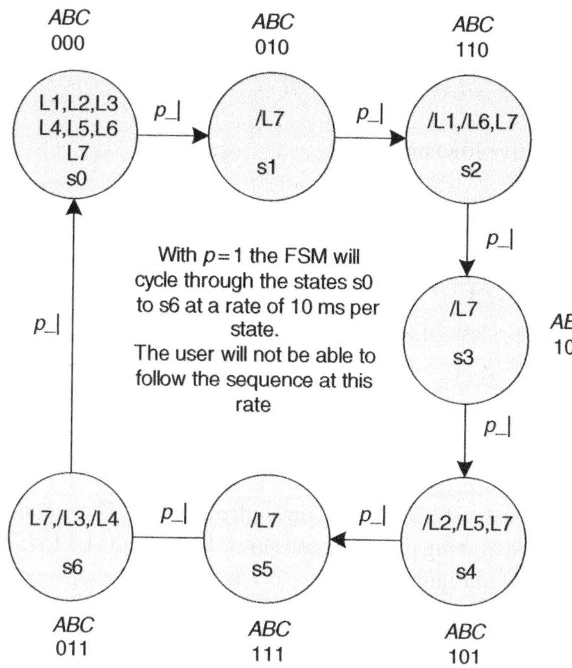

Figure 4.12 State diagram for the dice game.

slower clock frequency leads to a flicker effect and, thus, adds to the excitement of the game. Figure 4.12 shows the state diagram for the system.

4.5.1 Development of the Equations for the Dice Game

$$
\begin{aligned}
A \cdot d &= \mathrm{s1} \cdot p + \mathrm{s2} + \mathrm{s3} + \mathrm{s4} + \mathrm{s5} \cdot /p \\
&= /A \cdot B \cdot /C \cdot p + A \cdot B \cdot /C + A \cdot /B \cdot /C + A \cdot /B \cdot C + A \cdot B \cdot C \cdot /p.
\end{aligned}
$$

This can be reduced to

$$
\begin{aligned}
A \cdot d &= B \cdot /C \cdot p + A \cdot /C + A \cdot /p + A \cdot /B \\
B \cdot d &= \mathrm{s0} \cdot p + \mathrm{s1} + \mathrm{s2} \cdot /p + \mathrm{s4} \cdot p + \mathrm{s5} + \mathrm{s6} \cdot /p \\
&= /A \cdot /B \cdot /C \cdot p + /A \cdot B \cdot /C + A \cdot B \cdot /C \cdot /p + A \cdot /B \cdot C \cdot p \\
&\quad + A \cdot B \cdot C + /A \cdot B \cdot C \cdot /p,
\end{aligned}
$$

which reduces to

$$
\begin{aligned}
B \cdot d &= /A \cdot /C \cdot p + /A \cdot B \cdot /C + A \cdot C \cdot p + A \cdot B \cdot C + B \cdot /p \\
C \cdot d &= \mathrm{s3} \cdot p + \mathrm{s4} + \mathrm{s5} + \mathrm{s6} \cdot /p \\
&= A \cdot /B \cdot /C \cdot p + A \cdot /B \cdot C + A \cdot B \cdot C + /A \cdot B \cdot C \cdot /p,
\end{aligned}
$$

reducing to

$$C \cdot d = A \cdot /B \cdot p + B \cdot C \cdot /p + A \cdot C.$$

The outputs (LEDs are active low) are

L1 = (s0 + s1) = $(/A \cdot /B \cdot /C + /A \cdot B \cdot /C = /A \cdot /C)$ using active high in s0 and s1only.
L2 = (s0 + s1 + s2 + s3) = $/C$ using active high in these states only.
L3 = /s6(active low) = $/(/A \cdot B \cdot C)$.
L4 = /s6 = $/(/A \cdot B \cdot C)$ low in s6 only; hence invert.
L5 = /(s4 + s5 + s6) = $/(A \cdot C + B \cdot C)$ low in only these states; hence invert.
L6 = /(s2 + s3 + s4 + s5 + s6) or (s0 + s1) only high in s0 or s1 giving $(/A \cdot /C)$.
L7 = /(s1 + s3 + s5) = $/(/A \cdot B \cdot /C + A \cdot /B \cdot /C + A \cdot B \cdot C)$.

Figure 4.13 illustrates the dice FSM running through each state. The secondary state variables *a*, *b*, and *c* can be seen to be moving through each state. The outputs L1 to L7 are responding as expected and are illustrated in Figure 4.11.

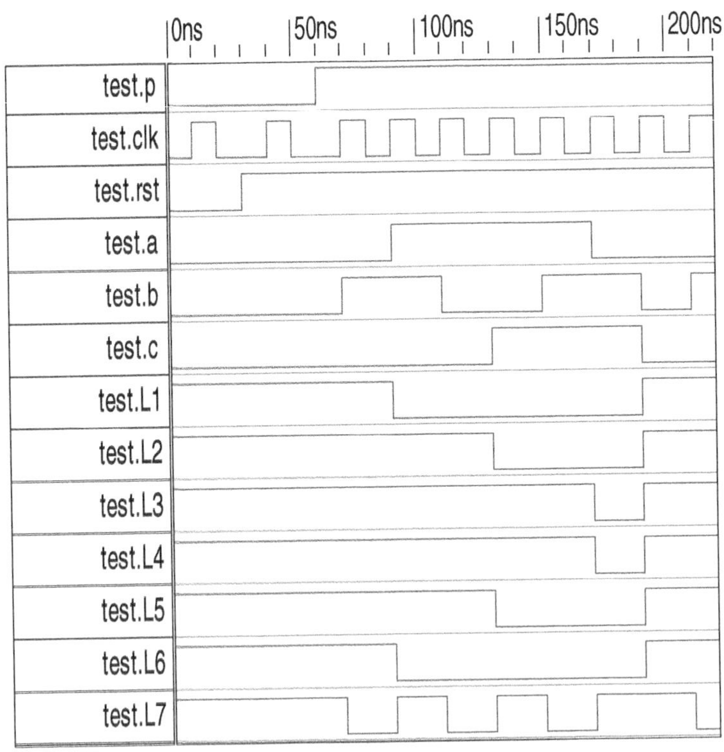

Figure 4.13 Simulation of the dice game.

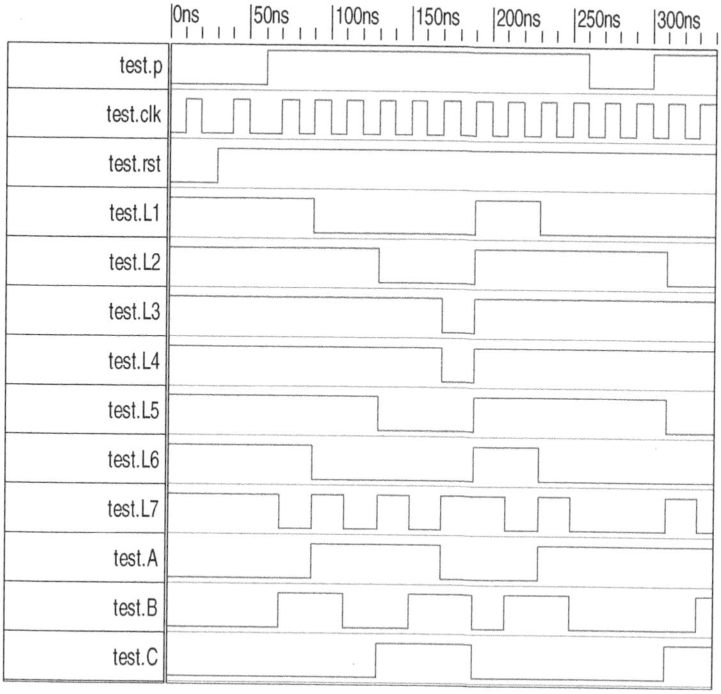

Figure 4.14 Dice game simulation with *p* input released showing FSM stopped in s3.

In Figure 4.14, the input *p* has been simulated as 'on' then 'off'. The FSM is seen to have stopped in state s3, then started again when *p* is set to logic 1.

Note that in both simulations the time-scale is in nanoseconds, but in practice the clock would be slowed down to a 10 ms period.

4.6 BINARY DATA SERIAL TRANSMITTER

The next example involves sending the 4-bit binary codes of a counter to a shift register to be serially shifted out over a serial transmission line.

Figure 4.15 shows the block diagram for a possible system. The FSM is used to control the operation of the Binary Counter and the Parallel Loading Shift Register. Both of these devices could be designed using the techniques described in Appendix B on counting methods. This leads to a Verilog description (module) for each device.

The system is started by raising the st input to logic 1. This is to cause the FSM to remove the reset from the Binary Counter and then load the current count value of the counter into the parallel inputs of the shift register. On releasing the parallel load input LD to logic 1, the shift register will clock the count value out over its transmit output (TX) at the baud rate dictated by the clock. When the shift register is empty its RE signal will go high and this

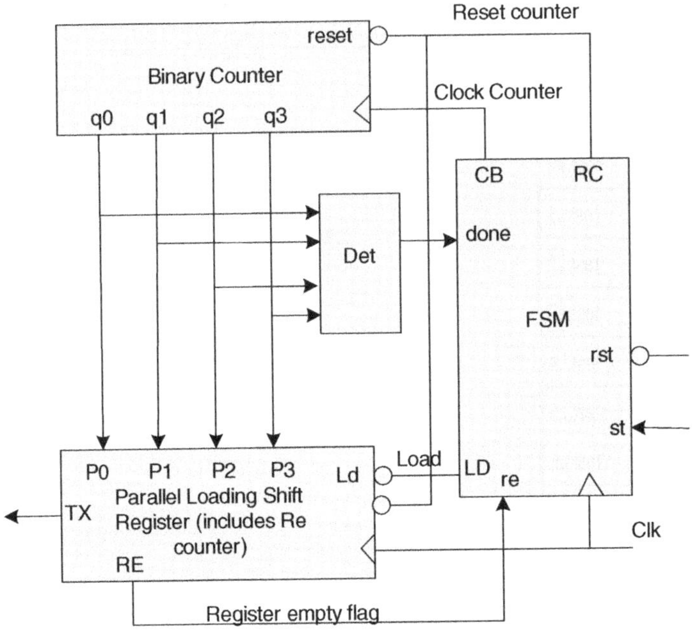

Figure 4.15 Block diagram of the binary data serial transmitter.

will be seen by the FSM, which will then determine whether the last count value has been sent. This is seen by the FSM when done = 1, detected by the detector block (an AND gate). If not the last counter value, then the next count value will be loaded into the shift register and the sequence repeated until all count values have been sent. At this point the system will stop and wait for st to be returned to its inactive state before returning the FSM to its s0 state.

From the above description, the state diagram in Figure 4.16 is developed. This state diagram is correct, but it is difficult to obtain a unit distance code for the secondary state variables. If a dummy state s7 is added, then a unit distance coding between s6 and s0 can be obtained for the secondary state variables A, B, and C. Note: it is not apparent from Figure 4.17, but the outputs in state s7 are the same as the state it is going to (s0), apart from the RC output. The s5 to s1 transition is not unit distance. If glitches are produced in any outputs, then dummy states could be introduced between s5 and s1 to establish unit distance coding. The reader might like to try to establish a unit distance code for the state diagram. This would require introducing an additional state variable (flip-flop), since all 2^3 states have been used in this design.

Using Figure 4.17, the equations for the FSM are obtained from the state diagram and implemented using D flip-flops:

$$A \cdot d = \text{s1} + \text{s2} + \text{s3} + \text{s4}$$
$$= /A \cdot B \cdot /C + A \cdot B \cdot /C + A \cdot /B \cdot /C + A \cdot /B \cdot C,$$

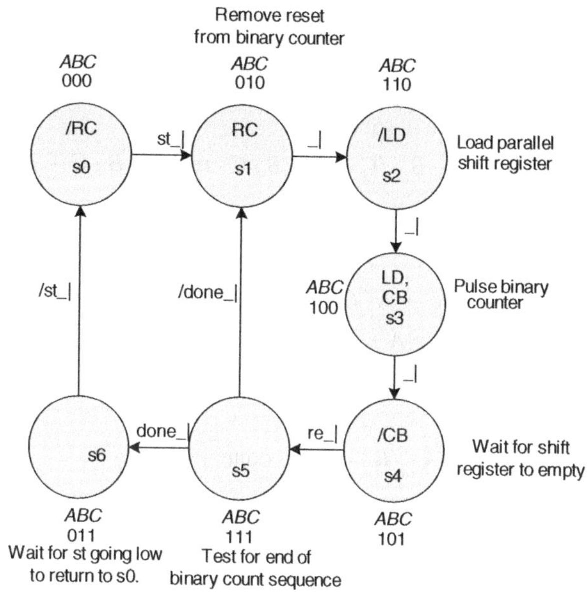

Figure 4.16 State diagram for the binary data serial transmitter.

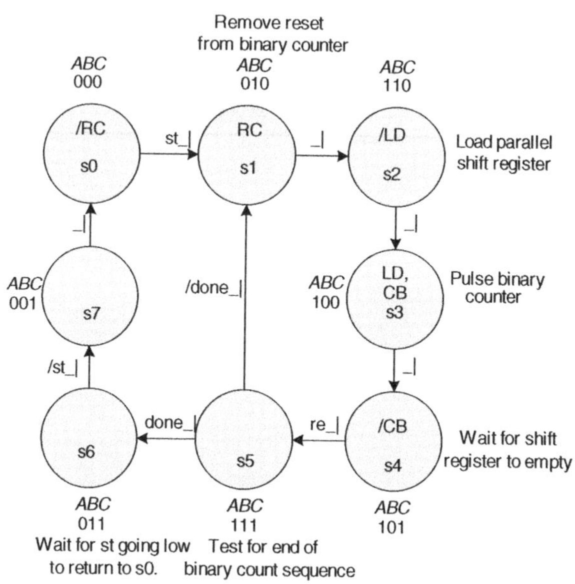

Figure 4.17 State diagram with additional dummy state s7 to obtain unit distance code for the secondary state variables.

reducing to

$$A \cdot d = B \cdot /C + A \cdot /B$$
$$B \cdot d = \text{s0} \cdot \text{st} + \text{s1} + \text{s4} \cdot \text{re} + \text{s5} + \text{s6} \cdot \text{st}$$
$$= /A \cdot /B \cdot /C \cdot \text{st} + /A \cdot B \cdot /C + A \cdot /B \cdot C \cdot \text{re} + A \cdot B \cdot C + /A \cdot B \cdot C \cdot \text{st},$$

reducing to

$$B \cdot d = /A \cdot /C \cdot st + /A \cdot B \cdot /C + A \cdot C \cdot \text{re} + B \cdot C \cdot \text{st} + A \cdot B \cdot C$$
$$C \cdot d = \text{s3} + \text{s4} + \text{s5} \cdot \text{done} + \text{s6} = A \cdot /B \cdot /C + A \cdot /B \cdot C + A \cdot B \cdot C \cdot \text{done} + /A \cdot B \cdot C,$$

reducing to

$$C \cdot d = A \cdot /B + B \cdot C \cdot \text{done} + /A \cdot B \cdot C.$$

The outputs (all Moore) are

$$\text{RC} = /\text{s0(active low)} = /(/A \cdot /B \cdot /C)$$
$$\text{LD} = /(\text{s2}) = /(AB/C)$$
$$\text{CB} = \text{s3(active high)} = A \cdot /B \cdot /C.$$

The serial transmitter simulation is shown in Figure 4.18. The state machine is tracked through its state sequence in the usual way by comparing the *A*, *B*, and *C* values in Figure 4.18 with the state diagram *A*, *B*, and *C* values in Figure 4.17.

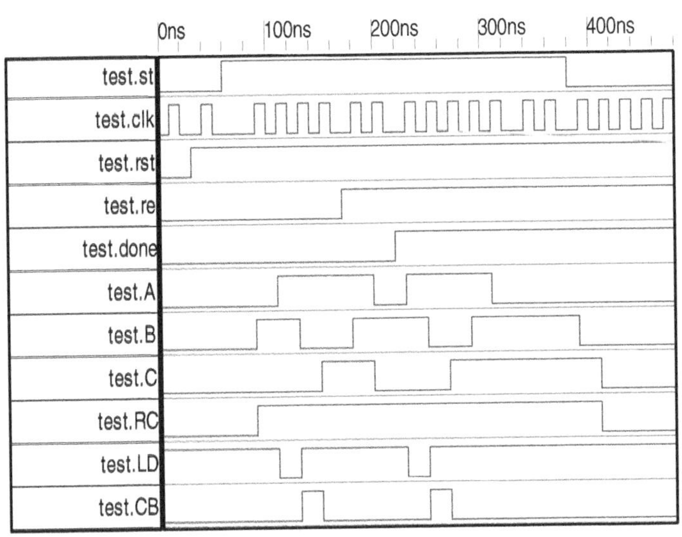

Figure 4.18 Simulation of the binary data serial transmitter.

4.6.1 The RE Counter Block in the Shift Register of Figure 4.15

The shift register in Figure 4.15 has an output RE to flag the point at which the register is empty. This can easily be obtained by using a four-stage Binary Counter that becomes enabled when the load input is disasserted (high). The counter can then be clocked with the same clock as the shift register; then, when it reaches its maximum count 1000, the most significant bit is used as the RE signal. Table 4.2 illustrates the effect.

From Table 4.2 it can be seen that when the counter reaches the eighth clock pulse the counter rolls over to set the most significant bit of the counter D to logic 1. This bit acts as the RE register empty bit. After shifting out the binary number, the FSM will return to its s0 state, where the RC output will once again go low and reset both the Binary Counter and the RE counter in the shift register. Note that in this particular design an additional flip-flop E could be added to the binary counter and this used as the RE output instead

The equations to describe the RE counter can be developed from the material in Appendix B on counting applications. The equations, using T-type flip-flops, are

$$A \cdot t = 1$$

$$B \cdot t = A$$

$$C \cdot t = A \cdot B$$

$$D \cdot t = A \cdot B \cdot C$$

$$\mathrm{RE} = D.$$

This last example has illustrated how a complete design can be developed in terms of Boolean equations that can be directly implemented in Verilog HDL (or any other HDL for that matter).

There are examples in Appendix B showing how a synchronous binary counter can be implemented using T flip-flops. Of course, the counter could be implemented as an asynchronous (ripple-through) counter if desired.

Table 4.2 Illustrating the effect of a binary counter used to determine shift register empty.

Binary counter					
RE D	C	B	A	Count value	
0	0	0	0	0	
0	0	0	1	1	
0	0	1	0	2	
0	0	1	1	3	
0	1	0	0	4	
0	1	0	1	5	
0	1	1	0	6	
0	1	1	1	7	
1	0	0	0	8	Shift register empty when $D = 1$
1	0	0	1	9	D output stays set

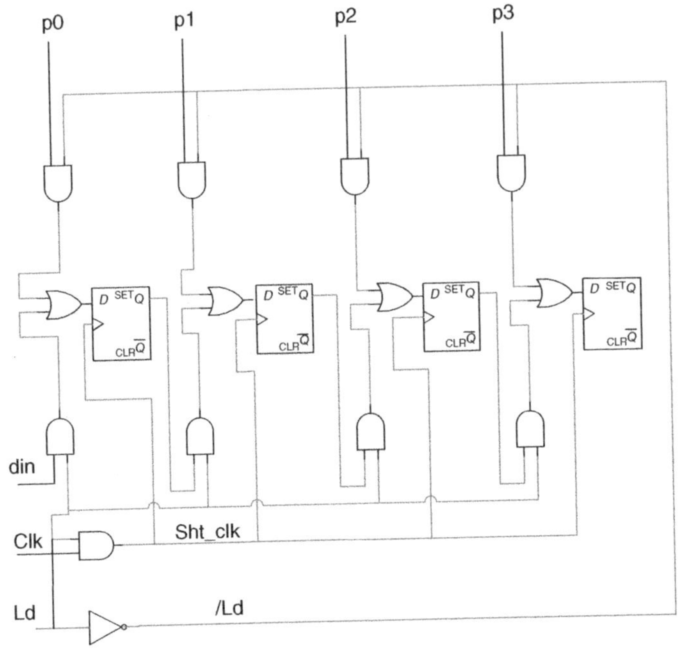

Figure 4.19 The 4-bit parallel loading shift register from Equations (B.9) to (B.12).

Also in Appendix B is an example of a parallel loading shift register using *D* flip-flops. The equations for a four-stage shift register are repeated below from Appendix B:

$$\mathrm{Q0} \cdot d = \mathrm{din} \cdot \mathrm{ld} + \mathrm{po} \cdot /\mathrm{ld} \tag{B.7}$$

$$\mathrm{Q1} \cdot d = \mathrm{q0} \cdot \mathrm{ld} + \mathrm{p1} \cdot /\mathrm{ld} \tag{B8}$$

$$\mathrm{Q2} \cdot d = \mathrm{q1} \cdot \mathrm{ld} + \mathrm{p2} \cdot /\mathrm{ld} \tag{B.9}$$

$$\mathrm{Q3} \cdot d = \mathrm{q2} \cdot \mathrm{ld} + \mathrm{p3} \cdot /\mathrm{ld} \tag{B.10}$$

$$\mathrm{Sft\_clk} = \mathrm{clk} \cdot \mathrm{ld}. \tag{B.11}$$

Figure 4.19 shows the schematic circuit for the 4-bit parallel loading shift register developed from Equations (B.7)–(B.11).

4.7 DEVELOPMENT OF A SERIAL ASYNCHRONOUS RECEIVER

Often, there is a requirement to use serial transmission and receiving of data in a digital system. Although there are lots of serial devices on the market, it is useful to be able to implement one's own design directly to incorporate into an FPGA device. The advantage of this approach is that the baud rate and protocols can be dictated by the designer, as can how the device will be controlled.

In this example, the serial data input is encapsulated into an asynchronous data packet with start (st) and stop (sp) protocol bits that have been added to the serial transmission packet. These

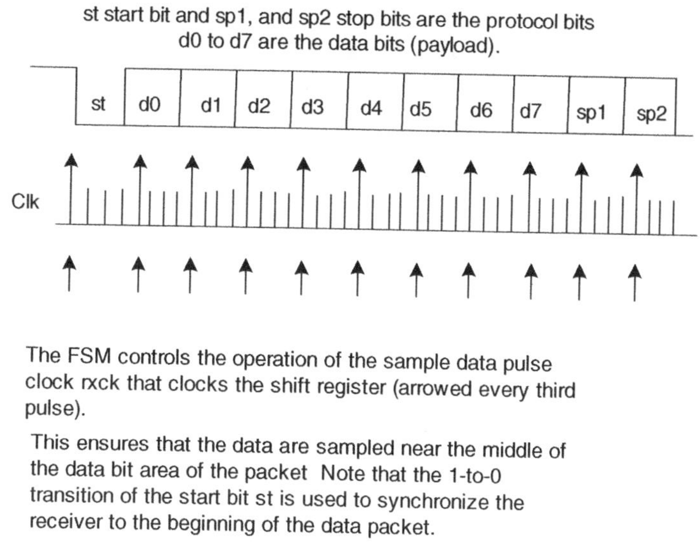

Figure 4.20 Protocol of the serial asynchronous receiver.

are used to provide a means of identifying the data packets as they arrive. This allows the data packets to arrive at any time and at any selected rate (dictated by the baud rate).

The problem with receiving data is that it is necessary to ensure that the shift register is clocked with correct data bits. To do this the FSM clock is used to drive an FSM to create a shift register clock RXCK in the middle of the data bit time period. This RXCK clock pulse can be seen in Figure 4.20 as the arrowed pulses occurring every third clk pulse. Thus, the clk signal runs four times faster than the RXCK signal generated by the FSM. Note, the FSM needs to detect the start of the data packet by looking for the 1-to-0 transition on the receiver input.

The block diagram for the serial asynchronous receiver is illustrated in Figure 4.21. The FSM is used to create the shift register clock, and to control the operation of the serial asynchronous receiver. The Divide by 11 Counter is used to count out the 11 bits that make up the protocol packet. This provides a shift register full signal rxf to indicate to the FSM that a complete data packet has arrived. The Data Latch is used for collecting the received data from the shift register to send to the outside world device controlling the asynchronous receiver.

The FSM must wait for start (by monitoring for the st bit change 1 to 0); this is just the first receive bit coming into the shift register. When detected, shift the data into the shift register. If the stop bit is not correct, then the FSM can issue an error via signal ERR. Note, in this version the start bit is tested along with the two stop bits via an AND gate (error detection signal ed) to ensure packet alignment after the complete packet is received, the receiver rx input is held at logic 1 by a pull-up resistor so that the start bit (active low) can be detected. The ack signal is available so that the outside world device using the system can respond to an error condition (no error means successful packet received). Healthy data packets will be latched into the data latch ready to be read by the controlling device.

The signal CDC is used to clear the shift register and set st to logic 1, i.e. the flip-flop representing the start bit of the shift register needs to be pre-set so that it can be cleared by the incoming start bit from the serial line.

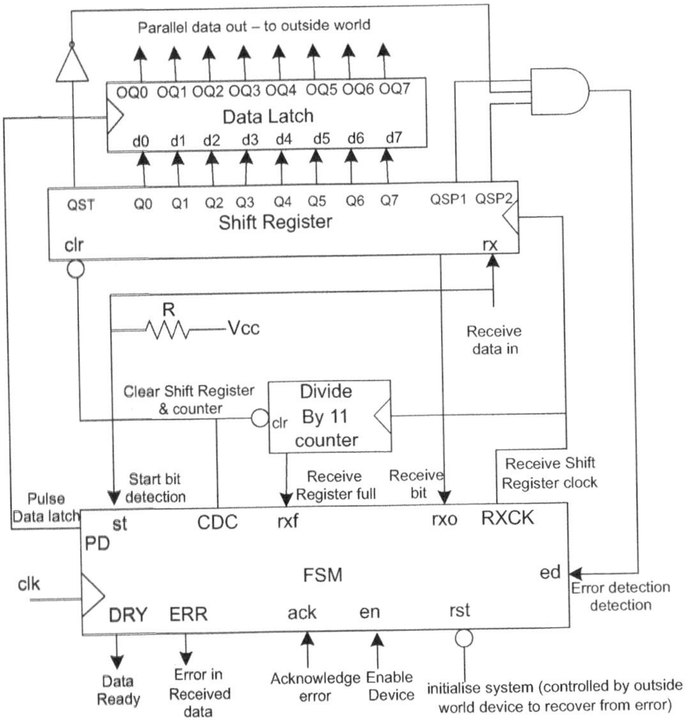

Figure 4.21 Block diagram of the serial asynchronous receiver.

The en signal is used to enable and start the asynchronous receiver. This is necessary to ensure that the system starts monitoring the clock so as to issue the shift register clock pulse (RXCK) at the right time (in the middle of the data bit period).

Figure 4.22 illustrates the state diagram for the system. In Figure 4.22, the FSM waits for the enable signal en going high and start signal st going low; it then moves through states s1, s2, and s4 and onto s5 to shift the start bit into the shift register. This is required in order to ensure that the start bit is detected and then shifted at the right time. In state s5, the shift register clock RXCK is pulsed to place the start bit into the shift register. It then falls into state s6, sending RXCK low and proceeds to cycle through the second loop consisting of states s5, s6, s7, and s8.

These states count out the clock cycles and produce a shift register clock pulse (RXCK) at the right time near the middle of each data bit. After all 11 bits have been clocked into the shift register the 11-bit counter will issue a receive register full signal rxf, and the FSM will now fall into state s9, where the start and stop bits are tested (ed should be logic 1). If ed = 0, then the FSM will move into s10 and issue the error signal.

The controlling device can then reset the asynchronous receiver and start again. If no error, then the FSM moves to s11 to latch the data in the shift register into the data latch ready for the controlling device to read (OQ0 to OQ7). It will also issue a data ready signal (DRY) to the controlling device, which will acknowledge this by raising an ack signal. The FSM can then move back to s0 via s12 (when ack goes low) to wait for the next data packet. The DRY and ack signals form a handshake mechanism between the FSM and the controlling device.

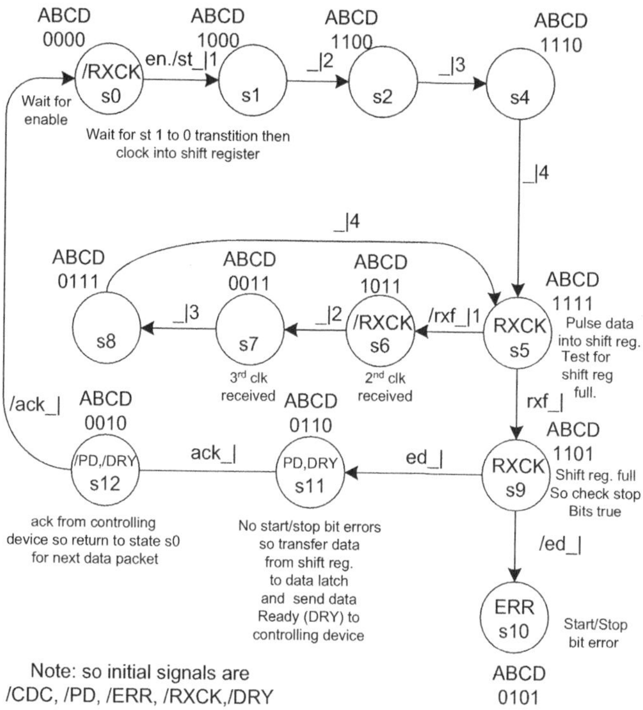

Figure 4.22 State diagrams for the serial asynchronous receiver.

The device enable signal en will be left high until all data packets have been received.

Note that the state assignments miss s3, which was removed from the state diagram during development when state s3 was no longer needed (owing to an error in the design at that time). State diagram development tends to be an iterative process.

4.7.1 Finite-State Machine Equations

$$A \cdot d = \text{s0} \cdot \text{en} \cdot /\text{st} + \text{s1} + \text{s2} + \text{s4} + \text{s5} + \text{s8}$$

$$B \cdot d = \text{s1} + \text{s2} + \text{s4} + \text{s5} \cdot \text{rxf} + \text{s7} + \text{s8} + \text{s9} + \text{s10} + \text{s11} \cdot /\text{ack}$$

$$C \cdot d = \text{s2} + \text{s4} + \text{s5} \cdot /\text{rxf} + \text{s6} + \text{s7} + \text{s8} + \text{s9} \cdot \text{ed} + \text{s11} + \text{s12} \cdot \text{ack}$$

$$D \cdot d = \text{s4} + \text{s5} + \text{s6} + \text{s7} + \text{s8} + \text{s9} \cdot /\text{ed} + \text{s10}$$

$$\text{RXCK} = \text{s5} = \text{ABCD}$$

$$\text{PD} = \text{dry} = \text{s11} = /\text{ABC}/\text{D}$$

$$\text{ERR} = \text{s10} = /\text{AB}/\text{CD}.$$

The reader may like to complete these to form the equations in terms of A, B, C, and D.

The complete asynchronous serial receiver block is simulated, together with all the modules in Figure 4.21, in Appendix B.

4.8 ADDING PARITY DETECTION TO THE SERIAL RECEIVER SYSTEM

The foregoing example could be improved upon by making the first stop bit sp1 into a parity bit. The parity bit would require combinational logic to check each bit of the protocol packet for either even parity or odd parity. This would require an exclusive OR block made up of the 11 bits of the packet.

For example, odd parity would require an odd parity output OP at the Transmitter of

$$\text{OP} = \text{bo}\wedge\text{b1}\wedge\text{b2}\wedge\text{b3}\wedge\text{b4}\wedge\text{b5}\wedge\text{b6}\wedge\text{b7}\wedge\text{b8}\wedge\text{b9}\wedge\text{b10}.$$

Or, including the protocol bits:

$$\text{OP}_{n+1} = \text{st}\wedge\text{d0}\wedge\text{d1}\wedge\text{d2}\wedge\text{d3}\wedge\text{d4}\wedge\text{d5}\wedge\text{d6}\wedge\text{d7}\wedge\text{OP}_n\wedge\text{sp}.$$

This output would be tested by the FSM for logic 1. If logic 0, this would indicate that one or more of the received bits was faulty.

Note that even parity EP can be detected by complementing the OP signal:

$$\text{EP}_n = /\text{OP}_n.$$

To implement the parity detector term, two input exclusive OR gates are cascaded with the last exclusive OR gate providing the OP_n signal. The output of the parity block at the receiver is *P*.

The inputs d0, d1, ..., d7 will be obtained from the output of the shift register in each case (see Figure 4.21).

4.8.1 To Incorporate the Parity

The parity detector inputs are connected to the outputs of the shift register and its output OP_n made available as an input to the FSM via the last two bit comparator comparing OP_n and OP_{n+1} in Figure 4.23.

Figure 4.24 shows the new protocol with the parity bit OP_n (shown in lower case) replacing sp1.

Figure 4.25 shows the additional parity block added to the block diagram. This version detects stop and parity bit errors at the output of the shift register; the start bit has not been tested (but could be included if desired).

Figure 4.26 illustrates the modified state diagram with ODD parity detection. Note that the input parity bit OP_{n+1} must be compared with the generated parity bit OP_n. If both are the same, then there is no parity error. This comparison can be made with a 2-bit exclusive NOR gate having an output *P* ($\text{OP}_n == \text{OP}_{n+1}$) being logic 1 if there is no parity error and logic 0 otherwise. This output is an input *p* to the state machine (see Figure 4.25).

In state s9, the bit sp is checked to find out whether the whole packet has been input, and s11 now tests for an odd parity error. In either case a failure will result in the FSM aborting the receive packet process and falling into state s10 to await a reset from the controlling device. The logic used in Figure 4.21 could be used to detect for start and stop bits if desired.

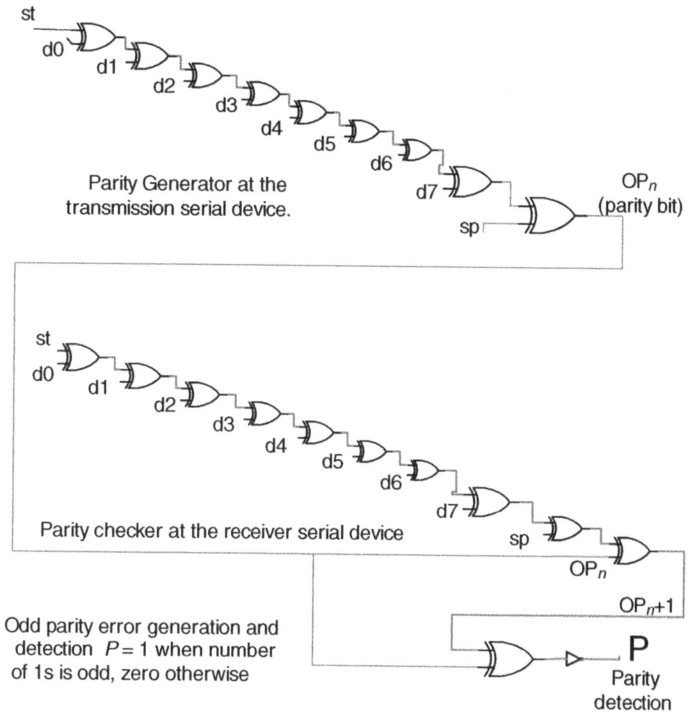

Figure 4.23 Arrangement of the parity generation and detection logic.

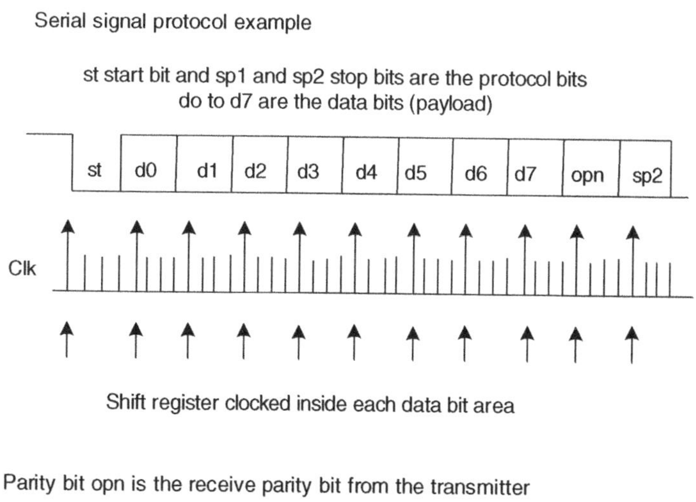

Figure 4.24 Protocol with parity detection bit added.

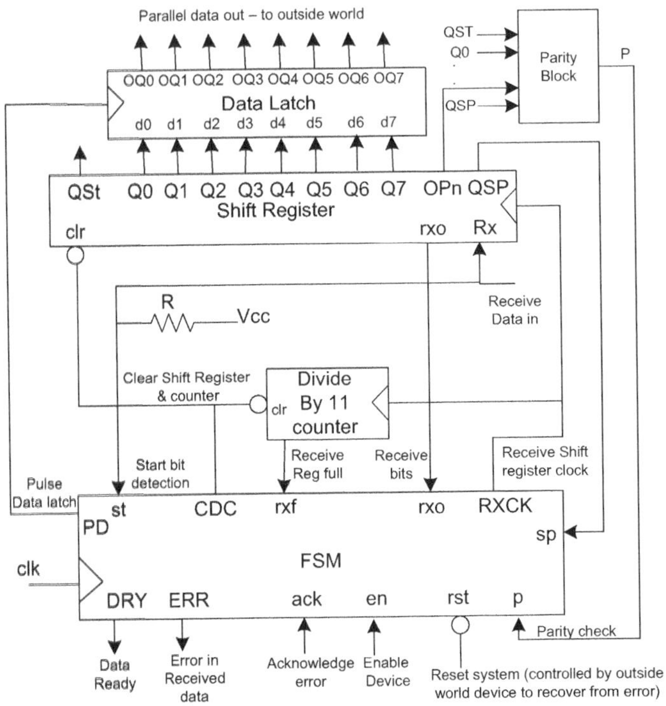

Figure 4.25 Block diagram with parity block added.

4.8.2 D-Type Equations for Figure 4.26

In the following equations, the variable P is the output of the parity check ($\mathrm{OP}_n = \mathrm{OP}_{n+1}$) connected to the input p of the FSM. See Figure 4.23.

$$A \cdot d = \mathrm{s0} \cdot \mathrm{en} \cdot /\mathrm{st} + \mathrm{s1} + \mathrm{s2} + \mathrm{s4} + \mathrm{s5} + \mathrm{s8} + \mathrm{s9} \cdot \mathrm{sp}$$
$$= /A/B/C/D \cdot \mathrm{en} \cdot /\mathrm{st} + A/B/C/D + AB/C/D + ABC/D + ABCD$$
$$+/ABCD + AB/CD \cdot \mathrm{sp}$$

$$B \cdot d = \mathrm{s1} + \mathrm{s2} + \mathrm{s4} + \mathrm{s5} \cdot \mathrm{rxf} + \mathrm{s7} + \mathrm{s8} + \mathrm{s9} \cdot /\mathrm{sp} + \mathrm{s10} + \mathrm{s11} + \mathrm{s12} \cdot /\mathrm{ack}$$
$$= A/B/C/D + AB/C/D + ABC/D + ABCD \cdot \mathrm{rxf} + /A/BCD + /ABCD$$
$$+AB/CD \cdot /\mathrm{sp} + /AB/CD + A/B/CD + /ABC/D \cdot /\mathrm{ack}$$

$$C \cdot d = \mathrm{s2} + \mathrm{s4} + \mathrm{s5} \cdot /\mathrm{rxf} + \mathrm{s6} + \mathrm{s7} + \mathrm{s8} + \mathrm{s11} \cdot p + \mathrm{s12} + \mathrm{s13} \cdot \mathrm{ack}$$
$$= AB/C/D + ABC/D + ABCD \cdot /\mathrm{rxf} + A/BCD + /A/BCD + /ABCD$$
$$+ A/B/CD \cdot p + /ABC/D + /A \cdot /B \cdot C \cdot /D \cdot \mathrm{ack}$$

$$D \cdot d = \mathrm{s4} + \mathrm{s5} + \mathrm{s6} + \mathrm{s7} + \mathrm{s8} + \mathrm{s9} + \mathrm{s10} + \mathrm{s11} \cdot /p$$
$$= ABC/D + ABCD + A/BCD + /A/BCD + /ABCD + AB/CD$$
$$+ /AB/CD + A/B/CD \cdot p.$$

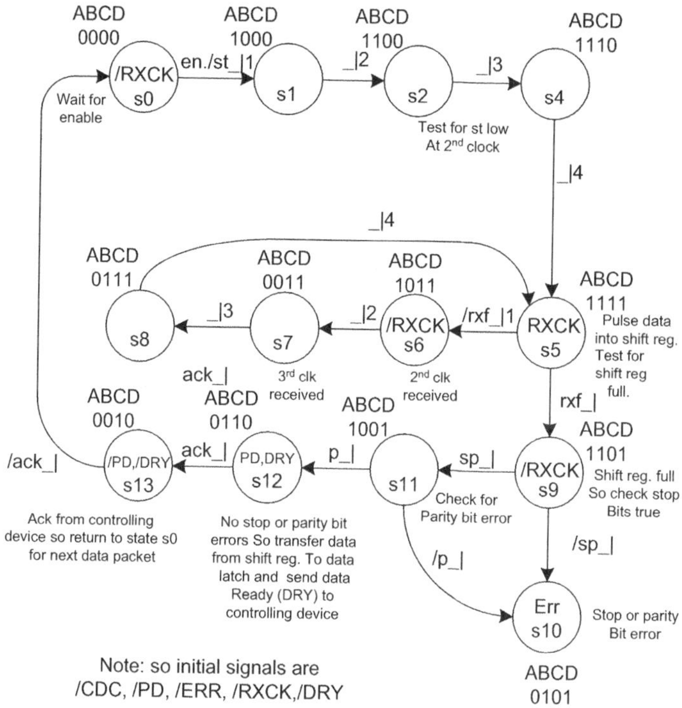

Figure 4.26 The state diagram with odd parity added to FSM.

The outputs are as they were in the state diagram of Figure 4.22, except for

$$\text{ERR} = \text{s10} = /AB/CD$$

$$\text{PD} = \text{dry} = \text{s12} = /ABC/D$$

$$\text{RXCK} = \text{s5} = A \cdot B \cdot C \cdot D.$$

The FSM part can be simulated, and this is illustrated in Figure 4.27. In this simulation, the test sequence is

s0, s1, s2, s4, s5, s6, s7, s8, s5, s9, s11, s12, s13, s0, s1, s2, s4, s5, s9, s10, s0, s1, s2, s4, s5, s9, s11, s10.

This ensures that all paths of the state diagram have been tested.

This should now be followed by a series of tests of all the other components, i.e. the shift register, the divide-by-11 counter, and the parity block, before going on to test the whole system.

4.9 AN ASYNCHRONOUS SERIAL TRANSMITTER SYSTEM

Having developed an asynchronous receiver module, an asynchronous transmitter is required to complete the serial device. Figure 4.28 shows the block diagram for an asynchronous serial transmitter.

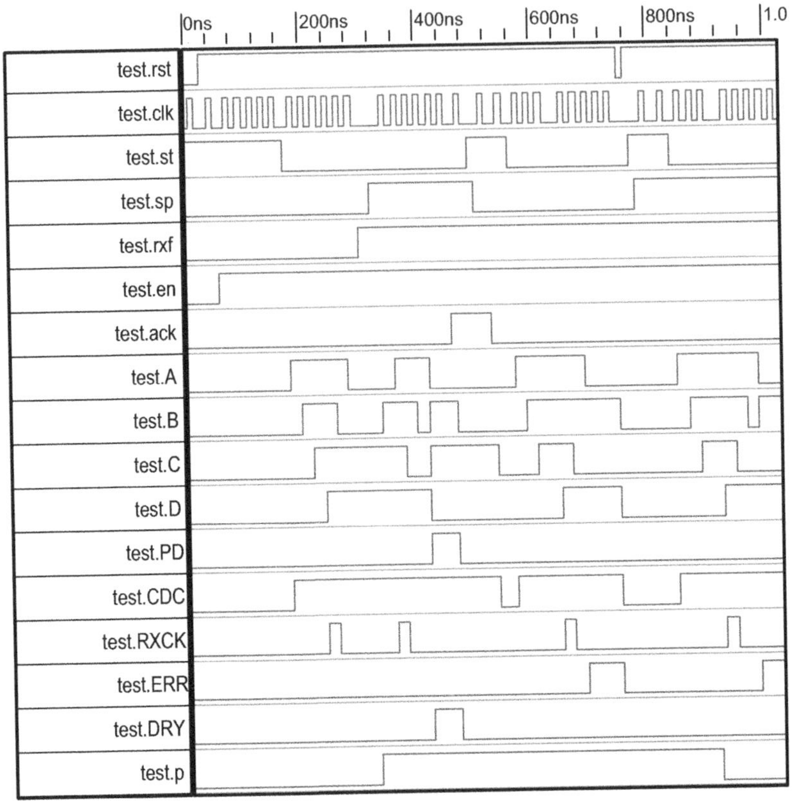

Figure 4.27 Simulation of the FSM for the serial receiver.

In Figure 4.28, the input Data Latch provides the data to be transmitted and the protocol bits st and sp are set to their expected values before being loaded into the Shift Register via the LD output from the FSM. Note that there is no need for a slower transmit clock, as the FSM can provide the shift register pulse at the right time.

The sequence is started by data being presented onto the parallel data inputs then the send input being sent high by the controlling device. The FSM then loads the data into the shift register and starts transmitting it out to line. The Divide by 11 Counter records the point at which the packet has been sent to line by raising the Transmit Register Empty (txe) signal high. The FSM can then send a Request To Send (RTS) signal to the controlling device to inform it that the data packet has been sent. The controlling device can set the ack signal high to say it has acknowledged this operation.

A possible solution is illustrated in Figure 4.29.

It is important to ensure that the clock signal to the shift register is the same frequency as the one used in the asynchronous receiver block. If it is not, then the receiver will not be able to receive the data packets. Even if the two clocks are different by only a small amount, a frame error could arise. This is when the difference in clock speeds produces a small difference in the total packet time and, hence, one or more data bits can be lost. In effect, start and stop bits must be sent and received correctly.

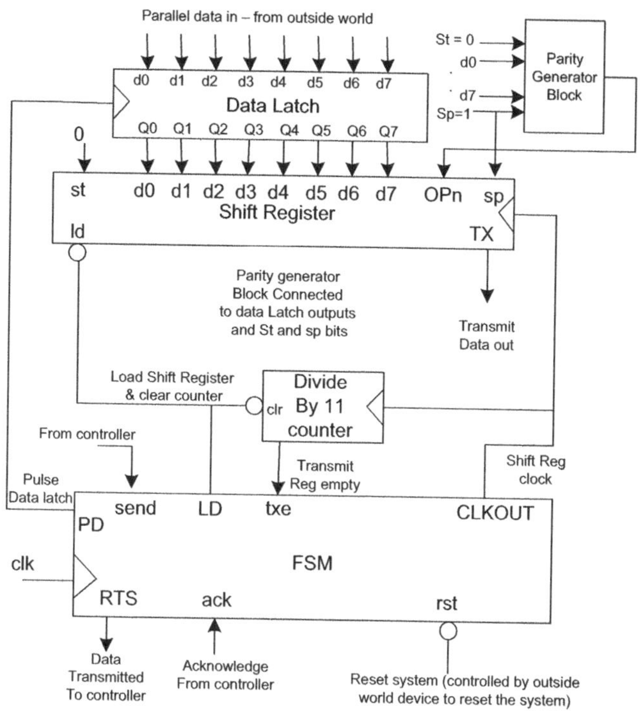

Figure 4.28 Block diagram for an asynchronous serial transmitter.

$$\text{Total packet time} = 11 \times 1/(\text{clock frequency}).$$

For example, if the transmitter shift register clock is 1 MHz (usually referred to as the baud rate), then

$$\text{Total packet time} = 11 \times 1/(1 \times 10^6) = 11 \times 1\,\mu\text{s} = 11\,\mu\text{s in duration.}$$

The receiver shift register clock does have a tolerance; this is a result of the fact that the data are sampled within a four-clock window (see Figure 4.20) and a small difference in the two packet lengths can be accommodated.

In some commercial Universal Asynchronous Receiver Transmitter (UART) devices, 16 (rather than 4) is used for the clk signal used to generate the shift register clock (RXCK), giving a greater resolution for detecting the logic value of the data bits.

Generally, if the clocks in both the transmitter and the receiver are of a high accuracy (as one would expect from crystal oscillators), then there is usually not a problem. It would be easy to restructure the receiver state diagrams of Figures 4.22 and 4.26 to accommodate a higher resolution shift register clock by adding more states in the loop comprising s5 to s8, and adding states between s1 to s5 for the start bit. However, such a design could make use of the One Hot method covered in Chapter 5.

Note that the FSM clock is four times that of the baud rate.

The state diagram for the asynchronous transmitter is illustrated in Figure 4.29. In this state diagram, the shift register is clocked every four FSM clock pulses as it moves between

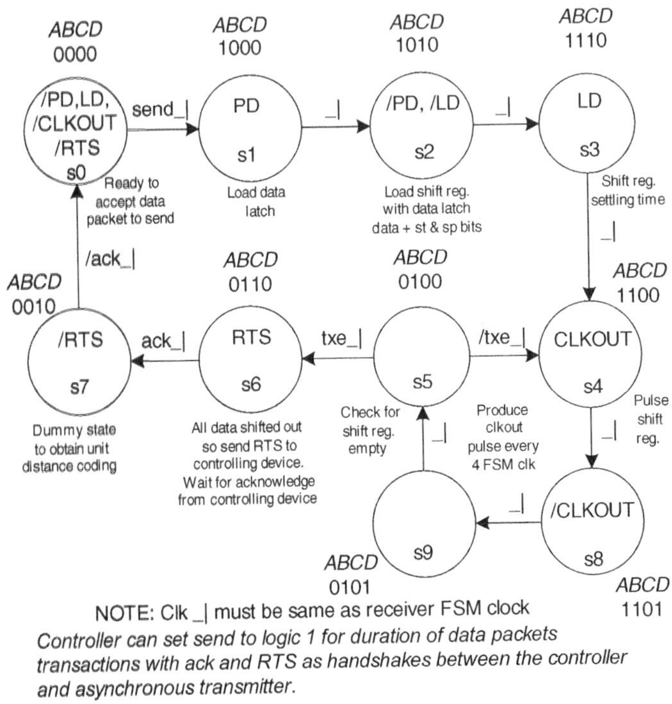

Figure 4.29 State diagram for the asynchronous serial transmitter.

s4, s8, s9, and s5. Note that for a 1 μs baud rate the transmitter FSM clock would need to be 4 MHz.

4.9.1 Equations for the Asynchronous Serial Transmitter

$$A \cdot d = \text{s0} \cdot \text{send} + \text{s1} + \text{s2} + \text{s3} + \text{s4} + \text{s5} \cdot /\text{txe}$$
$$= /B \cdot /C \cdot /D \cdot \text{send} + A \cdot /B \cdot /D + A \cdot C \cdot /D + A \cdot B \cdot /D + B \cdot /C \cdot /D \cdot /\text{txe}$$
$$B \cdot d = \text{s2} + \text{s3} + \text{s4} + \text{s5} + \text{s8} + \text{s9} + \text{s6} \cdot /\text{ack}$$
$$= A \cdot C \cdot /D + B \cdot /C + /A \cdot B \cdot /D \cdot /\text{ack}$$

$$C \cdot d = \text{s1} + \text{s2} + \text{s5} \cdot \text{txe} + \text{s6} + \text{s7} \cdot \text{ack}$$
$$= A \cdot /B \cdot /D + /A \cdot B \cdot /D \cdot \text{txe} + /A \cdot C \cdot /D$$
$$D \cdot d = \text{s4} + \text{s8}$$
$$= A \cdot B \cdot /C \cdot /D + A \cdot B \cdot /C \cdot D$$
$$= A \cdot B \cdot /C$$
$$\text{PD} = \text{s1} = A \cdot /B \cdot /C \cdot /D$$
$$\text{CLKOUT} = \text{s4} = A \cdot B \cdot /C \cdot /D$$
$$\text{LD} = /\text{s2} = /(A \cdot /B \cdot C \cdot /D)$$
$$\text{RTS} = \text{s6} = /A \cdot B \cdot C \cdot /D.$$

A simulation of the FSM results in the waveforms of Figure 4.30. In this simulation, the test sequence is s0, s1, s2, s3, s4, s8, s9, s5, s4, s8, s9, s5, s6, s7, s0.

Using the asynchronous transmitter and receiver FSMs just described, it would be possible with modern FPGAs to run at quite high baud rates, as illustrated below.

FSM clock	Receiver RXCK	Transmitter clock CLKOUT	Baud rate
4 MHz	1 MHz	1 MHz	1 mega baud
8 MHz	2 MHz	2 MHz	2 mega baud
16 MHz	4 MHz	4 MHz	4 mega baud
32 MHz	8 MHz	8 MHz	8 mega baud
80 MHz	20 MHz	20 MHz	20 mega baud

Both transmit and receiver units use the same FSM clock frequency generated with their own clock circuits.

The higher baud rates would need to use twisted-pair cables over relatively short transmission distances up to around 1 m. Transmission line effects would need to be taken into account, but this is beyond the scope of this book.

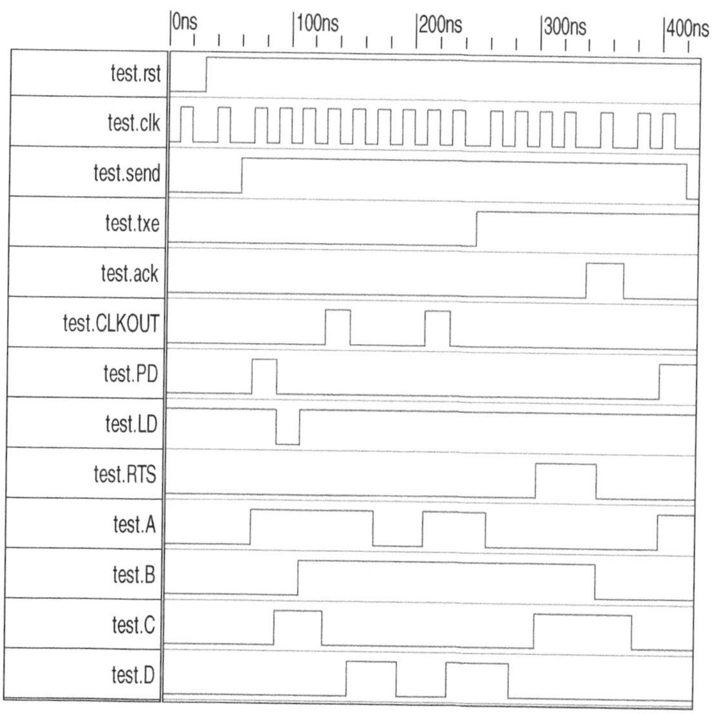

Figure 4.30 Simulation of the serial transmitter FSM.

4.10 CLOCKED WATCHDOG TIMER

Most microcontrollers these days have a built in watchdog timer (WDT). The WDT is an addressable device that can be written to on a regular basis. The idea is that the timer (usually a down counter) is regularly written to reinitialize it to a known count value. Between writes, the counter will be clocked towards zero. If the microcontroller does not write to the WDT between countdown periods, then the counter will reset to zero and this action can be used to reset the microcontroller.

The WDT thus acts as a safeguard to prevent the microcontroller from running out of control (jumping to an instruction that is not part of the program sequence), perhaps due to a transient in the power system.

Another use is in a microprocessor-based system where the operating system (perhaps a real-time operating system) can regularly reset the WDT and, hence, provide a means of determining a microprocessor system failure.

The application program running on the microcontroller needs to write regularly to the WDT to prevent it from reaching the reset state.

Although most microcontrollers have this feature, a lot of microprocessor systems do not. Therefore, a circuit would need to be designed for this purpose.

The clocked FSM system shown in Figure 4.31 is a basic system designed to perform the action of a WDT. The system needs to be designed around the specific memory/IO cycle timing of the microprocessor. In Figure 4.31 the memory/IO write cycle is based around a four-clock pulse cycle time T1 to T4.

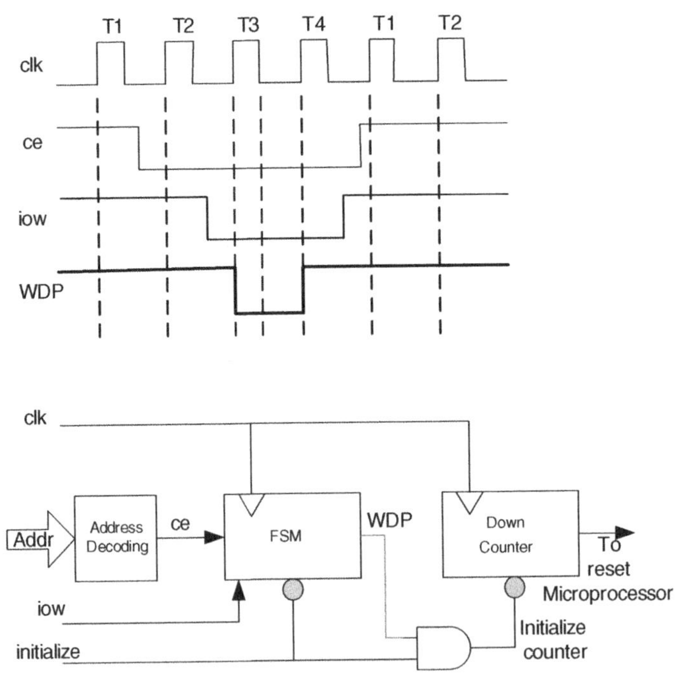

Figure 4.31 Block diagram for a WDT for a microprocessor system.

The system is controlled by an FSM that monitors the chip enable ce controlled by the address decoding logic. This can respond to a particular address from the microprocessor. In addition, the iow signal controlled by the microprocessor is also monitored by the FSM. When the microprocessor addresses the WDT, ce goes low, followed by iow in the T2 clock period. On the rising edge of the T3 clock period, the WDT pulse is generated. The FSM must produce this watchdog pulse (WDP) at exactly the right time in the write cycle (T3 period). Both the FSM and the down counter are clocked by the same microprocessor clock clk.

In Appendix B, the design of a down binary counter is described and Section B.1 shows how this can be done. To provide this counter with a fixed starting value (to count down from), the flip flips of the counter can be preset to a known value, using a parallel loading counter (see Section B.3). This is the purpose of the initialize input in Figure 4.31 (essentially a parallel load input to the down counter).

Note that this same input provides the initial state for the FSM (which will be state zero). The WDP will provide frequent reinitialization pulses to the down counter and, thus, prevent it from reaching its zero state (which would otherwise cause a microprocessor reset).

A suitable state diagram is illustrated in Figure 4.32, wherein the FSM waits in state s0 for the microprocessor to write to the address of the WDT. This will cause ce to go low during the T1 state of the memory/IO cycle (see Figure 4.31) so that on the T2 rising clock edge the FSM will move into s1. Here, it waits for the microprocessor to lower iow; then, on the next clock pulse (T3), the FSM will move into state s2, where it will lower the WDP output signal. On the next clock pulse (T4), the FSM will move to s3, raising the WDP, and wait for the ce signal to go high. This will occur at the end of the memory/IO write cycle and will be seen by the FSM on the rising edge of T1.

The equations for the FSM that follow are from Figure 4.32.

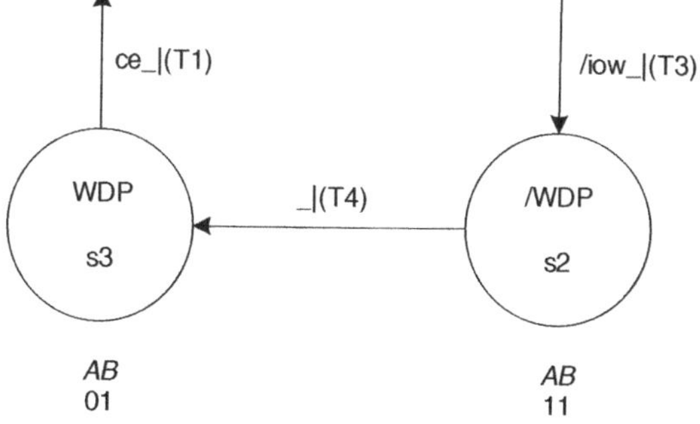

Each clock pulse corresponds to a *T* state

Figure 4.32 State diagram for the WDT.

4.10.1 D Flip-Flop Equations

$$\begin{aligned} A \cdot d &= \mathrm{s0} \cdot /\mathrm{ce} + \mathrm{s1} \\ &= /A \cdot /B \cdot /\mathrm{ce} + A \cdot /B \\ &= /B \cdot \mathrm{ce} + A \cdot /B \\ B \cdot d &= \mathrm{s1} \cdot /\mathrm{iow} + \mathrm{s2} + \mathrm{s3} \cdot /\mathrm{ce} \\ &= A \cdot /B \cdot /\mathrm{iow} + A \cdot B + /A \cdot B \cdot /\mathrm{ce} \\ &= A \cdot /\mathrm{iow} + A \cdot B + B \cdot /\mathrm{ce}. \end{aligned}$$

4.10.2 Output Equation

$$\mathrm{WDP} = /(\mathrm{s2}) = /(A \cdot B).$$

The equation for ce would depend upon the desired address assigned to the WDT. For example, if the address assigned was 300h (11 0000 0000 binary), then the equation would result in

$$\mathrm{ce} = /(\mathrm{a9} \cdot \mathrm{a8} \cdot /\mathrm{a7} \cdot /\mathrm{a6} \cdot /\mathrm{a5} \cdot /\mathrm{a4} \cdot /\mathrm{a3} \cdot /\mathrm{a2} \cdot /\mathrm{a1} \cdot /\mathrm{a0}).$$

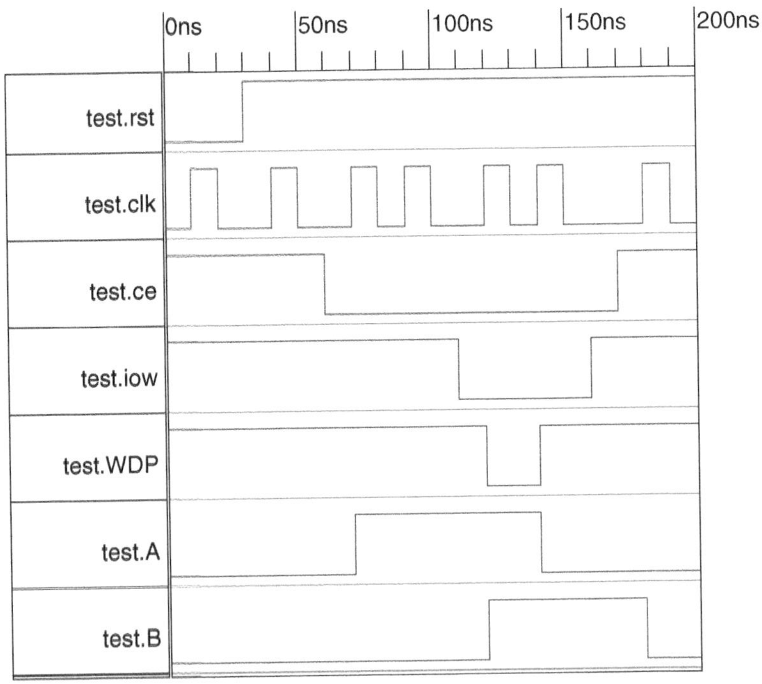

Figure 4.33 The WDT FSM simulation.

There could be additional qualifier signals, i.e. in a PC using the IO memory map the signal /aen would be required in order to distinguish between dynamic memory access (DMA) cycles and IO cycles (see Chapter 5 for DMA). Also, the /iow signal would be needed to identify a write cycle.

The above equation for ce would then be

$$\text{ce} = /(\text{a9} \cdot \text{a8} \cdot /\text{a7} \cdot /\text{a6} \cdot /\text{a5} \cdot /\text{a4} \cdot /\text{a3} \cdot /\text{a2} \cdot /\text{a1} \cdot /\text{a0} \cdot /\text{aen} \cdot /\text{iow}).$$

The equations to describe the down counter are repeated below from Appendix B for convenience.

$$\text{Q}n \cdot t = \prod_{p=1}^{p=n} (/\text{q}p) \quad \text{for an } n\text{-stage counter, with the first } T \text{ } mflip\text{-flop q0} \cdot t \text{ input} = 1.$$

This equation expands to

$$\begin{aligned}
\text{Q0} \cdot t &= 1 \\
\text{Q1} \cdot t &= /\text{q0} \\
\text{Q2} \cdot t &= /\text{q0} \cdot /\text{q1} \\
\text{Q3} \cdot t &= /\text{q0} \cdot /\text{q1} \cdot /\text{q2} \\
\text{Q4} \cdot t &= /\text{q0} \cdot /\text{q1} \cdot /\text{q2} \cdot /\text{q3}.
\end{aligned}$$

for a four-stage down counter.

Note that the counter needs an asynchronous initialization signal connected to each *T* flip-flop to form the parallel loading input logic (see Equation (B.4) and Figure B.4).

Figure 4.33 shows the FSM in action. The output WDP goes low during state s2 after the address-decoding ce and iow have been detected going low in sequence. The FSM state transitions are clearly seen in the flip-flop *A* and *B* outputs.

Note that in the above simulation there are additional clock pulses. These have been generated by the test bench generator to test for the FSM remaining in states s0 and s1 until changes in the ce and iow signals occur. This would not happen in practice, since the microprocessor has control of iow and the address-decoding logic ce.

4.11 SUMMARY

In this chapter, a number of practical examples have been developed using the block diagram and state diagram approach developed in the Chapters 1–3. These have then been implemented in terms of *D*-type flip-flops. You may well decide to use some of these examples in your own designs, or expand upon them to make them fit your own requirements.

In the next chapter, the idea of having a state for each *D*-type flip-flop will be introduced, leading to systems that do not need secondary state variables.

5

The One Hot Technique in Finite-State Machine Design

5.1 THE ONE HOT TECHNIQUE

The FSMs designed up to now have used secondary state variables to identify each state. This requires the use of unit distance assignment, where possible, to try to avoid potential glitches in output signals.

An alternative would be to assign a flip-flop for each state. Although this may be considered wasteful, it has the advantage that it would in theory avoid the generation of output glitches, since each state would have its own flip-flop. At any one time, only one flip-flop would be set, i.e. the one corresponding to the state the FSM was currently in.

This idea is called 'One Hotting' and is much used in FSM designs that are targeted to FPGAs. This is because FPGAs have an architecture that consists of many cells that can be programmed to be flip-flops, or gates. So a large number of flip-flops is not difficult to achieve. A PLD, on the other hand, has an architecture with only a limited number of flip-flops controlled from AND/OR 'sum of product' terms.

Another feature of the One Hot technique is that it can require fewer logic levels because there is no required logic from other state variables apart from the primary inputs and previous state(s). This can result in faster logic speeds.

The method of implementing a 'One Hot' FSM will now be described.

Consider Figure 5.1. In this example of the use of the One Hot technique, the single-pulse generator with memory problem is revisited. It uses three states (rather than the four-state FSM used in the original design). This is possible because one does not have to consider unit distance coding and, hence, there are no secondary state variables.

The equations on the right in Figure 5.1 are the equations necessary to synthesize the FSM. To understand where these come from, consider the One Hot state diagram.

Initially, the FSM should be in state s0. This can be arranged via an initialization input so that the flip-flop representing state s0 (called FFS0) is set, and all other flip-flops (FFS1 and FFS2) are reset.

FSM based Digital Design using Verilog HDL Peter Minns and Ian Elliott

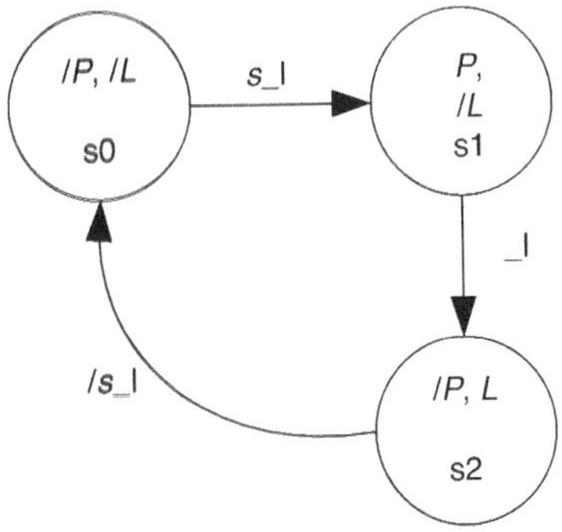

Design equations:

$s0 \cdot d = s2 \cdot /s + s0 \cdot /s$

$s1 \cdot d = s0 \cdot s$

$s2 \cdot d = s1 + s2 \cdot s$

Output is $P = s1$

$L = s2$

The state diagram does not need any secondary state variables since each state is represented by a *D*-type flip-flop.

At initialization, the flip-flops representing s1 and s2 are reset, while that representing state s0 is set.

Figure 5.1 An example of the use of the One Hot technique.

Consider state s0. Here, the FSM should remain in state s0 until the condition to exit s0 occurs. This is, of course, when the primary input signal *s* becomes logic 1.

However, the flip-flop FFS0 needs a signal on its *D* input that will keep it in the set state. The required signal is

$$s0 \cdot /s.$$

This is obtained from the fact that the FSM is in state s0 and the 'leaving condition' from state s0 is *s*, so that while *s* is not true, i.e. $s = 0$, or /*s*, the flip-flop should remain set.

This term s0 · /*s* is known as a 'hold term' because it holds the FFS0 set until it is required to change to the next state, s1.

Also, when the FSM reaches state s2 it will only return to state s0 when the signal *s* is logic 0. So there is another term:

$$s2 \cdot /s.$$

This is known as the 'set term', or 'turn on' term, for the flip-flop.

The complete equation for the state s0 flip-flop FFS0 is

$$s0 \cdot d = \underbrace{s2 \cdot /s}_{\text{set term}} + \underbrace{s0 \cdot /s}_{\text{hold term}}.$$

Now consider state s1. The condition to enter state s1 is when the FSM is in state s0 and $s = 1$. So, the equation for flip-flop FFS1 is

$$s1 \cdot d = s0 \cdot s.$$

Note that the 'leaving condition' from s1 is a simple clock pulse. There is no input condition along the transitional line between s1 and s2; therefore, when the FSM reaches state s1, it will naturally exit state s1 on the next clock pulse, so a 'hold term' is not needed.

Now consider the final state s2.

The condition to enter state s2 is s1, since there is no input condition along the transitional line between states s1 and s2. There will, however, be a holding term between s2 and s0, which is

$$\mathrm{s2} \cdot s.$$

While $s = 1$ the FSM must remain in state s2. So the equation for FFS2 will be

$$\mathrm{s2} \cdot d = \mathrm{s1} + \mathrm{s2} \cdot s.$$

Finally, the output signal is

$$P = \mathrm{s1},$$

since only in state s1 will the output P be logic 1; L will only be active in state s2:

$$L = \mathrm{s2}.$$

The circuit for this FSM is illustrated in Figure 5.2. Note in Figure 5.2 the initialization logic is fitted retrospectively. In a One Hot system, one of the flip-flops, representing the initial state in the FSM, needs to be set, while all other flip-flops need to be cleared. If flip-flops without preset

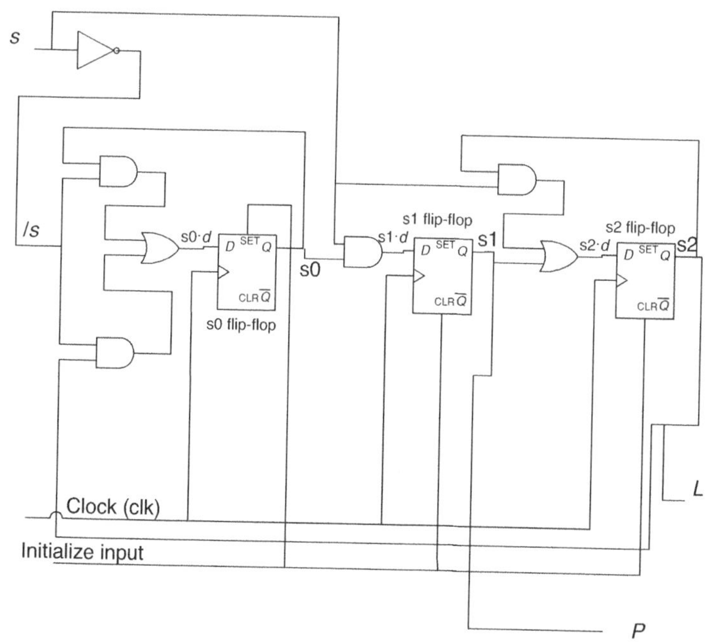

Figure 5.2 Circuit for the One Hot version of the single-pulse FSM.

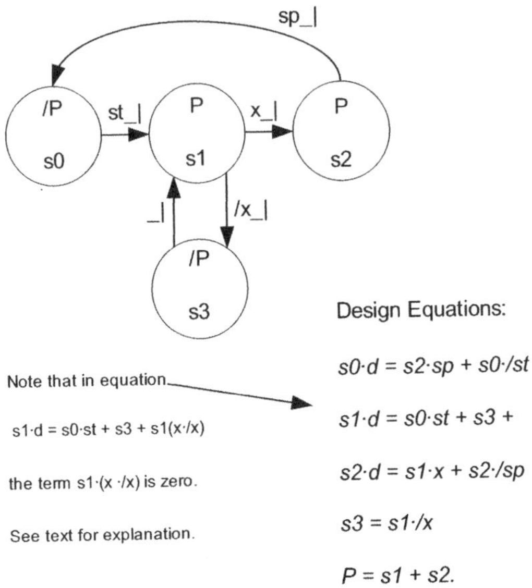

Figure 5.3 A second example with two-way branch.

and clear inputs are used, then a synchronous reset scheme needs to be adopted (as seen in Chapter 3, Frames 3.16 and 3.19).

Now consider the two-way branch FSM design in Figure 5.3. In this example, the equation for FFS0 follows the rules already explained for the first example. In the equation for FFS1, however, note that there is a term for entering state s1 via s0 (s0 · st) and a term to enter via s3.

The two-way branch leaving state s1 is via s1 · x (to state s2) and s1 · $/x$ (to state s3), and the combined terms result in

$$\mathrm{s1} \cdot d = \mathrm{s0} \cdot \mathrm{st} + \mathrm{s3} + \mathrm{s1} \cdot x \cdot /x,$$

which reduces to

$$\mathrm{s1} \cdot d = \mathrm{s0} \cdot \mathrm{st} + \mathrm{s3}$$

because the s1 · x · $/x$ terms would reduce to zero:

$$\mathrm{s1}(/x \cdot x) = 0.$$

The FSM is held in s1 by complementing the inputs such that the leaving term between s1 and s2 (x) is complemented ($/x$) and the leaving term between s1 and s3 ($/x$) is also complemented (x) so as to imply a hold in s1. Of course, this leads to

$$\mathrm{s1}(/x \cdot x) \text{ as } \mathrm{s1}(1 \cdot 0) \text{ or } \mathrm{s1}(0 \cdot 1) \quad \text{resulting in the term s1 being zero.}$$

Looking at the state diagram of Figure 5.3, it can be seen that once the FSM reaches state s1 it should leave this state either via the transition to state s2 or via the transition to state s3 on the next clock pulse. There is no reason to hold it in state s1.

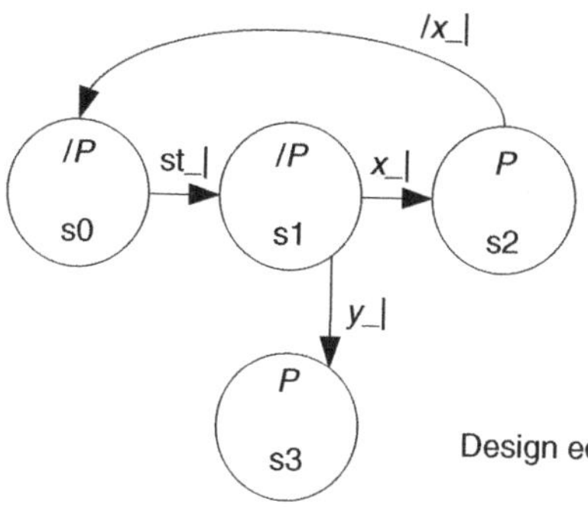

Design equations:

$$s0 \cdot d = s2 \cdot /x + s0 \cdot /st$$

$$s1 \cdot d = s0 \cdot st + s1 \cdot (/x \cdot /y)$$

$$s2 \cdot d = s1 \cdot x + s2 \cdot x$$

$$s3 = s1 \cdot y + s3$$

$$P = s2 + s3$$

Figure 5.4 An example with a two-way branch with noncomplementary inputs.

Therefore, the above interpretation for s1 is correct. Hence, the equation

$$s1 \cdot d = s0 \cdot st + s3$$

is the correct one.

Note: in a state diagram with a two-way branch transition with complementary inputs (in this case x and $/x$), the two-way branch term is dropped.

The other equations in Figure 5.3 follow in the usual way.

Now consider the following FSM shown in Figure 5.4. In this example there is again a two-way branch, but this time the exit from each branch path is not complementary. Notice how the equation for s1 $\cdot$ d contains a term

$$s1 \cdot (/x \cdot /y).$$

This is the required holding term that will hold the FSM in state s1 until either x becomes logic 1 or y becomes logic 1, i.e. the FSM will remain in state s1 while both x and y are logic 0.

Note: when using a two-way branch with different inputs along each transitional line (like x and y), the two inputs (x and y) must be mutually exclusive.

Continuing with example of Figure 5.4, the invariant state s3 is entered from state s1, but once it is entered there is no transition from this state. The FSM will remain in state s3 until the FSM is reinitialized to its initial state of s0. For this reason, the s3 term on the right-hand side of the equation for s3 is needed.

Figure 5.5 shows an example you might like to attempt on your own. Do not look at the solution below the figure until you have attempted to do it yourself.

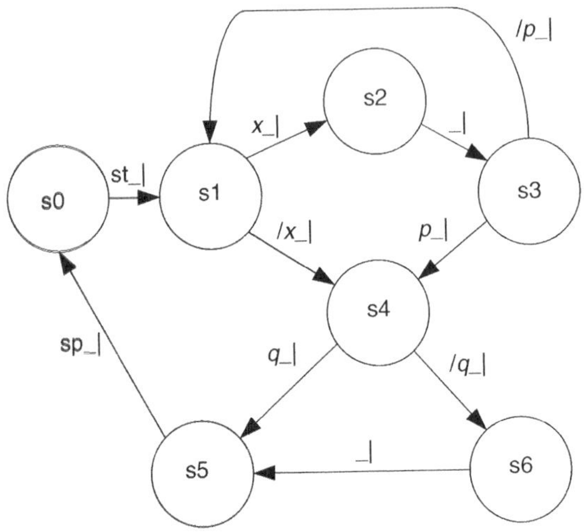

Solution:

$$s0 \cdot d = s5 \cdot \mathrm{sp} + s0/\mathrm{st}$$

$$s1 \cdot d = s0 \cdot \mathrm{st} + s3 \cdot /p$$

$$s2 \cdot d = s1 \cdot x$$

$$s3 \cdot d = s2$$

$$s4 \cdot d = s1 \cdot /x + s3 \cdot p$$

$$s5 \cdot d = s4 \cdot q + s6 + s5 \cdot /\mathrm{sp}$$

$$s6 \cdot d = s4 \cdot /q$$

Figure 5.5 Example for the reader. Do not look at the solution below until you have attempted to do it yourself.

The One Hot technique is ideal for large state machines to be implemented using FPGA devices, since an FPGA can accommodate a large number of flip-flops. Also, the development of the equations is very easy for a design developed at the logic gate level.

The rest of this chapter looks at a number of more complex FSM examples making use of the One Hot technique. The following examples illustrate how an FSM can be used to implement typical design problems where perhaps a microcontroller might have been used. Each example features ideas that you might wish to incorporate into your own designs.

5.2 A DATA ACQUISITION SYSTEM

Usually, a microcontroller, or digital signal processor (DSP), is used to implement a DAS. In the case of the microcontroller the ADC is built into the microcontroller chip. For applications using a microcontroller with built-in ADC, the system will usually make use of integer data values from the ADC. For DASs requiring high-speed data calculations, a DSP may be used. These can

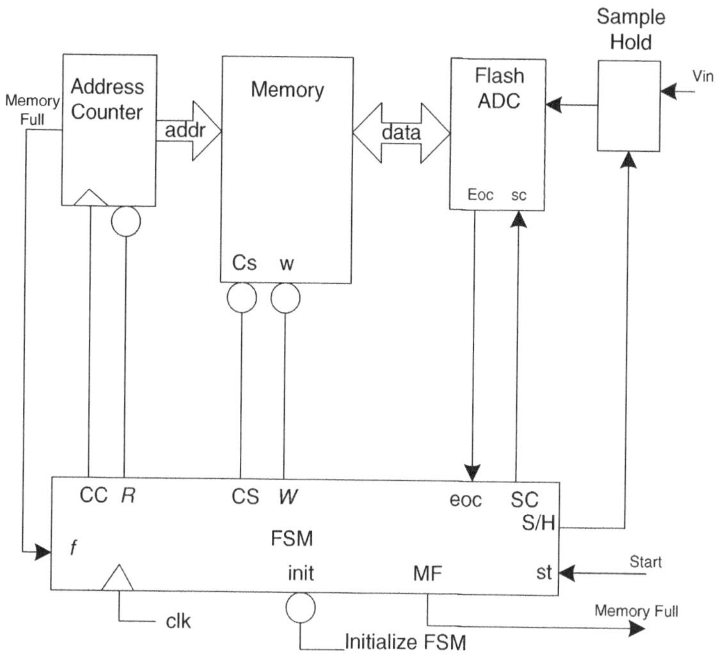

Figure 5.6 Basic high-speed DAS.

be obtained using either integer arithmetic circuits or a built-in floating-point processor to carry out the processing with 'real' numbers.

One problem with all DASs is that they have finite processing speed limitations, usually due to the processing limitations of the microprocessor used. To some extent this can be overcome by using parallel processing and hardware arithmetic circuits.

A totally hardware arrangement could be designed around an FSM controlling hardware adder/subtractor/multiplier/divider subsystems. This could increase the throughput of such systems. Alternatively, the FSM could be used to 'gather' the data and store it for subsequent processing by a microprocessor or DSP in situations where 'real-time' processing is not required.

This next example illustrates a much simpler system looked at in Chapter 2 and illustrated in Figure 5.6. This basic system could use a flash ADC to allow very fast conversion times. The overall system makes use of high-speed static RAM to store the converted digital values. The system is designed to interact with another system. This other system starts the process off by asserting the st input, and the FSM sends a memory full (*f*) response in due course.

For now, a state diagram can be developed for this basic system as illustrated in Figure 5.7. This is much along the lines of the one developed in Frames 2.4–2.10. In this state diagram, the sequence of control is clear. Once the external system sends a request for the system to start filling the memory with data (st = 1), the following occurs:

- The sample-and-hold circuit is placed into hold mode ready for the ADC (s1).
- The flash ADC is placed into conversion mode and the FSM waits for the end of conversion eoc signal to go high, signifying that a conversion has taken place (s2).
- In s3, the FSM selects the memory device by asserting (low) its chip select input CS. The FSM will move to s4 only when the ADC eoc signal returns to logic 0.

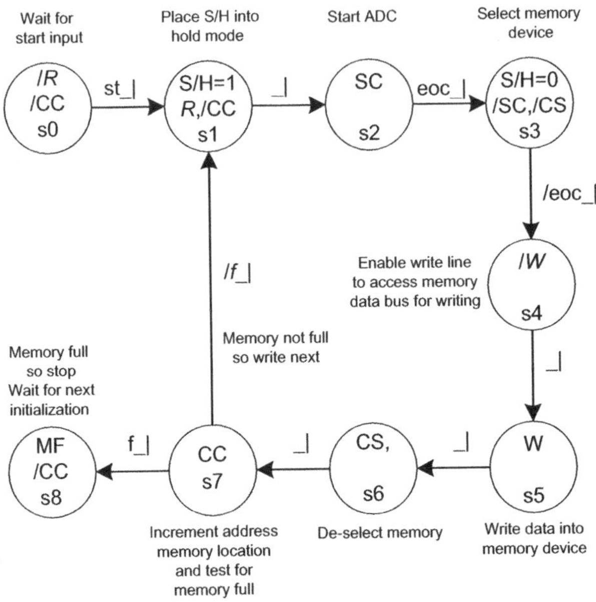

Figure 5.7 State diagram for the DAS.

- In s4, the FSM activates the memory chips write enable signal *W* (low).
- In s5, the memory write signal is taken high to write the data into the memory device.
- In s6, the chip select is taken high to deselect the memory. This ensures that the memory chip is deselected before the address is changed.
- In s7, the address counter is pulsed by making CC $= 1$; the address counter is pulsed on the rising edge of this signal. In s7, a check is made to see whether the last memory location has been used (f); if not, the FSM moves around the loop comprising s1 to s7 again.
- This will continue until all the available memory has been filled with data, at which point the FSM will fall into s8 and assert the mf output to the external device.

Note that the MF signal could be connected to the interrupt input of the remote device so that it could start the process with st $= 1$ and be interrupted when the task is complete.

The One Hot equations now follow:

$\mathrm{s0} \cdot d = /\mathrm{st}$ flip-flop s0 will be set during initialization and held until st $= 1$

$\mathrm{s1} \cdot d = \mathrm{s0} \cdot \mathrm{st} + \mathrm{s7} \cdot /f$

$\mathrm{s2} \cdot d = \mathrm{s1} + \mathrm{s2} \cdot /\mathrm{eoc}$

$\mathrm{s3} \cdot d = \mathrm{s2} \cdot \mathrm{eoc} + \mathrm{s3} \cdot \mathrm{eoc}$

$\mathrm{s4} \cdot d = \mathrm{s3}$

$\mathrm{s5} \cdot d = \mathrm{s4}$

$\mathrm{s6} \cdot d = \mathrm{s5}$

$\mathrm{s7} \cdot d = \mathrm{s6}$

$\mathrm{s8} \cdot d = \mathrm{s7} \cdot f + \mathrm{s8}$ will hold in this state until reset.

The outputs are

$S/H = s1 + s2$

$SC = s2$

$CS = /(s3 + s4 + s5)$ an active-low signal in states s3 to s5

$W = /s4$ an active-low signal in state s4 only

$MF = s8$

$R = /s0$ an active-low signal in state s0 only

$CC = s7$ pulsing CC high as s7 is entered; CC goes low on leaving state s7.

These signals can be used to construct a Verilog file and simulated, as illustrated in Figure 5.8. From Figure 5.8, it can be seen that the FSM loops four times, ending up in s8 at the end of the third loop. Note the control of the memory chip select and write signals and the address counter pulses. Also, at the end of the simulation the memory full mf signal goes high in state s8. The reset is applied to return the FSM to s0.

The system developed in Figure 5.6 allows digitized data to be stored into the memory, but it does not provide any way of getting access to the data once it has been saved. The reader

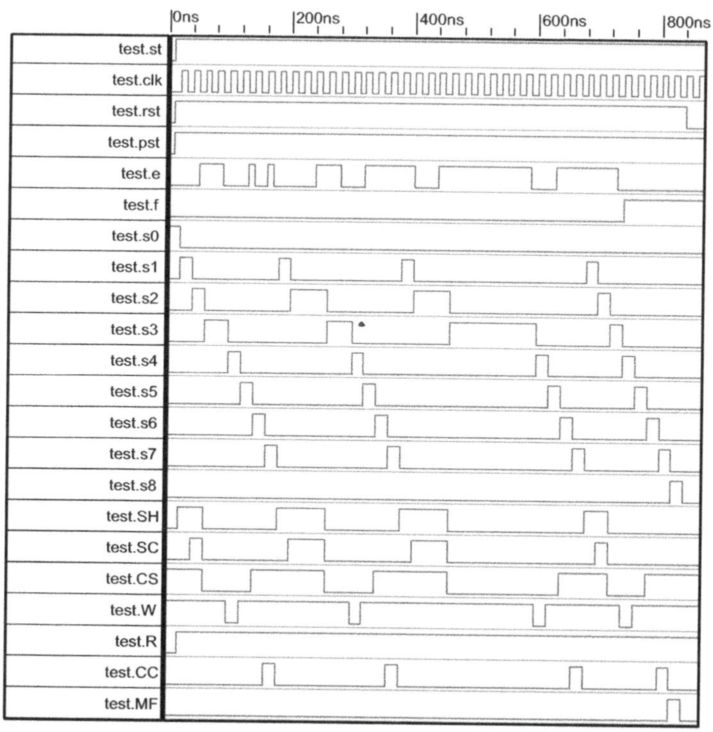

Figure 5.8 Simulation of the data acquisition FSM controller.

might like to modify the system to allow this to happen, but some thought needs to be given to what device is to be used to perform this operation.

The next example illustrates how memory can be controlled in this way.

5.3 A SHARED MEMORY SYSTEM

It is often required to be able to access the data stored in memory via some other controlling device. For example, this could be an external microprocessor to process the stored data in the memory. The example in Figure 5.9 illustrates how this might be done. In this system the memory can be accessed by either the FSM or the external system (which could be a microprocessor or DSP system). The memory is, in effect, 'shared'. The idea is that during the data-gathering phase, the FSM has sole access to the memory and deposits digitized samples of data under its own control. During the data delivery phase the external device can access the memory, but only when there is data to be read.

The external device must wait for the RMA (read memory available) signal going high, for only when this signal is high will the FSM have disconnected itself from the memory device. Also, when the external device has completed the read memory transaction, and disconnected itself from the memory device, it must send an acknowledge signal ack to the FSM so that the FSM can revert to its initial state. The FSM in this system is the master device. Signals RMA and ack form a handshake mechanism.

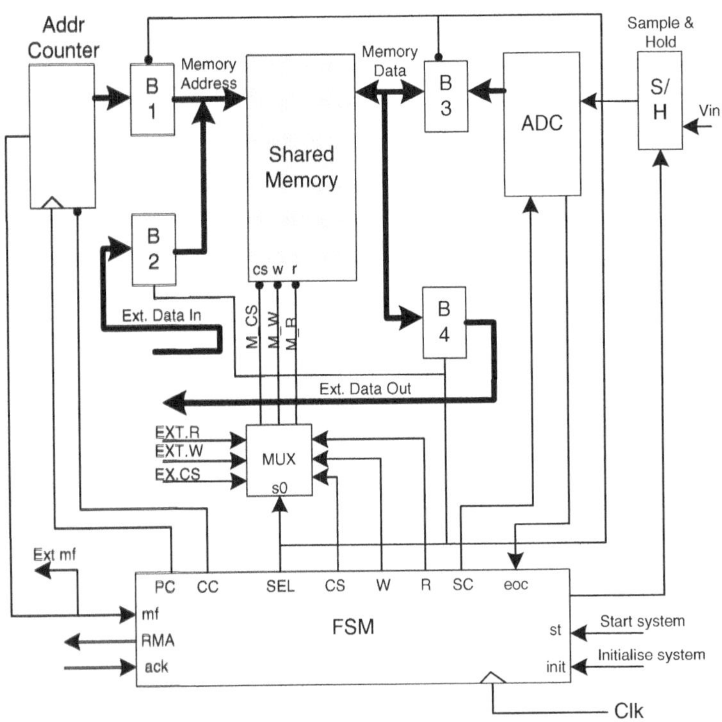

Figure 5.9 Block diagram of a shared memory system.

Note that the FSM uses its SEL signal to control the selection of the tri-state buffers B1 to B4, so that buffers B1 and B3 are selected when SEL = 0. Buffers B2 and B4 are selected by making SEL = 1 to allow the external device to control the memory.

The 'tri-state' devices are thus connected to the memory device to allow it to be 'shared'.

- The tri-state buffers B1 to B4 control the connection of the address and data buses.
  The two-way Multiplexer M is used to control the memory device from the two sources (FSM and external device).
- When its control input s0 = 0, the CS, *W*, and *R* control lines from the FSM are connected to the memory device. Otherwise, the external device has control of these three signals when s0 = 1.

The following equations describe the behaviour of the multiplexer:

$$\mathrm{M\_CS} = \mathrm{CS} \cdot /\mathrm{SEL} + \mathrm{EXT\_CS} \cdot \mathrm{SEL}$$
$$\mathrm{M\_W} = W \cdot /\mathrm{SEL} + \mathrm{EXT\_W} \cdot \mathrm{SEL}$$
$$\mathrm{M\_R} = R \cdot /\mathrm{SEL} + \mathrm{EXT\_R} \cdot \mathrm{SEL}.$$

Note that the handshake signals RMA and ack are mandatory for this system to work, since the external device must not have access to the memory unless it receives the RMA = 1 from the FSM. Likewise, only when the external device has disconnected itself from the memory can it send the ack = 1 signal to the FSM.

The state diagram for this system (Figure 5.10) is very similar to that in Figure 5.7, but has signals to control the memory device connection.

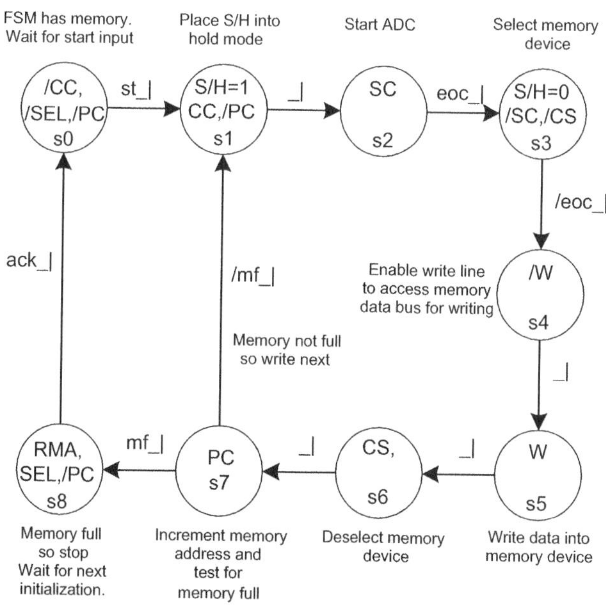

Figure 5.10 State diagram for the shared memory FSM system.

Note that in the state diagram in Figure 5.10 it is assumed that the ADC is slower than the time for the FSM to move from state s3 back round to state s2 and in s3 it waits for eoc to return low before moving to s4.

The equations for this design can be obtained from the state diagram as follows.

D flip-flop *d* inputs:

$$\begin{aligned}
\mathrm{s0} \cdot d &= \mathrm{s8} \cdot \mathrm{ack} + \mathrm{s0} \cdot /\mathrm{st} \\
\mathrm{s1} \cdot d &= \mathrm{s0} \cdot \mathrm{st} + \mathrm{s7} \cdot /\mathrm{mf} \\
\mathrm{s2} \cdot d &= \mathrm{s1} + \mathrm{s2} \cdot /\mathrm{eoc} \\
\mathrm{s3} \cdot d &= \mathrm{s2} \cdot \mathrm{eoc} + \mathrm{s3} \cdot \mathrm{eoc} \\
\mathrm{s4} \cdot d &= \mathrm{s3} \cdot /\mathrm{eoc} \\
\mathrm{s5} \cdot d &= \mathrm{s4} \\
\mathrm{s6} \cdot d &= \mathrm{s5} \\
\mathrm{s7} \cdot d &= \mathrm{s6} \\
\mathrm{s8} \cdot d &= \mathrm{s7} \cdot \mathrm{mf} + \mathrm{s8} \cdot /\mathrm{ack}.
\end{aligned}$$

Output equations:

$\mathrm{CC} = /\mathrm{s0}$ active-low output

$\mathrm{SEL} = \mathrm{s8}$

$\mathrm{RMA} = \mathrm{s8}$

$\mathrm{S/H} = \mathrm{s1} + \mathrm{s2}$

$\mathrm{SC} = \mathrm{s2}$

$\mathrm{CS} = /(\mathrm{s3} + \mathrm{s4} + \mathrm{s5})$ active-low output

$W = /\mathrm{s4}$ active-low output

$\mathrm{PC} = \mathrm{s7}$ and assumes that the address counter is positive-edge triggered. PC reverts to its inactive (PC = 0) state on leaving s7.

5.4 FAST WAVEFORM SYNTHESIZER

A number of design issues will be covered in this example, including some aspects of interfacing to a microprocessor or microcontroller to an FSM-based design.

A frequency synthesizer is to be developed based around an FSM. The idea here is to be able to transfer a set of data from a microprocessor/microcontroller via a parallel portal into a memory device. Once this is done, the FSM is to read consecutive memory locations and output them to a DAC. A block diagram of the system is illustrated in Figure 5.11.

Note that the waveform data may be any number of data samples in the memory, depending upon the waveform period and sampling frequency. Therefore, the memory full signal mf is actually an ‘end of waveform’ signal, generated by comparing the address bus value with an ‘Address Limit Value’ sent by the controlling device.

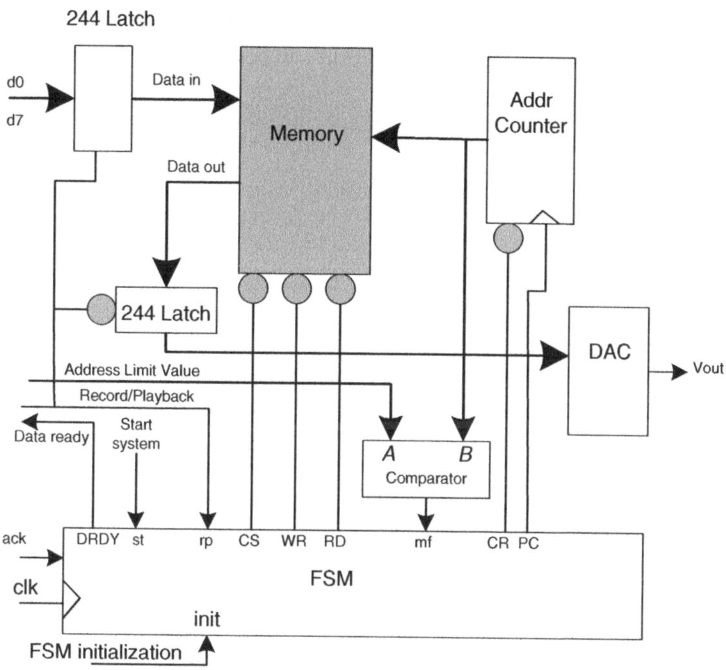

Figure 5.11 The fast waveform synthesizer block diagram.

Of course, the total number of waveform samples must be able to fit into the memory device, but the end of waveform must be detected so that when the FSM cycles through to memory location zero the waveform at the DAC output looks continuous and starts at the correct point in the waveform.

In this diagram, the parallel ports to/from a microcontroller, say, are used to provide waveform data to the memory. st is the start input and rp is an input to define record mode (logic 1) or playback mode (logic 0). These two inputs could be from the microcontroller or simply provided as user-activated switches.

5.4.1 Specification

On power up, the FSM looks for st asserted. Then, if the rp is logic 1, it will assert its DRDY output high to let the microcontroller know that it is expecting a data byte. The micro-controller puts a data byte onto the parallel port outputs d0 to d7. The FSM then writes a data byte to the memory device and then lowers its DRDY signal, to let the microcontroller know it has dealt with the data byte. On seeing the DRDY signal go low, the microcontroller lowers its ack signal line to let the FSM know that the transfer is complete. This process continues until the memory is full. Note that memory full depends upon the number of waveform samples placed into the memory device. The microcontroller places a limit value onto the data lines, so that the FSM has a memory limit value to reach. At this point the memory full signal mf will go to logic 1.

If the input rp is turned to the play position, then the FSM will start to send the data in the memory repeatedly to the ADC so that the waveform will be displayed until such times as the st input is disasserted.

A state diagram will be created based upon the specification and then implemented using One Hot equations.

5.4.2 A Possible Solution

This is a relatively complex design making use of a program running on the microcontroller to control the system via the parallel ports.

The state diagram needs two main loop paths: one for record mode and the other for playback mode. By making use of Mealy outputs, it is possible to produce a state diagram using 13 states. This is illustrated in Figure 5.12.

There are, of course other possible solutions, some of which will contain more states (particularly if the outputs are all Moore). This solution makes use of Mealy outputs so that the main part of the loop can be used for both write and read operations. The *R* and *W* signals are active-low signals and are dealt with in the manner discussed in Frame 3.26.

A brief description of the state diagram is now given.

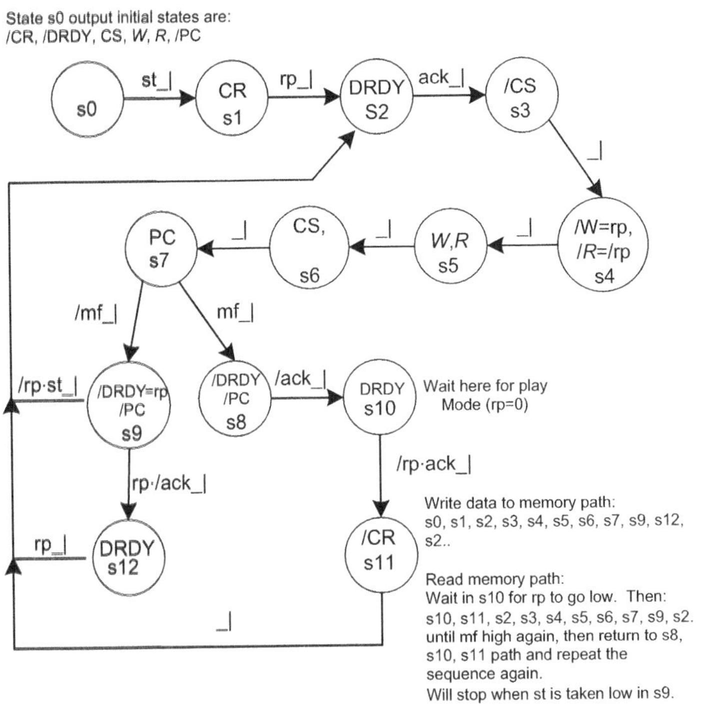

Figure 5.12 One Hot State diagram for the waveform synthesizer FSM.

On operation of the start input st the state machine will leave state s0 to s1 where it will remove the address counter reset CR before moving, on the next clock pulse, to s2 to raise its ready flag DRDY. On receiving the DRDY signal from the FSM, the microcontroller (via its parallel port) will enable the tri-state data buffer connecting the parallel port to the memory data bus so that data can be written to the latter – this by making rp = 1. This will also disable the other tri-state buffer used for reading the memory data. The microcontroller will raise its ack signal to allow the FSM to move to state s3, the memory chip select will be activated (CS = 0) to enable the memory device, and on moving to s4 the memory write W will be lowered, since rp = 1 (write mode). Note that in memory play mode rp = 0 it will be the read signal line that will be lowered in state s4. On moving to s5, the CS and W (or R) will be raised to perform the memory write (or memory read) of that particular memory location.

The FSM will, on the next clock pulse, move to s6 to deselect the memory chip before moving on to s7, where it will raise the PC signal to pulse the address counter. A test will be performed to see whether the memory is full. If the memory is not full, then the state machine will follow the path s7 to s9, where it will lower the DRDY flag if in record mode (rp = 1) and wait for an ack from the microcontroller (this allows the microcontroller to prepare the next data byte to be sent to the memory). On reaching state s12 the state machine will move on to state s2 to repeat the operation for the next memory location. Note, as usual, PC is lowered on leaving s7.

This will continue until all of the memory is full. When this happens, the transition from s7 will be to s8, not s9, and the state machine will send its usual DRDY to zero and wait for acknowledgement from the microcontroller. On receiving the acknowledgement flag ack, it will wait in s10 for the user to set the rp input to zero (indicating that the system is now in playback mode).

In playback mode, the state machine will move to state s11 to reset the address counter and thereby back to s2 to repeat the loop s2, s3, s4, s5, s6, s7, s9, and s2 repeatedly while rp = 0 and st = 1. In this loop, the memory is being read, but now, since rp = 0, the address counter will continue to roll over to zero after running through the memory up to the memory limit value until the start input st = 0.

Note that the FSM waits for ack to be disasserted in states s8 and s9 to complete the handshakes.

A reset can be added to the system to force it back to state s0 at any point in the state sequence. Development of the One Hot equations from the state diagram can now be undertaken.

5.4.3 Equations for the d Inputs to D Flip-Flops

$\mathrm{s0} \cdot d = \mathrm{s0} \cdot /\mathrm{st}$ hold term only

$\mathrm{s1} \cdot d = \mathrm{s0} \cdot \mathrm{st} + \mathrm{s1} \cdot /\mathrm{rp}$

$\mathrm{s2} \cdot d = \mathrm{s1} \cdot \mathrm{rp} + \mathrm{s11} + \mathrm{s12} \cdot \mathrm{rp} + \mathrm{s9} \cdot /\mathrm{rp} \cdot \mathrm{st} + \mathrm{s2} \cdot /\mathrm{ack}$

$\mathrm{s3} \cdot d = \mathrm{s2} \cdot \mathrm{ack}$

$\mathrm{s4} \cdot d = \mathrm{s3}$

$\mathrm{s5} \cdot d = \mathrm{s4}$

$\mathrm{s6} \cdot d = \mathrm{s5}$

$s7 \cdot d = s6$

$s8 \cdot d = s7 \cdot mf + s8 \cdot ack$

$s9 \cdot d = s7 \cdot /mf + s9 \cdot /(rp \cdot /ack) \cdot /(/rp \cdot st)$ note hold term for two-way branch

$s10 \cdot d = s8 \cdot /ack + s10 \cdot /(/rp \cdot ack)$

$s11 \cdot d = s10 \cdot /rp \cdot ack$

$s12 \cdot d = s9 \cdot rp \cdot /ack + s12 \cdot /(rp).$

The output equations follow.

5.4.4 Output Equations

$CR = /(s0 + s11)$

$DRDY = s2 + s3 + s4 + s5 + s6 + s7 + s10 + s11 + s12$ alternatively, $DRDY = /(s8 + s9 \cdot rp)$ as an active-low signal

$CS = /(s3 + s4 + s5)$

$W = /(s4 \cdot rp)$

$R = /(s4 \cdot /rp)$

$PC = s7.$

These can all be implemented in Verilog HDL directly.

5.5 CONTROLLING THE FINITE-STATE MACHINE FROM A MICROPROCESSOR/MICROCONTROLLER

In order to develop the program, one needs a programmer's model to illustrate the connection interface between the FSM and the microcontroller.

From Figure 5.13 it can see that the microcontroller needs to use a byte-wide output port to send waveform data to the memory, and two additional bits to form a handshake between the microcontroller and the FSM. There is also a need for a byte-wide output port to send the memory limit value. The main purpose of the microcontroller is to generate the waveform data to be used by the FSM-based synthesizer. It is beyond the scope of this book to go into how this might be done, but the individual digital values could be computed by the microcontroller to be sent to the memory device.

Listing 5.1 illustrates a program fragment for possible execution on the microcontroller. The program is written in C, which is very common for microcontroller programming.

```
//-- includes needed by the program -------------------------
#include <microcontroller.h> // standard C header file for the particular
microcontroller.
//-----printer port register addresses -----------------------
#define dataport 0x300 // address for port data outputs (change to suit
microcontroller)
#define ackdryrp 0x301 // address for handshake bits and rp (change to suit
microcontroller)
```

```
#define memlim 0x302 // address for the memory limit portal.
#define MAX 1024 // Limit of memory size - can be
          // changed to suit your requirements. Not used in this example.
unsigned char mem_limit_value; // location to save limit value in.
// C Fuction prototypes used by the program.
void get_data(void); // used to get the data from the FSM.
void Send_data_to_FSM(void); // Use to send data to the parallel port.
int i;
unsigned char inbyte, outbyte;
unsigned char array[MAX] ;
//--main program function------------------------------
int main(void)
{
    get_data(); // a C function that deals with the data you want to send.
    Send_data_to_FSM(); // see below.
    // could do other things here.
return (0); terminate the C program here.
} // end of main program.
// The C functions now follow.
void Send_data_to_FSM(void)
{
    mem_limit_value = 255; //get the memory limit value to send.
    MemLim = mem_limit_value; // send limit value to its portal.
    for(i = 0; i < sizeof(array); i++)
      {
        do{ // wait for data ready flag to go low from FSM.
          inbyte = ackdryrp; // input from the ackdry port register.
          inbyte &= 0x01; //mask all bits except the drdy bit.
      } while(inbyte != 0x00); //keep on looping until data ready flag set
from FSM (active-low).
//---------------------------------------------------
      outbyte = array[i] ; //get next data byte to send to FSM from array.
      dataport = outbyte; // send it to FSM.
      ackdryrp |= 0x02; //set ack bit to tell FSM
      do{ // wait for drdy to go high again.
        inbyte = acktryrp;
        inbyte &= 0x01;
      } while(inbyte != 0x01);
      }         // end of for loop.
}               // end of C function to send data to FSM.
void get_data(void)
{       // just generate data for a ramp waveform. Simple example.
    for (i = 0; i < mem_limit_value; i++)
      {
        array[i] = i;
      }
}       // end of get_data;
```

Listing 5.1 Example C code to control the waveform synthesizer.

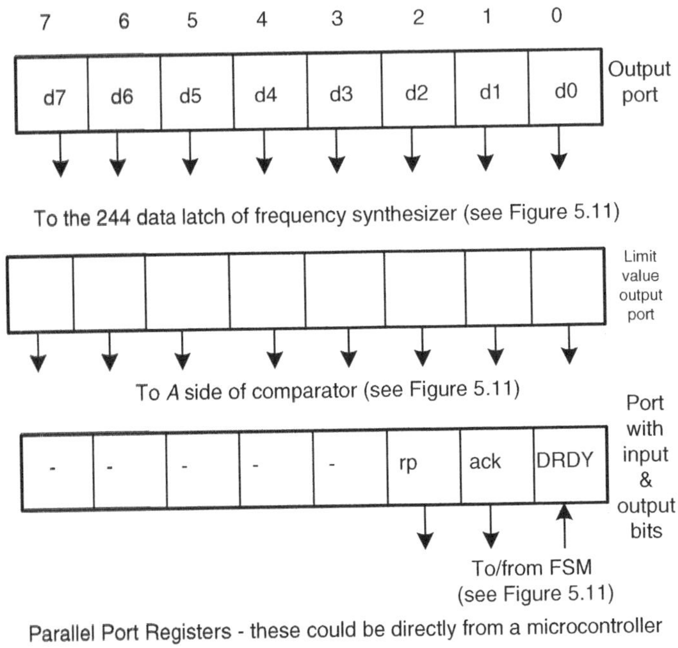

Figure 5.13 Parallel port registers and their bit functions.

Listing 5.1 is very generic and would need to be tailored to a particular microcontroller. It is made up of a main program function `main` () which calls two C functions.

In this example, the first of these functions, get_data (), is used to create a simple ramp waveform by writing bytes to an array with the line

```
array[i] =i;
```

up to a memory limit value. The **`for`** loop simply increments the i value from 0 up to `mem_limit_val` and stores it into consecutive elements of the array. Note, `mem_limit_val` would be the value sent to the Comparator *A* inputs in Figure 5.11 to activate the mf signal when the address inputs from the counter were the same as the 'Address Limit Value'.

The second C function takes the content of the array and sends it to the FSM memory, via the dataport of the microcontroller:

```
outbyte=array[i]; //get next data byte to send to FSM from array.
dataport=outbyte; // send it to FSM.
ackdry=0x02; //set ack bit to tell FSM.
```

To control this operation, and to synchronize the FSM to the microcontroller, the `dry` and `ack` signals are used as handshake signals. The microcontroller uses **`do-while`** loop constructs to perform these operations.

```
do{ // wait for data ready flag to go low from FSM.
  inbyte = ackdry; // input from the ackdry port register.
  inbyte &= 0x01; //mask all bits except b1, the dry bit.
} while (inbyte != 0x00); //keep on looping until data ready flag
cleared from FSM.
```

The **do-while** loop is used to read in the status of the drdy bit `(inbyte = ackdry)`. This is then stripped of all bits except the bit b0 dry with the instruction inbyte `&= 0x01`. This is compared with `0x00`, and if not equal (!=) causes the **do-while** loop to repeat until dry is set to zero, making the **while** `(inbyte != 0x00)` false and causing the program to fall out of the **do-while** loop. In this way, the program cannot get past the first **do-while** loop until `dry = 0`. The second **do-while** loop looks for drdy to go high before getting the next data value from the array to send to the FSM.

The program continues to repeat the actions again until all the data in the array have been sent to the FSM memory.

This short description should give you an insight into how the waveform data can be sent to the FSM. For the generation of more complex data, e.g. sine waves and exponentially decaying sine waves, a more complex `get_data()` function would need to be developed.

5.6 A MEMORY-CHIP TESTER

An FSM-based test system can be used to test memory chips prior to fitting them onto a circuit board. Fitting memory chips direct from the manufacture can be expensive if a faulty memory device is discovered at the final testing stage of production and the defective memory has to be removed, particularly if the device is soldered directly onto the printed circuit board.

The memory tester could typically be used in the Goods Inward Department of a factory that was using a large number of memory chips. This would allow each memory chip to be tested and could form the basis of a quality control on overall quality of the memory chips received from a particular manufacturer. The memory tester should be easy to use by an unskilled operator and function as a 'go–no-go' tester.

The basic idea is to write some data into the memory chip and read the data back to check that they are the same. In such a test, any location found to be faulty would deem the memory chip to be faulty and it would, therefore, be rejected.

Figure 5.14 illustrates the block diagram for the memory tester. In this version, the data 55 hex (0101 0101 binary) is written into each consecutive memory location, then read back and compared using the digital bitwise comparator. The bitwise comparison follows the Boolean equation

$$\text{Bit}n = /(\text{A}n \,\hat{}\, \text{B}n),$$

where ^ is the exclusive OR operator. This operation is (with the NOT operator /) the exclusive NOR i.e. exclusive OR negated. *n* represents the bit being ex-NORed. The ex-NOR operation is shown below for completeness.

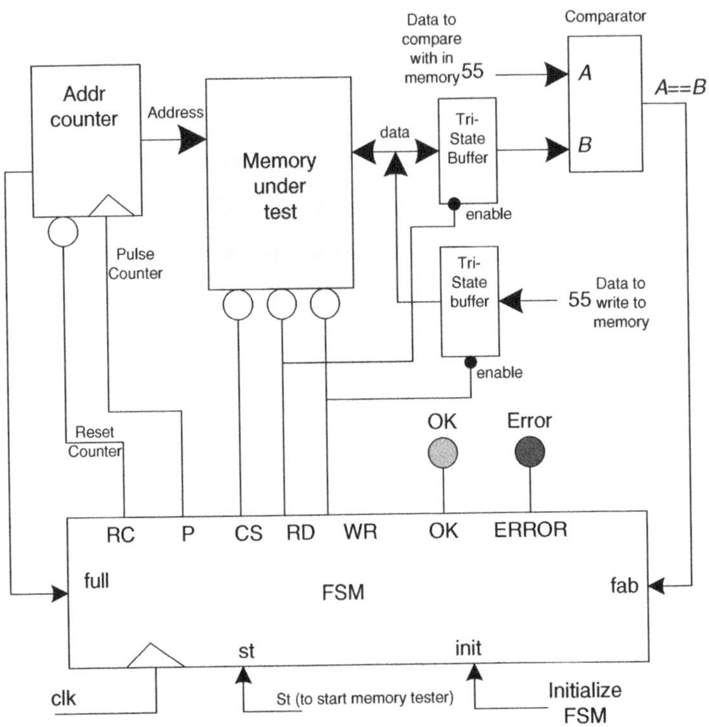

Figure 5.14 Block diagram for the memory tester.

An	^	Bn	Bit$n$
0		0	1
1		0	0
0		1	0
1		1	1

The system can be started by raising input st, the start input. The FSM will control the memory operations and test the fab input to determine whether what was written is the same as what is read.

A more sophisticated version could be developed in which each memory location is tested with the data 55 hex, then retested with the data AA hex to check for adjacent stuck at 1 or 0 faults. Other tests, such as checking adjacent memory locations to test for inter-memory location faults, could also be included; however, for this simple tester the 55 hex data will suffice.

The output '$A = B$' connected to the fab input of the FSM is the logical product of all 8-bit comparisons bit0–bit7; so, if all exclusive NOR outputs are at logic 1, then the '$A = B$' output will be logic 1. This is expressed mathematically as

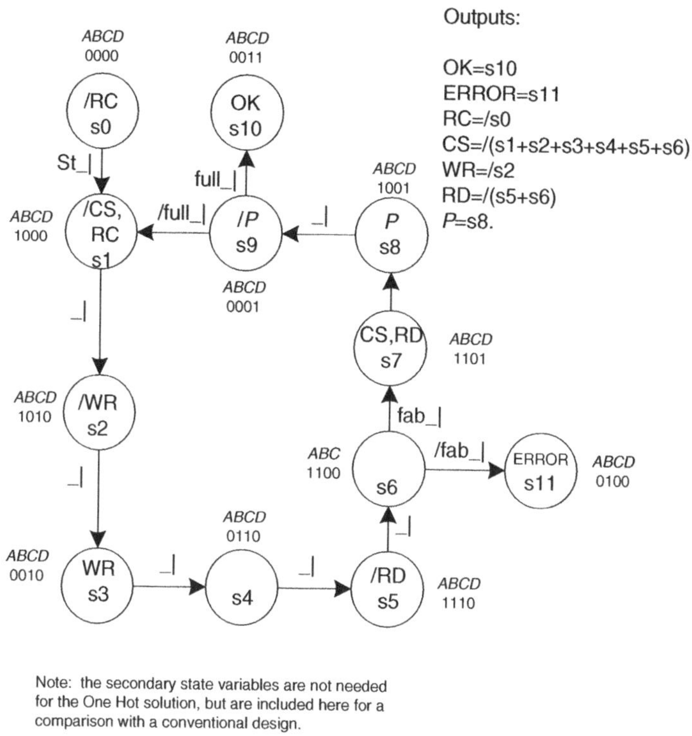

Figure 5.15 State diagram for the memory tester.

$$\text{`}A == B\text{'} = \text{fab} = \prod_{n=0}^{n=7} /(\text{A}n \,\hat{}\, \text{B}n),$$

where An and Bn are bit n on each A and B input and Π indicates that each /(An ^ Bn) is ANDed (i.e. product).

The state diagram for the memory tester is illustrated in Figure 5.15. In this state diagram, the initial states of the outputs have not been shown, but they can, of course, be deduced from the state diagram, since each state shows the change of outputs. So, for example, RC = 0 in s0, then in s1 it becomes RC = 1, and remains so for all other states in the diagram. Likewise, CS = 0 in s1, so it must be CS = 1 in s0. Following on, the other initial values in s0 are $P = 0$, ERROR = 0, OK = 0, $W = 1$, RD = 1. Note that the state diagram has been allocated a set of secondary state variables *ABCD*. These are not needed in the One Hot design, but they are used later on when a comparison with the more conventional method used in Chapter 4 is made.

In states s1, s2, and s3, the data 55 hex is written into the current memory location pointed to by the address counter. States s4, s5, and s6 are used to read the memory location and in state s6 the FSM tests fab. If fab = 1, then the memory location is OK and the FSM proceeds to pulse the address counter in s8 and checks to see whether all memory locations have been tested in state s9. If not, the whole process is repeated.

One Hot Design Equations:
s0·d = s0·/st
s1·d = s0·st + s9·/full
s2·d = s1
s3·d = s2
s4·d = s3
s5·d = s4
s6·d = s5 (no hold term since two-way branch)
s7·d = s6·fab
s8·d = s7
s9·d = s8 (no hold term since two-way branch)
s10·d = s9·full + s10
s11·d = s6·/fab + s11

Outputs:

OK=s10
ERROR=s11
RC=/s0
CS=/(s1+s2+s3+s4+s5+s6)
WR=/s2
RD=/(s5+s6)
P=s8.

Figure 5.16 The One Hot equations for the memory tester.

In the case of a good memory chip the FSM will loop around the states s1 to s9 repeatedly until the memory full indicator forces the FSM into state s10. The only way out of this state is via a system reset. This ensures that, after a memory test, the system waits for operator intervention.

At any time a memory location is found to be faulty, the FSM will drop into s11 and stop. The only way out of s11 is via a system reset.

The One Hot equations for the memory tester are given in Figure 5.16.

The state diagram of Figure 5.15 has a Moore output *P*. The rising edge of *P* will clock the address counter on entering state s8, *P* being lowered on leaving s8. The memory chip enable is disasserted in s7 prior to this action. The address counter only responds to the rising edge of *P*, so that on the next clock pulse the state of full can be tested in state s9.

5.7 COMPARING ONE HOT WITH THE MORE CONVENTIONAL DESIGN METHOD OF CHAPTER 4

In Figure 5.15, a set of secondary state variables has been provided so that this example could be implemented with four flip-flops. If this was done, the *D*-type equations would be as shown in Figure 5.17.

This, of course, uses the same technique used in Chapter 4, not the One Hot method. You might like to complete the equations and minimize to compare with the One Hot solution above.

It is useful at this stage to do a comparison between the One Hot method and the method that uses secondary state variables in the last example.

D Flip Flop Design Equations:

A·d = s0·st + s1 + s4 + s5 + s6·fab+ s7 + s9·/full.

B·d = s3 + s4 + s5 + s6 + s11.

C·d = s1 + s2 + s3 + s4 + s9·full.

D·d = s6·fab + s7 + s8 + s9·full.

Outputs:

OK=s10
ERROR=s11
RC=/s0
CS=/(s1+s2+s3+s4+s5+s6)
WR=/s2
RD=/(s5+s6)
P=s8.

Figure 5.17 Memory tester design implemented with four flip-flops.

	One Hot	Secondary state
Complexity	Simple	Need to define the state
Number of flip-flops	12	4
Combinational logic	Simple	Complex

The One Hot design is simple, uses more flip-flops but has simple combinational logic. The design using secondary state variables needs to be assigned a unique secondary state coding and has more complex combinational logic. However, it requires only four flip-flops. The One Hot arrangement needs 12 flip-flops and 15 gates, whereas the secondary state implementation needs four flip-flops and 13 gates. A hidden advantage of the One Hot design is that it makes more efficient use of the space on an FPGA device.

5.8 A DYNAMIC MEMORY ACCESS CONTROLLER

DMA controllers are used in some computer systems in order to allow data to be moved from one part of the memory system to another or from memory to a peripheral device (such as a printer or disk drive for example). If these data moves were done by the computing microprocessor, this would tie the microprocessor up and slow down the computing system. The PC has a special chip called the DMA controller, the 8257 (now largely integrated into an ASIC device), that performs these tasks.

This next example gives some idea of how a simple DMA controller could be developed around an FSM. The design could be integrated into an FPGA.

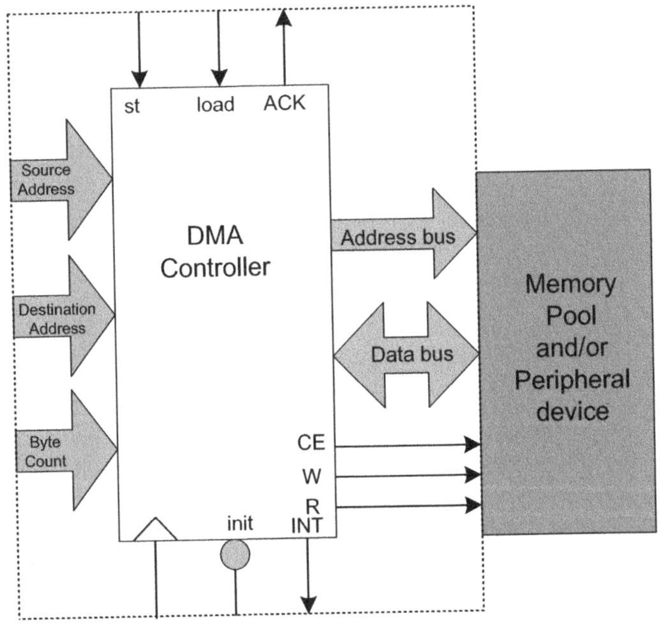

Figure 5.18 Block diagram of a possible DMA controller.

Figure 5.18 shows a possible arrangement for a DMA controller. The source and destination addresses need to be supplied by the microprocessor, as well as the number of words to be transferred (Byte Counter). The size of the data could be bytes (8 bits), words (16 bits) or even double words (32 bits), since the design can be scalable. In this design, it is assumed that these are delivered via an input port, but registers could be provided with address decoding for a memory-mapped DMA controller.

The dashed line marks the boundary of the DMA controller. The Memory Pool/Peripheral Device is external.

A DMA controller must be able to isolate itself from the memory/peripheral device when not being used, and this is achieved using tri-state devices.

Essentially, the DMA controller is designed to respond to an input st. At this point it should accept the source, destination addresses, and the number of words/bytes to be transferred. Then it should interrupt the microprocessor to let it know it is about to take over the memory/peripheral. The microprocessor will then isolate itself from these devices and send the load signal high to let the DMA controller know this has been done, and also provide it with the source/destination addresses and the byte count. At this point, the DMA controller will load the source, destination counters, and the byte counter.

Note that the registers are clocked synchronously with the system clock (on negative edge of clk) but enabled via the FSM output ec. The DMA controller now has enough information to carry out the transaction. This involves:

1. Selecting the source address and reading its content into a buffer.
2. Selecting the destination address and depositing the buffer content into this address.

3. Decrementing the byte counter and advancing the source and destination address counters.
4. Repeating 1 to 3 until all data transactions are completed (indicated by the byte counter reaching zero).

The DMA controller can now be developed in more detail. Clearly, a parallel-loading up counter is needed for both the source address and the destination address. Also, a parallel-loading down counter is required for the byte counter. Appendix B describes how these can be simply designed in detail.

Since the source and destination counter outputs need to be connected to the address bus, they should have tri-state buffers to isolate them from the memory/peripheral address bus when the DMA controller is not in use. The source or the destination address counters are used one at a time to avoid bus contention. The DMA controller will also need a data register and buffer connected so that the data read from one memory location can be fed to another memory location. This data buffer acts as a holding register within the DMA controller. The buffer needs to be isolated from the memory/peripheral data bus when not being used. Finally, all these internal devices need to be controlled by the FSM.

Figure 5.19 illustrates a possible block diagram for the DMA controller. Figure 5.19 shows a lot of detail and contains internal signals used by the FSM to control the operation of the DMA controller.

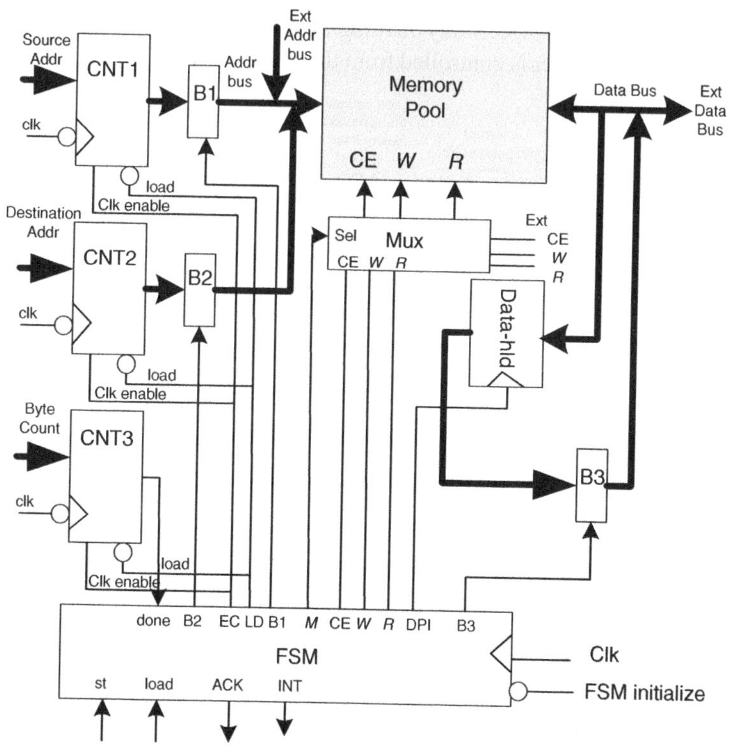

Figure 5.19 Detailed block diagram for the DMA controller system.

The FSM must carry out the transactions 1 to 4 detailed above. These, in turn, need to be defined in terms of the actions required to control the hardware in Figure 5.19. These actions will involve:

1. Waiting for the start signal st.
2. Providing an interrupt to the microprocessor to get it to isolate itself from the memory.
3. Waiting for a load signal from the microcontroller; when obtained, loading the source, destination, and byte count into the relevant counters.

Then:

4. The source memory needs to be selected and data read from the memory into the data holding register.
5. The source address needs to be isolated from the memory and the destination memory selected.
6. The data in the holding register needs to be transferred into the output buffer B3 and stored into the memory destination address.

After all this:

7. The byte counter needs to be decremented and checked to see whether all bytes of data have been transferred.
8. If there are more bytes to transfer, then the FSM needs to repeat 1 to 7 again. This is to continue until all bytes are transferred, indicated by the byte counter being decremented to zero.

The state diagram for the DMA controller can now be developed. The final form of this state diagram is illustrated in Figure 5.20. Study this diagram together with the diagram of Figure 5.19 to see how the DMA controller is controlled from the FSM.

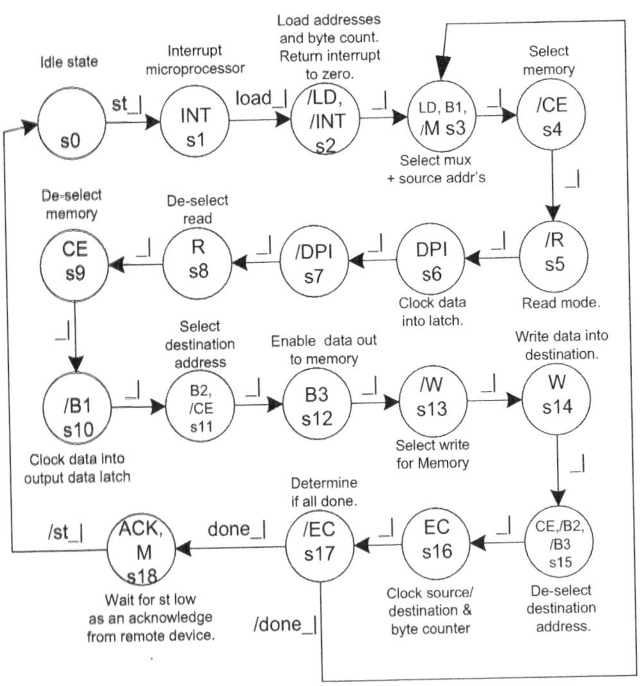

Figure 5.20 The state diagram for the DMA controller FSM.

A number of points need to be considered:

1. When reading the source memory location (states s4 to s7), the chip select and read signals CE and R need to be kept active while data is transferred into the holding register (s6 and s7) before they are disasserted (to their high state in states s8 and s9 respectively). This is different to the way in which memory read cycles have been done in other examples.
2. Writing the data from the output buffer follows the more usual arrangement, whereby the chip is selected (s11), then write is selected (s13), and finally both CE and W are deselected (s14 for W, s15 for CE) to write the data into the memory destination location.
3. Note that the source, destination, and byte count registers are enabled via the EC output from the FSM in state s16, and that the system clock clk clocks the data on the negative clock edge.

The One Hot equations can now be determined.

5.8.1 Flip-Flop Equations

$$\mathrm{s0} \cdot d = \mathrm{s18} \cdot /\mathrm{st} + \mathrm{s0} \cdot /\mathrm{st} \qquad \mathrm{s10} \cdot d = \mathrm{s9}$$
$$\mathrm{s1} \cdot d = \mathrm{s0} \cdot \mathrm{st} + \mathrm{s1} \cdot /\mathrm{load} \qquad \mathrm{s11} \cdot d = \mathrm{s10}$$
$$\mathrm{s2} \cdot d = \mathrm{s1} \cdot \mathrm{load} \qquad \mathrm{s12} \cdot d = \mathrm{s11}$$
$$\mathrm{s3} \cdot d = \mathrm{s2} + \mathrm{s17} \cdot /\mathrm{done} \qquad \mathrm{s13} \cdot d = \mathrm{s12}$$
$$\mathrm{s4} \cdot d = \mathrm{s3} \qquad \mathrm{s14} \cdot d = \mathrm{s13}$$
$$\mathrm{s5} \cdot d = \mathrm{s4} \qquad \mathrm{s15} \cdot d = \mathrm{s14}$$
$$\mathrm{s6} \cdot d = \mathrm{s5} \qquad \mathrm{s16} \cdot d = \mathrm{s15}$$
$$\mathrm{s7} \cdot d = \mathrm{s6} \qquad \mathrm{s17} \cdot d = \mathrm{s16}$$
$$\mathrm{s8} \cdot d = \mathrm{s7} \qquad \mathrm{s18} \cdot d = \mathrm{s17} \cdot \mathrm{done} + \mathrm{s18} \cdot \mathrm{st}.$$
$$\mathrm{s9} \cdot d = \mathrm{s8}$$

5.8.2 Output Equations

$\mathrm{INT} = \mathrm{s1}$

$\mathrm{LD} = /\mathrm{s2}$ active-low signal

$\mathrm{B1} = \mathrm{s3} + \mathrm{s4} + \mathrm{s5} + \mathrm{s6} + \mathrm{s7} + \mathrm{s8} + \mathrm{s9}$

$\mathrm{B2} = \mathrm{s11} + \mathrm{s12} + \mathrm{s13} + \mathrm{s14}$

$\mathrm{B3} = \mathrm{s12} + \mathrm{s13} + \mathrm{s14}$

$\mathrm{CE} = /(\mathrm{s4} + \mathrm{s5} + \mathrm{s6} + \mathrm{s7} + \mathrm{s8} + \mathrm{s11} + \mathrm{s12} + \mathrm{s13} + \mathrm{s14})$ active-low signal

$R = /(\mathrm{s5} + \mathrm{s6} + \mathrm{s7})$ active-low signal

$W = /\mathrm{s13}$ active-low signal

$\mathrm{EC} = \mathrm{s16}$

$\mathrm{DPI} = \mathrm{s6}$

$M = (\mathrm{s0} + \mathrm{s1} + \mathrm{s2} + \mathrm{s18});$ considering the high signal levels instead of low signal levels

$\mathrm{ACK} = \mathrm{s18}.$

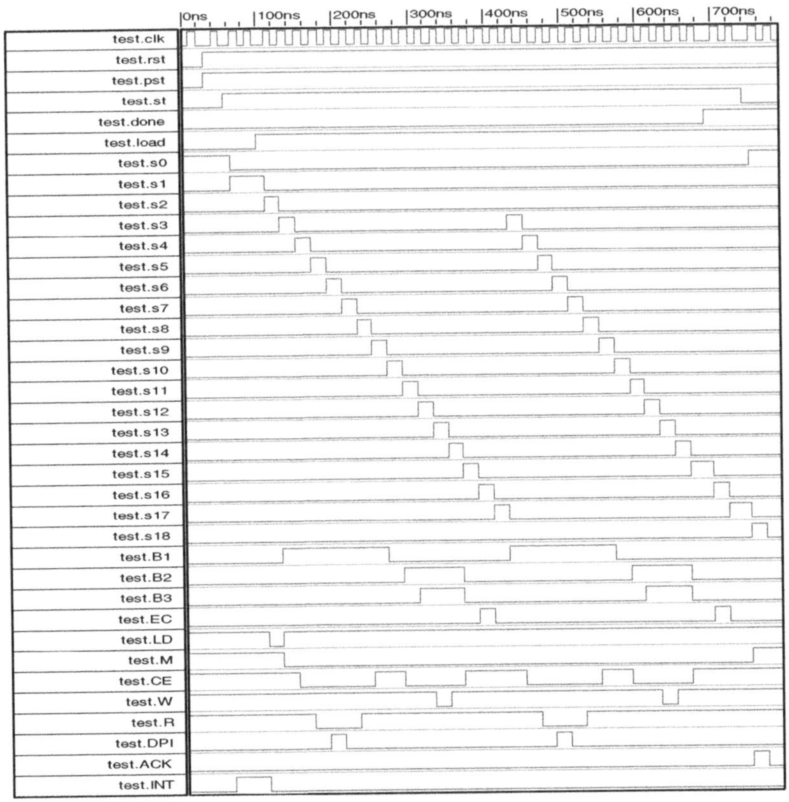

Figure 5.21 Simulation of the DMA FSM block.

The FSM block is simulated in Figure 5.21. In this simulation, the main loop comprising of states s3 to s17 is traversed twice. On the second loop, the done input is true (logic 1) and the FSM moves to s18 before returning back to state s0. This proves the operation of the FSM.

5.9 HOW TO CONTROL THE DYNAMIC MEMORY ACCESS FROM A MICROPROCESSOR

The DMA system is started with the start input, which in the previous design would need to be via an output port from the microprocessor. This is sometimes useful, since it avoids the need for address decoding logic.

A more appropriate way would be to have this signal via the memory (or I/O map) of the microprocessor. Normally, this would require using a byte-wide port.

In Figure 5.22, the start signal st is generated by a microprocessor using a spare address location. The address used here is 380 hex or 11 1000 0000 binary for this purpose. A typical memory or Input/Output access cycle is illustrated, from which it is clear that when the chip enable Ce and the IOW are both low (as would be generated by the microprocessor) the output from the address decoding logic corresponding to the address 380 hex would go high. The next

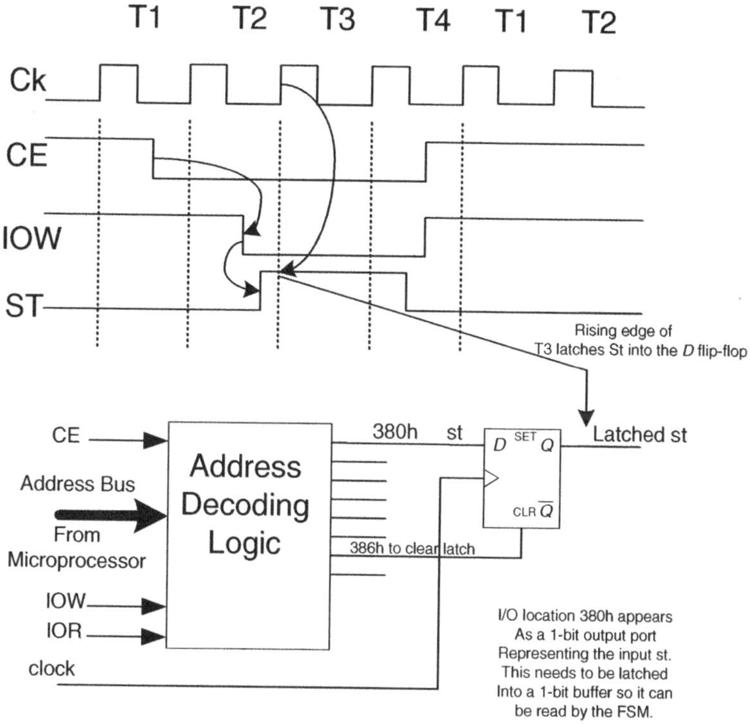

Figure 5.22 Generating a start signal from a microprocessor for the FSM.

clock pulse from the microprocessor clock (T3 rising edge) would clock the st value into the *D*-type flip-flop.

The microprocessor would need to use an address (386 hex in this case) to reset the *D* flip-flop at an appropriate time. However, before this, the microprocessor would need to wait for the ACK response from the FSM.

Figure 5.23 illustrates how this could be done, together with the generation of the st signal. In Figure 5.23, the additional data latch is used to store the state of the FSM output ACK. The FSM raises the ACK signal line and clocks it into the data latch with the pak signal (added to the FSM for this purpose). The microprocessor can read the ACK signal by addressing 381 hex, which takes the tri-state buffer out of its tri-state, thus setting bit d0 of the data bus to that of the ACK signal stored in the *D*-type flip-flop during the memory or IO read cycle of the microprocessor.

The ACK signal will be read by the microprocessor in the T4 state on the rising edge of CE and IOR signals during the read cycle (see Figure 5.24) into an appropriate internal register within the microprocessor.

The state diagram fragment shown in Figure 5.25 shows the relevant state sequence needed to use the microprocessor memory or IO mapped control. Note that this can be used with the other states of Figure 5.20 in the DMA controller.

Of course, this example is based upon hypothetical microprocessor bus timing, but it does illustrate a possible method.

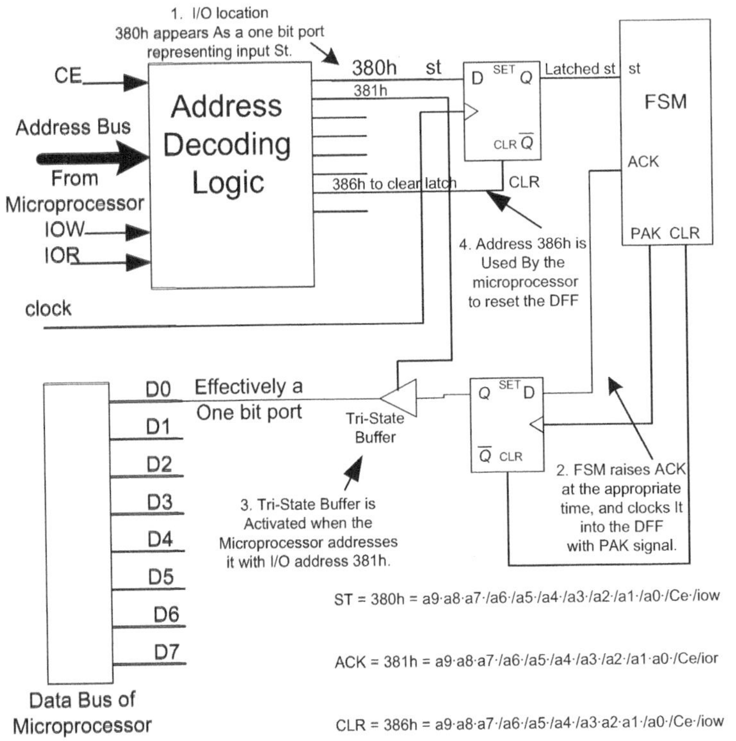

Figure 5.23 The whole arrangement for writing to and reading from the FSM.

5.10 DETECTING SEQUENTIAL BINARY SEQUENCES USING A FINITE-STATE MACHINE

A very common requirement in communication and computer network systems is to detect binary sequences. The following example illustrates the idea and can be scaled up and changed to detect other sequences.

One common approach is to insert a shift register into the transmission line and use digital comparators to detect the incoming binary stream after the number of bits corresponding to the binary code have been shifted into the shift register. This, of course, introduces an n-bit delay. So, to detect a 4-bit code introduces a 4-bit delay. If the code to be detected is longer (e.g. 8 or 16 bits), or other devices are to be added to the line to detect other codes, then the delay increases.

An alternative approach is to monitor the transmission line passively in real time and process the binary bits in an FSM. This will not introduce any delay.

Consider a system such as the one shown in Figure 5.26. In this system, the FSM monitors the input binary sequence continuously looking for the sequence $d = 1101$ (this could be any sequence in practice, but this one will suffice).

The FSM needs to synchronize to 4-bit data streams; in Figure 5.26, the first data stream is 1011, then the next 1101 (the required sequence), followed by the sequence 0011. The output M should go high at the end of the sequence 1101.

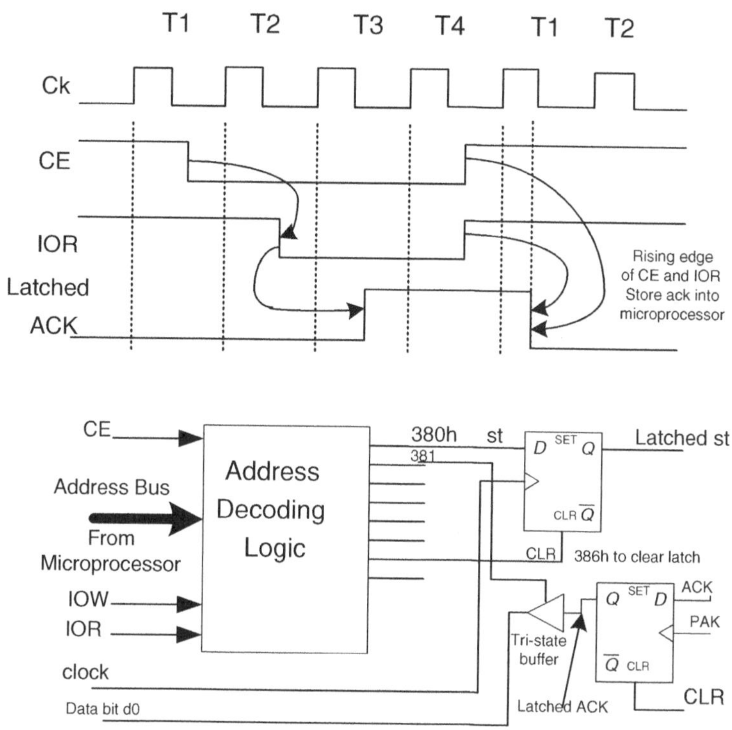

Figure 5.24 The arrangement for reading the ACK signal from the FSM.

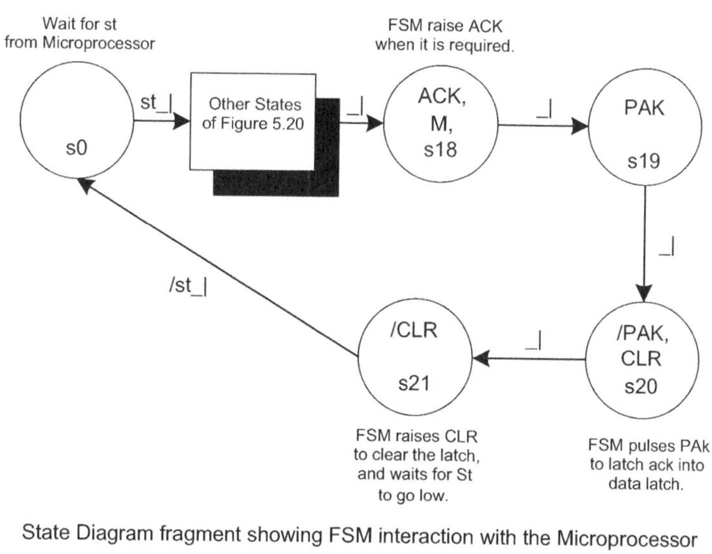

Figure 5.25 Illustration of the state sequence needed for using the microprocessor memory or IO mapped control.

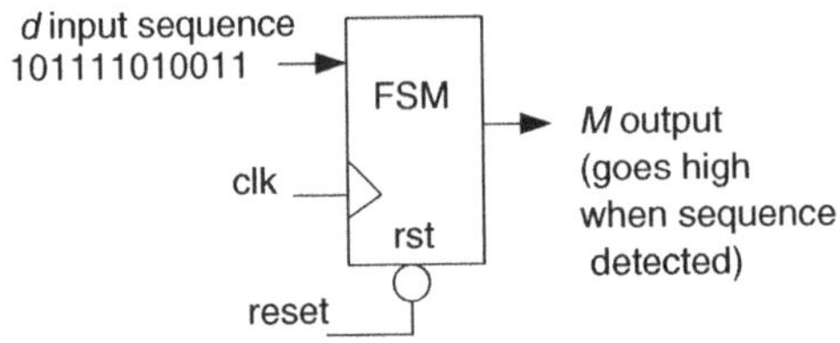

Figure 5.26 Binary sequence detector.

The best way to develop the state diagram for this application is to start with a state diagram that follows the required sequence; see Figure 5.27, where the required sequence $d = 1101$ is detected in state s4, where the FSM stops.

However, it is necessary to go through the 4-bit sequence and return to state s0 if the required sequence is not detected. This is shown in Figure 5.28, where the state diagram is seen to cater for all possible combinations.

For example, an input sequence of $d = 1100$ would follow states s0, s1, s2, s3, s0. An input sequence 1111 would follow states s0, s1, s2, s7, s0; and so on. In this way, the FSM keeps in step with the incoming binary sequence.

Once the correct sequence is found, the FSM will stop in state s4.

The FSM clock needs to synchronize with the middle of the data bits being monitored; this could be done using the same technique used in the asynchronous receiver design of Figure 4.20 in Section 4.7.

The design can be implemented using the One Hot method, resulting in the following equations:

$$\mathrm{s0} \cdot d = \mathrm{s3} \cdot /d + \mathrm{s7}$$
$$\mathrm{s1} \cdot d = \mathrm{s0} \cdot d$$
$$\mathrm{s2} \cdot d = \mathrm{s1} \cdot d$$
$$\mathrm{s3} \cdot d = \mathrm{s2} \cdot /d$$

$$\mathrm{s4} \cdot d = \mathrm{s3} \cdot d + \mathrm{s4}$$
$$\mathrm{s5} \cdot d = \mathrm{s0} \cdot /d$$
$$\mathrm{s6} \cdot d = \mathrm{s5} + \mathrm{s1} \cdot /d$$
$$\mathrm{s7} \cdot d = \mathrm{s6} + \mathrm{s2} \cdot d$$

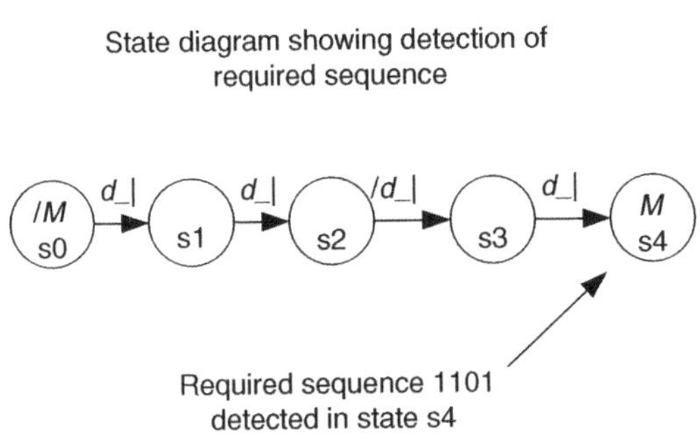

Figure 5.27 State diagram segment to detect required sequence.

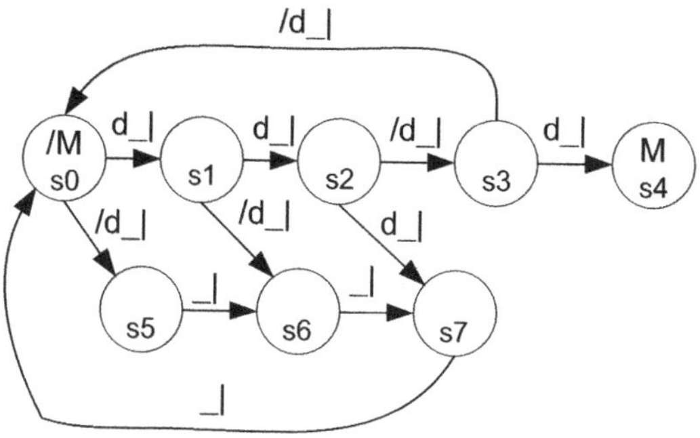

Figure 5.28 State diagram completed for all possible input combinations.

and output

$$M = s4.$$

This design can be built up in Verilog and simulated as illustrated in Figure 5.29. This simulation runs through all possible paths of the state diagram in order to test out the FSM logic.

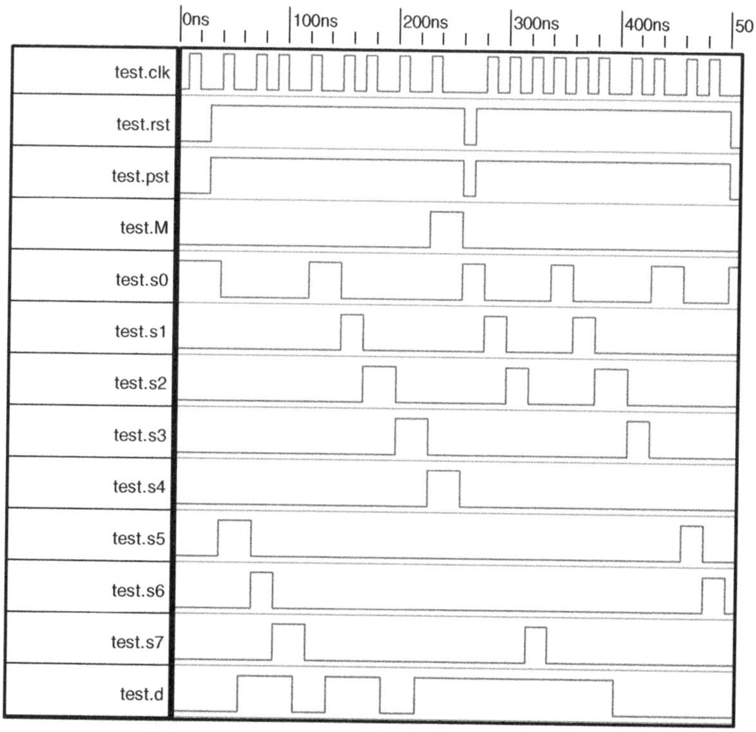

Figure 5.29 Simulation of the sequence detector.

In the first sequence, the simulation is seen to follow the sequence s0, s5, s6, s7, s0. This is followed by the sequence s0, s1, s2, s3, s4, with $M = 1$. After this, the FSM is reinitialized back to s0 for another sequence by lowering rst and pst (asynchronous initialization). Then, the sequence s0, s1, s2, s7, s0 occurs. This is followed by other sequences to complete the testing of all paths through the state diagram. Note that during the last sequence, i.e. s0, s5, s6, the asynchronous initialization forces the FSM back to s0.

The system could be modified so that it continues indefinitely to monitor the incoming sequence, providing an $M = 1$ output whenever the correct sequence is detected. This can easily be done by making M a Mealy output in state s3, so that

$$M = s3 \cdot d.$$

If d is not 1 in state s3, neither is M. Of course, state s4 is no longer needed in this case.

Figure 5.30 shows the final state diagram In Figure 5.30, the M output is made a Mealy output in s3 so that the FSM can return to s0 after any sequence. In this way the FSM can continue to monitor incoming sequences forever and remain synchronized to the 4-bit pattern.

In Figure 5.31, the sequence detector can be seen to return to s0 after detecting the 1101 sequence. Note: the output M is only 1 when $d = 1$.

The same technique could be applied for longer sequences, making use of more states and more flip-flops.

One limitation of the sequence detector of Figure 5.30 is that it is limited to detecting one particular binary sequence, in this case 1101. It would be more useful to have an FSM that could accept any binary sequence without having to redesign the state diagram.

In Figure 5.30, the FSM looks at the line bits with the same variable d. Instead, the d input could be compared bit by bit with a number of digital 1-bit comparators (exclusive NOR gates), each one having a bit of the code to compare the incoming bits with. Figure 5.32 illustrates a possible arrangement. In this case, a more realistic 8-bit code is to be detected.

Also note that the code to be detected can be stored into a data latch prior to starting the detection process. This system can be used to detect up to 255 different codes (assuming the code 0000 0000 is not used).

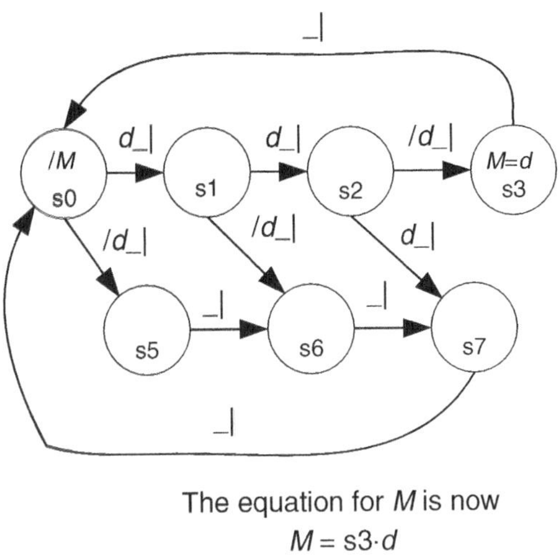

Figure 5.30 Final state diagram for continuous monitoring for the $d = 1101$ sequence.

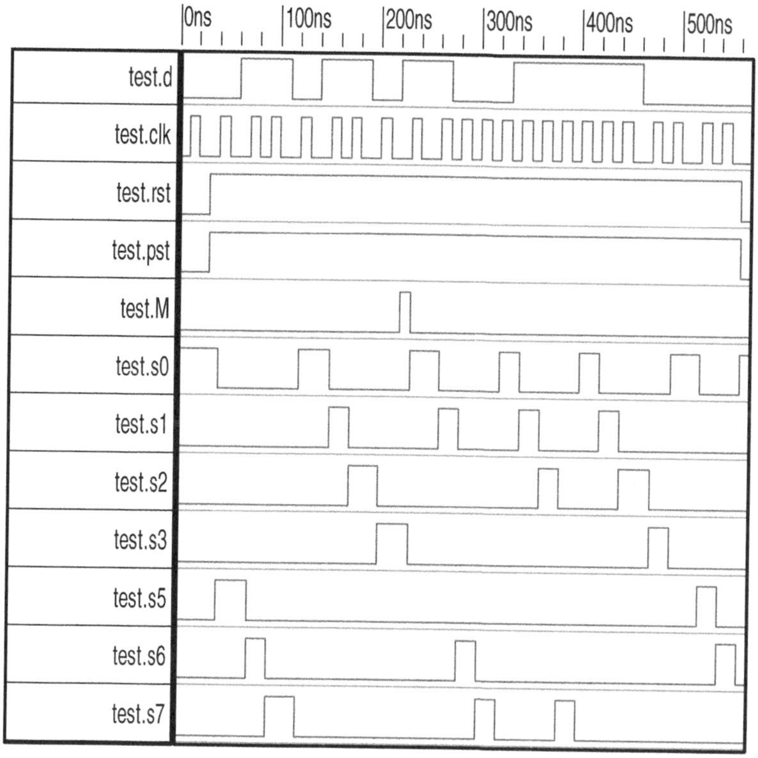

Figure 5.31 Final simulation of the sequence detector.

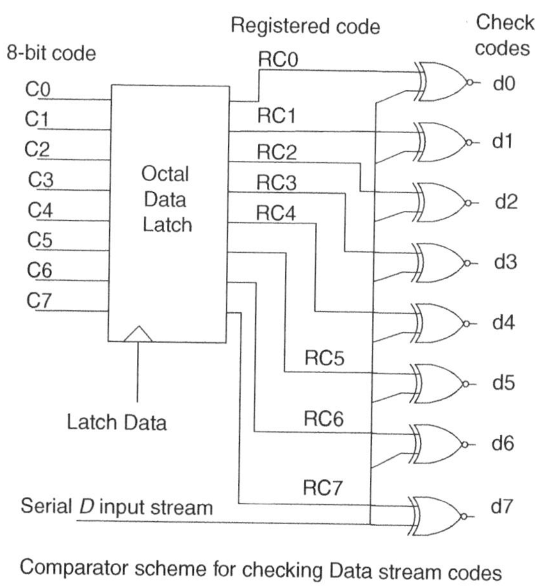

Figure 5.32 Comparators used to compare each bit with a pre-stored code.

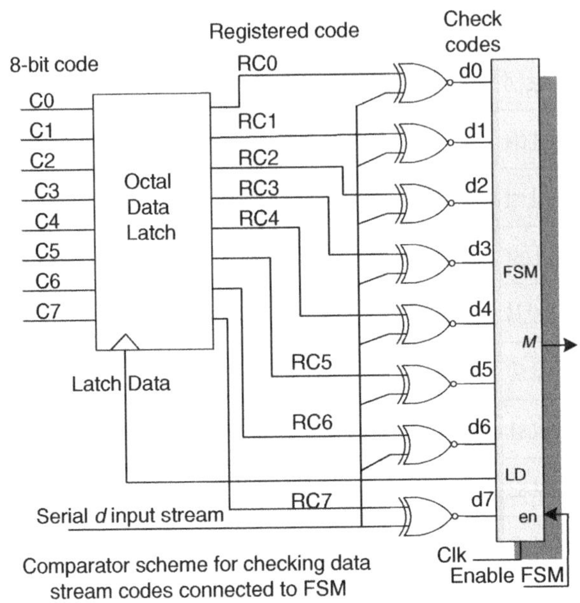

Figure 5.33 Full system of the general 8-bit binary code detector.

The code C0 to C7 is loaded into the data latch and is presented as a registered code RC0 to RC7 and connected to one input of the single-bit digital comparators.

The input bits from line *d* are all connected to the other comparator inputs, so that eight compared bits d0 to d7 are available to the FSM.

Figure 5.33 shows the full system: an additional input to the FSM en is used to start the system and an additional output LD is used to latch the code value to be detected.

The state diagram for the FSM is illustrated in Figure 5.34. The state diagram of Figure 5.34 follows the same basic idea developed in the state diagram of Figure 5.30, but for a byte-wide code. Note that rather than compare each *d* bit at the line, the FSM now compares each bit after it has been compared with the desired code with the 1-bit comparators, first bit d0, then d1, through to bit d7.

Now the FSM is a fixed sequence that can detect any possible 8-bit code. All that needs to be done is load the required code into the data latch before starting the detector. The system can be disabled at any time by disasserting the input en. This will cause the system to stop at the end of the current sequence then return to state s0.

A little thought shows that the same FSM could be used to detect a number of different codes one after the other by simply changing the codes in sequence.

One aspect of the system not yet discussed is how to synchronize the system to the line bit stream. One way to do this would be to start the system off with a synchronization bit stream code, say 10101010*xx*, prior to starting the code detection process, where *x* is an additional bit that could be either 0 or 1. This could be broadcast by the sender.

The additional bits are needed to allow the FSM to load the data latch with the desired code to be detected. The same FSM could be used for this, since all that needs to be done is to load up the synchronization code. Once the synchronization byte is detected (via *M*) the code to be detected would be loaded into the data latch and the code detection sequence started.

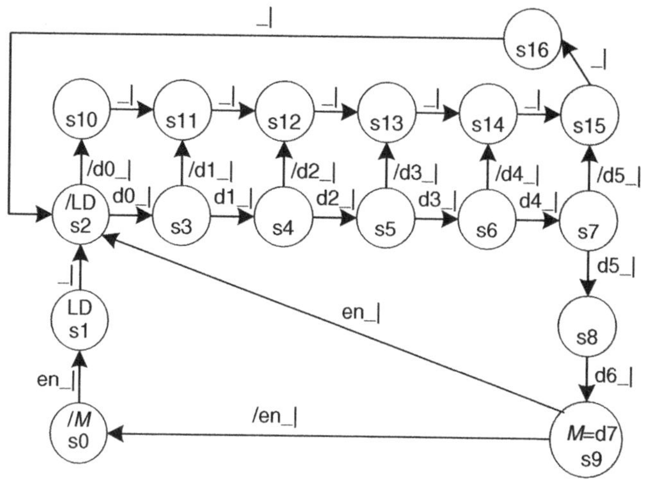

Initialize input starts FSM in s0. Controlling device loads code to be detected into data latch. Then enables detector with En=1. FSM latches this code into the data latch with Ld signal.
Thereafter, FSM cycles through states according to data input received. Controlling device can stop the detector at the end of any sequence by lowering En to 0 and stopping the FSM in state s0.

Figure 5.34 The state diagram for the FSM-based byte-wide code detector.

The One Hot equations can be obtained directly from the state diagram of Figure 5.34:

$$\mathrm{s0} \cdot d = \mathrm{s9} \cdot /\mathrm{en} + \mathrm{s0} \cdot /\mathrm{en}$$
$$\mathrm{s1} \cdot d = \mathrm{s0} \cdot \mathrm{en}$$
$$\mathrm{s2} \cdot d = \mathrm{s1} + \mathrm{s9} \cdot \mathrm{en} + \mathrm{s16}$$
$$\mathrm{s3} \cdot d = \mathrm{s2} \cdot \mathrm{d0}$$
$$\mathrm{s4} \cdot d = \mathrm{s3} \cdot \mathrm{d1}$$
$$\mathrm{s5} \cdot d = \mathrm{s4} \cdot \mathrm{d2}$$
$$\mathrm{s6} \cdot d = \mathrm{s5} \cdot \mathrm{d3}$$
$$\mathrm{s7} \cdot d = \mathrm{s6} \cdot \mathrm{d4}$$
$$\mathrm{s8} \cdot d = \mathrm{s7} \cdot \mathrm{d5}$$
$$\mathrm{s9} \cdot d = \mathrm{s8} \cdot \mathrm{d6}$$
$$\mathrm{s10} \cdot d = \mathrm{s2} \cdot /\mathrm{d0}$$
$$\mathrm{s11} \cdot d = \mathrm{s3} \cdot /\mathrm{d1} + \mathrm{s10}$$
$$\mathrm{s12} \cdot d = \mathrm{s4} \cdot /\mathrm{d2} + \mathrm{s11}$$
$$\mathrm{s13} \cdot d = \mathrm{s5} \cdot /\mathrm{d3} + \mathrm{s12}$$
$$\mathrm{s14} \cdot d = \mathrm{s6} \cdot /\mathrm{d4} + \mathrm{s13}$$
$$\mathrm{s15} \cdot d = \mathrm{s7} \cdot /\mathrm{d5} + \mathrm{s14}$$
$$\mathrm{s16} \cdot d = \mathrm{s15}$$

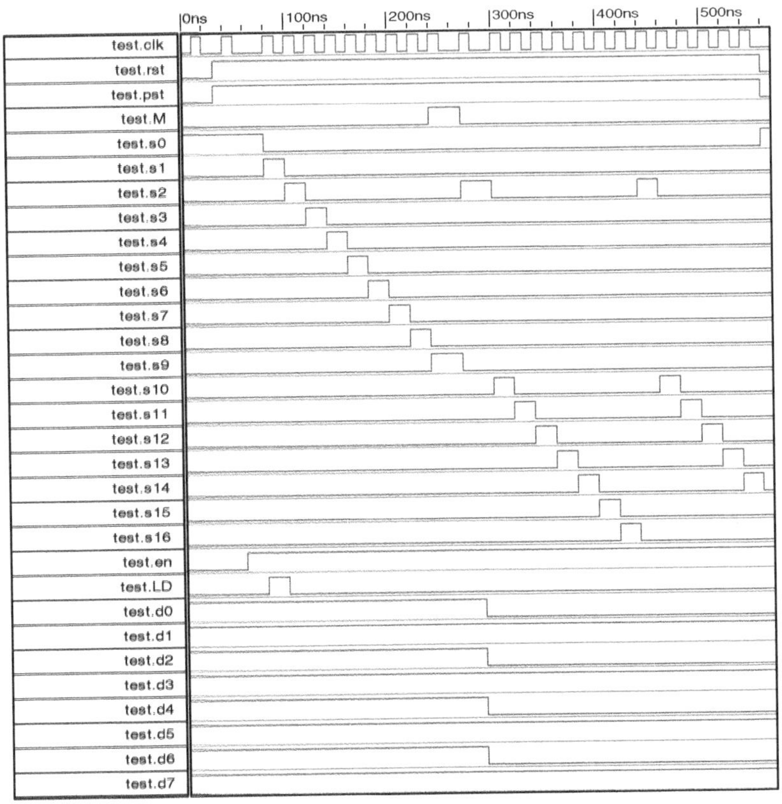

Figure 5.35 Simulation of the FSM sequence detector using a code 11001011.

with outputs

$$M = \mathrm{s9} \cdot \mathrm{d7}$$

$$\mathrm{LD} = \mathrm{s1}.$$

Of course, the code to be detected could be any length, since the state diagram could be developed for any particular length following the same basic idea.

The simulation in Figure 5.35 shows the system in which the code to be identified is 11001011. This code is first loaded into the latch via the C0 to C7 inputs. The simulation then presents a number of serial *d* input sequences, with the last one being the one the system is trying to detect. The *M* output goes high at the end of this sequence.

The complete system, comparator, octal latch and FSM as connected up in Figure 5.33 is simulated and illustrated in Figure 5.36. Only the system inputs and output signals are visible here (see block diagram in Figure 5.33), along with the FSM state sequence so that the state sequence of the state machine in Figure 5.34 can be followed. Note that the sequence to be detected in this simulation is C[7:0] = 11001011. This sequence is detected at the end of the simulation at around 700 ns into the simulation, and can be clearly seen in the bottom two signals (*d* input and *M* output).

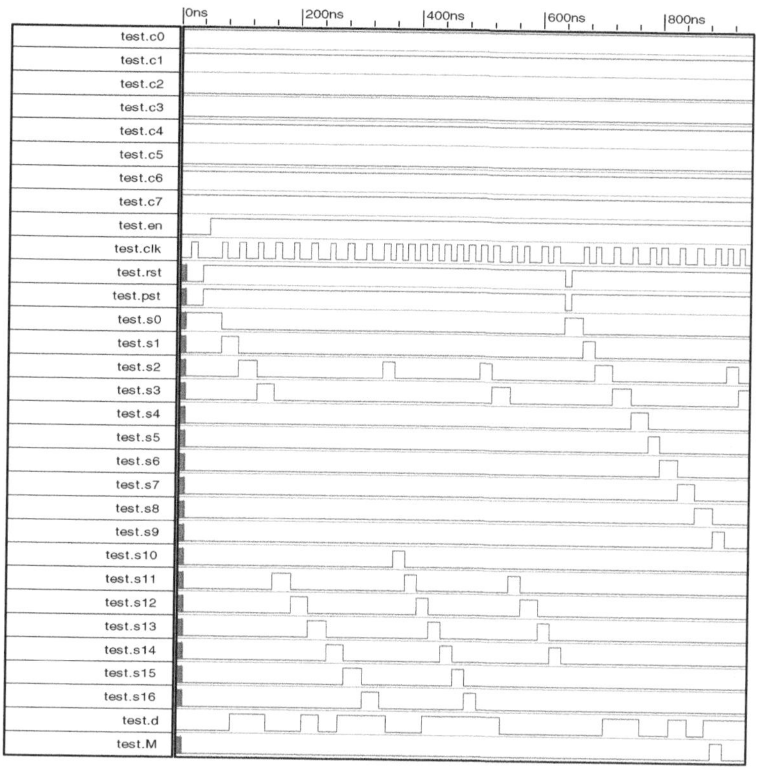

Figure 5.36 Simulation of the complete 8-bit sequence detector.

5.11 SUMMARY

This chapter has explored the use of the One Hot technique to implement FSMs. These are particularly useful for implementation in FPGA devices and have the advantage of not requiring secondary state variables. The hand calculations are much easier to perform and can be converted into Verilog HDL easily. Also, owing to the large size of FPGAs, large FSM designs can be implemented without the need to consider secondary state variable assignment.

6

Introduction to Verilog HDL

6.1 A BRIEF BACKGROUND TO HARDWARE DESCRIPTION LANGUAGES

This chapter will introduce the fundamental aspects of what has become an essential tool for the modern digital system designer, namely the HDL. There are many different HDLs used for a variety of purposes. Some are best suited to low-level design, making use of logic gates and Boolean equations (e.g. ABEL [1]), while other so-called *system-level* languages are intended to aid the design and verification of entire systems comprising both hardware and software (examples are SystemC [2] and SystemVerilog [3]).

In addition to the support for digital systems, in which events and values are modelled in *discrete* terms, HDLs have evolved to encompass the realm of *continuous time* or *analogue* behaviour. Apart from mentioning these languages in passing, this book will not consider the details of this category of HDL.

The HDL described in this book is the very popular, and relatively easy-to-learn, Verilog HDL [4], often referred to as 'Verilog' or 'HDL' (the names 'Verilog' or Verilog HDL' are used interchangeably throughout this book). The language has a considerable user base among the digital design communities within both industry and academia across the globe. Verilog HDL is unique with regard to the breadth of support it provides for describing and simulating digital systems. Using built-in models of metal oxide–semiconductor (MOS) transistors, the language allows digital circuits to be described at the so-called *switch level*, where individual switches can exhibit detailed timing and signal strength behaviour. The switch level is very close, in representative terms, to the actual physical implementation of the digital integrated circuit, this makes Verilog HDL the first choice of language used to verify designs beyond the circuit level. At the other extreme, the high-level language constructs contained within the language facilitate the use of a more abstract and, therefore, powerful representation known as *behavioural* or *register transfer level* (RTL) in which the design is represented by storage registers and operations involving the movement and processing of information stored in them. It is perhaps the latter capability that makes Verilog HDL and other similar languages the only effective way of dealing with the complexities of contemporary digital design.

The Verilog HDL started out as a proprietary tool in the 1980s, but soon gained widespread popularity as digital integrated circuits and systems became more complex. Consequently, it was introduced into the public domain and subject to standardization by the IEEE in the mid 1990s. The majority of the examples used in this book make use of the Verilog HDL defined by

FSM based Digital Design using Verilog HDL Peter Minns and Ian Elliott

the IEEE Standard 1364-2001 released in 2002. This version of Verilog HDL introduced many new powerful features, along with some cosmetic changes, bringing it in line with one of the other most popular HDLs, namely VHDL (Very High Speed Integrated Circuit Hardware Description Language) [5].

The two most widely used HDLs, i.e. Verilog HDL and VHDL, despite sharing the same acronym, differ in terms of syntax and general appearance, the latter being similar to the Ada programming language [6] and the former having some C-like features. Despite these cosmetic differences, the two HDLs share very similar semantics and tend to be used in the same manner towards achieving the same eventual goal of designing and implementing a cost-efficient digital system that meets the specification in terms of performance and economics.

In addition to their use in design simulation and verification, both Verilog and VHDL can be used as the input language to the automated process of hardware creation known as *logic synthesis* [7]. The vast majority of digital circuits implemented in actual hardware have been *synthesized* from a design description written in one of these languages. Modern logic-synthesis software tools are highly reliable, producing optimum and efficient logic circuits often implemented in the form of programmable logic. It should be noted that the role of the digital designer is no less important, however, despite the availability of such tools. What has happened is that the designer is now able to work at a higher level of abstraction, making use of the expressive power of the HDL to create ever more complex designs, while the detailed issues and processes surrounding implementation have been largely automated.

The use of design languages is now well established and the modern electronic designer needs a working knowledge of at least one of the popular HDLs to compete in the employment market. Migrating designs between one particular HDL and another is a relatively straightforward task, once the fundamentals have been mastered. It is far more challenging to learn and master an HDL from scratch for the first time, and apply it to a real-world design problem, than it is to convert a given design into an alternative language, having already mastered an HDL.

As mentioned previously, the huge growth in the use of HDLs such as Verilog HDL and VHDL, along with the constant increase in complexity and integration of hardware and software, brought about by the advances in microelectronic technology, has resulted in the development of what are referred to as *system-level* languages such as SystemC [2] and SystemVerilog [3].

While SystemC has been developed around the popular C++ language and, therefore, lacks support for low-level digital design, SystemVerilog is a superset of Verilog HDL and, therefore, possesses all of the digital hardware modelling capabilities of Verilog in addition to the higher level data abstraction and software integration needed by today's system-on-a-chip designers.

By learning Verilog HDL, therefore, the digital designer is setting down the foundations for a long and prosperous career, with the comfort of knowing that support exists within the design tools and languages for the ever more complex designs of the future.

To summarize this section, here are some of the key advantages of using an HDL such as Verilog HDL:

- Technology independence – designs written in an HDL are largely independent of the target technology and, therefore, future-proof.
- Textual descriptions are concise, unambiguous and self-documenting.
- Standard language promotes design reuse and portability between design tools.

- Textual descriptions replace or augment schematics.
- Automated design – logic synthesis tools accept designs written using an HDL.
- High-level design – the designer is freed from the tedium of gate-level design to concentrate on system-level aspects.

6.2 HARDWARE MODELLING WITH VERILOG HDL: THE MODULE

In Verilog HDL, the basic unit of hardware is known as a *module*. In common with C-language functions, modules are free standing and cannot, therefore, contain definitions of other modules. A module can be *instantiated* within another module, however, in a similar manner to which a C function can be *called* from another C function; this provides the basic mechanism for the creation of design hierarchy in a Verilog description.

Listing 6.1 shows the basic layout of a module:

```
module module-name(list-of-ports);
   local wire/reg declarations
    parallel statements
endmodule
```

Listing 6.1 Basic layout of a module.

Note that in this and subsequent listings all keywords are shown in bold. The hardware description is enclosed by the keywords **`module`** and **`endmodule`**, the former being immediately followed by the name of the module and a list of ports enclosed in parentheses. (Some modules do not require ports; therefore, the `list-of-ports` is empty.)

Note that the semicolon at the end of the first line (the module header) is always required, but no semicolon is required after the bracketing keyword **`endmodule`**.

Within the module header, the `list-of-ports` enclosed between the parentheses fully specifies the size (number of bits) and direction of the ports (input or output, etc.), along with the name of the port.

In this manner, the first line of a module contains all of the details of the module that are visible from outside, i.e. the module header represents the *interface specification* or module *prototype*.

Immediately below the module header, items that are to be used within the confines of the module are declared. The second line of Listing 6.1 shows the most common local objects to be **`reg`** and **`wire`**; these represent internal storage and/or connections used within the module. Consistent with other languages, Verilog requires that all objects must be declared before they are referenced; therefore, this means that they tend to be located at the top of the module body. The local **`wire`** and **`reg`** objects represent signals used within the module to link together the logical elements described by the so-called *parallel statements*. The term 'parallel statements' refers to the manner in which this group of statements executes during a simulation, i.e. concurrently, in a manner similar to that of real digital hardware. The parallel statements describe the behaviour, structure and/or data flow of the design encapsulated by the module. They can take a variety of forms; among these are *primitive gates*, module *instantiations* and *continuous assignments*, all of which will be described in detail in due course.

```
1 module myxor(output y, input a, b);

2 assign y = a ^ b;

3 endmodule
```

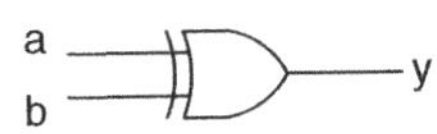

Figure 6.1 A simple module.

A Verilog module description consists of case-sensitive ASCII text, the file containing the text of a module with a given *module-name* is conventionally stored under the filename 'module-name.v'. An example of a very simple module is shown in Figure 6.1. The listing in the figure includes line numbers that are for reference purposes only; they must not appear in the actual source text.

As shown in Figure 6.1, the module describes a two-input exclusive-OR gate named `myxor` having single-bit inputs `a` and `b` and an output `y`. The names and direction of the module ports are specified by the comma-separated list enclosed within parentheses on line 1. The ports of a Verilog module have a default width of 1 bit, and in some cases a port may need to be both an input and an output, i.e. *bidirectional*. Verilog uses the reserved word **`inout`** to specify a bidirectional port.

The functionality of the module given in Figure 6.1 is defined by the so-called *continuous assignment* statement on line 2, assigning the output `y` the expression `a ^ b` (where ^ is bit-wise exclusive-OR in Verilog). The keyword **`assign`** is used to indicate a continuous assignment. Such a statement creates a static binding between the expressions on the left- and right-hand sides of the = operator; it is most commonly used to describe combinational logic.

Despite the similarity with an assignment used in the C language, the continuous assignment on line 2 in Figure 6.1 is a *parallel statement*; this means that it is constantly active and awaiting events on either of the input signals `a` and `b` to trigger its execution. Such events would depend on the activity of external sources applied to the module inputs.

Specifically, whenever a change in value occurs on either or both of the inputs `a` and `b`, the expression on the right-hand side of the assignment on line 2 is evaluated and the result is assigned to the target of the continuous assignment on the left-hand side of the = operator (output `y`) at the start of the next simulation cycle.

A module may contain any number of continuous assignment statements, all of which act in parallel and, therefore, may be written in any order.

Figure 6.2 shows an example of a module containing multiple continuous assignments. Such a description is sometimes referred to as a *dataflow* style description. The Verilog source describes the logic diagram shown below the text in the figure. In this example, each gate is modelled by a separate continuous assignment on lines 7, 8 and 9. An alternative would have been to describe the logic using a single statement such as

```
assign F = ~((A & B) | (C & D));
```

The above assignment illustrates the similarity between Verilog and the C language in terms of the bit-wise logical operators: inversion `(~)`, logical AND `(&)` and logical OR `(|)`. Also, note

```
1   //Note - Comments are written in the same
2   //style as C++ (block comments use /* */)
3   //Verilog description of a AND-OR-INVERT gate
4
5     module AOI(input A, B, C, D, output F);

6     wire  W1, W2;

7     assign W1 = A & B;
8     assign W2 = C & D;
9     assign F = ~( W1 | W2);

10   endmodule
```

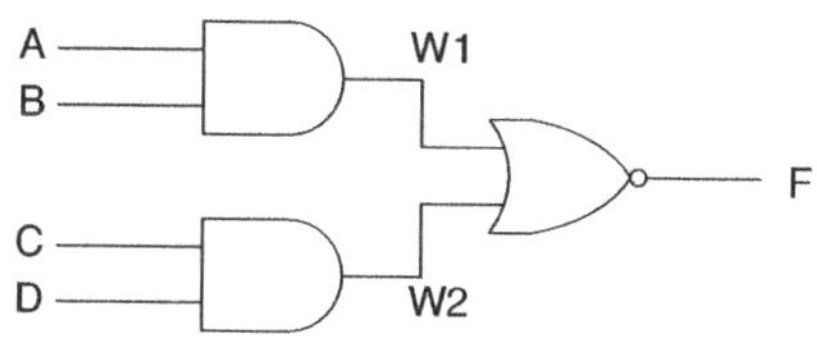

Figure 6.2 A Verilog AND-OR-INVERT module.

the use of parentheses in the above assignment, these force the order of operator evaluation to reflect the logical structure being described.

Note that the continuous assignment statements on lines 7, 8 and 9 could have been written in any order without changing the behaviour of the logic, internal single-bit **wires** (declared on line 6) are used to connect the outputs of the two AND assignments (lines 7 and 8) to the inputs of the two-input NOR assignment. The order of execution of the continuous assignments on lines 7, 8 and 9 is determined by events on the primary inputs A, B, C, D and internal **wires** W1 and W2.

For example, if input A changed from logic 0 to logic 1, this event on A would cause the assignment on line 7 to execute. This, in turn, would cause the value on **wire** W1 to change from logic 0 to logic 1, assuming input B was already at logic 1. It should be noted that the event on W1 occurs at the same time as the event on input A, since the continuous assignment does not specify any propagation delay. However, the simulator updates signals using a mechanism that involves discrete cycles known as *simulation cycles*, in which signals are updated as a result of assignment execution.

An infinitesimally small delay, sometimes referred to as *delta delay*, elapses each time the simulation cycle advances. So, if the event on input A occurred at a time of 10 ns, the resulting change in **wire** W1 would occur at a time of 10 ns + 1*d*, where *d* represents 'delta'.

Referring back to Figure 6.2, an event on W1 has the effect of triggering the continuous assignment on line 9, which, depending on the value of W2, may or may not result in a change in the module output F. If a change in F were to occur, it would be at a time of 10 ns + 2*d*, due to the one-delta introduced by the assignment execution.

A **wire** is a particular case of the more general category of Verilog objects known as *nets*, all of which share the common requirement of having to be driven continuously, either by a continuous assignment or by virtue of being connected to the output of a primitive gate or module instantiation.

Note: the left-hand side, or target, of a continuous assignment statement must be a **`wire`**.

The ports of a module (`A`, `B`, `C`, `D` and `F` on line 5 of the example shown in Figure 6.2) are also **`wires`** by default; as such, they may appear on the left- or right-hand sides of continuous assignments, depending on whether they are outputs or inputs respectively. Unlike some HDLs, the Verilog language allows ports that have been defined as outputs to appear on the right-hand side of an assignment. This flexibility is included to reflect a common situation in hardware, where a module output signal is internally fed back into an input within the same module.

Figure 6.2 also shows the format used in Verilog for adding comments to a description. Lines 1, 2 and 3 illustrate the similarity between Verilog comment delimiters and those used by the C and C++ languages. Comments are a useful tool for adding documentation to a design description.

The next example of a Verilog module illustrates several additional aspects of the language not yet mentioned.

Figure 6.3 shows a very simple Verilog description of a 4-bit binary adder along with its corresponding symbolic representation. The module header, spanning lines 1, 2, 3 and 4, shows how multi-bit ports are defined. In this case, the inputs `a` and `b` and the output `sum` are all 4 bits wide, being represented as buses on the symbol.

```
1 module add4( output [3:0] sum,
2              output c_out,
3              input [3:0] a, b,
4              input c_in);

5 assign #15 {c_out, sum} = a + b + c_in;

6 endmodule
```

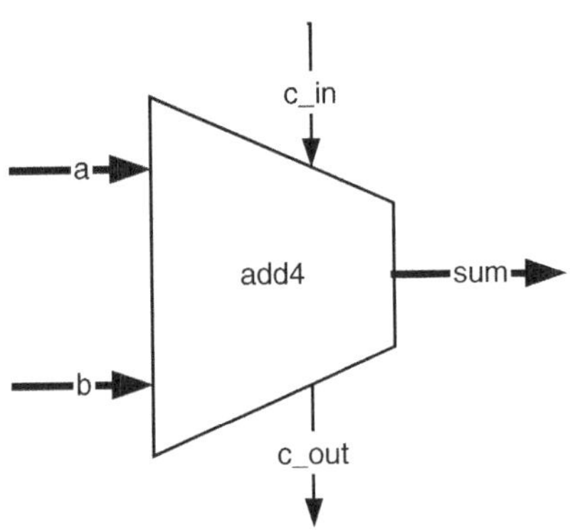

Figure 6.3 Verilog module and symbol for a 4-bit adder.

For example, line 3 of the listing given in Figure 6.3 defines two input ports each having 4 bits ordered 3 down to 0:

```
input [3:0] a, b,
```

Ports having the same direction and width can be listed together or on separate lines, whichever is preferred. The expression [3:0] is the *bit range* of the port; for mathematical purposes, the left-hand bit (in this case bit 3) is always assumed to be the most significant bit.

The module presented in Figure 6.3 is described by a single continuous assignment statement situated on line 5:

```
assign #15 {c_out, sum} = a + b + c_in;
```

The above assignment gives some indication of the expressive power of an HDL such as Verilog. To describe an adder, it is simply a case of adding the three input port values together using the built-in + operator whenever any of the inputs change, and continuously assigning the result to the outputs. If required, the adder could also have been described in terms of Boolean equations, logic gates or even individual MOS transistor switches, such is the flexibility of Verilog. There are a couple of important points concerning the above assignment that are worth highlighting at this point:

- The expression on the right-hand side of the assignment operator performs an *unsigned* addition by default.
- Since `a` and `b` are referred to without specifying a bit range, their entire 4-bit values are added along with the single-bit carry input `c_in`.
- The carry input `c_in` is automatically added to the least significant bits of `a` and `b` (`a[0]` and `b[0]`).
- The result produced by adding the three inputs is potentially 5 bits in length; therefore, the target of the continuous assignment is the *concatenation* of the outputs `c_out` and `sum` (using the { } operator), with `c_out` occupying the most significant bit position (bit 4).
- The inclusion of `#15` after the keyword **assign** indicates a delay of 15 time-units between any input change and the resulting change in the outputs. Time delays are described in more detail in Chapter 7.

In all of the above examples of Verilog modules, all the objects representing digital signals are of type **wire**. This includes both the internal signals and the module ports. This is due to the simple combinational nature of the examples considered thus far: each module has defined a set of simple combinatorial relationships between the inputs and outputs; there is no need to store any values. Unconnected **wire**s are effectively undriven and, therefore, are assigned the high-impedance value `z`.

In addition to **wire**s, Verilog provides the **reg** (short for register) type variable to describe signals that have the ability to retain, or store, the last value assigned to them.

In common with **wire**s, the **reg**-type signal defaults to 1 bit, but it can also be defined as having multiple bits using the same notation as **wire**s, as illustrated by the following example:

```
reg [7:0] count; //an 8-bit register variable
```

The use of the **reg** object will be considered in detail in Chapter 7.

6.3 MODULES WITHIN MODULES: CREATING HIERARCHY

An important tool used by software engineers is so-called *top-down design*. Simply described, this involves breaking a complex problem into a set of clearly defined sub-problems, which may in turn be further subdivided into yet simpler problems. In the C/C++ languages, and others, the basic unit of execution is the *function* or *procedure*; these self-contained blocks of code are intended to perform a relatively simple task. The software engineer will create the functions required to implement the low-level tasks and make use of them in higher level functions by means of the function or procedure *calling* mechanism. In this manner, a complex software application can be implemented as a hierarchy of functions nested to any required depth. In digital hardware design, a similar hierarchical approach can be applied to complex design problems by means of module *instantiation*.

As stated earlier, modules can instantiate, or create an occurrence of, other user-defined modules as well as predefined gates and switches. In this manner, Verilog provides support for the fundamental tools used in the creation of complex digital systems, namely hierarchy, modularity and regularity [8].

Creating hierarchical designs in Verilog is quite straightforward. Having defined a module and stored it in a text file, it may be *compiled* into a library (or, in some tools, a project database) and referenced in other modules using the following syntax:

```
module-name instance-name(list-of-connections);
```

In the above, the `module_name` is the name of the module as defined by the module header, the `instance-name` is a unique name assigned to this particular instance or occurrence of the module. The `list-of-connections` defines the details of how the instanced module's ports are connected within the enclosing, or *parent*, module.

Figure 6.4 shows the block diagram of a digital system described by a Verilog module named `modT`. As shown in the figure, the so-called *parent* module, `modT`, contains three

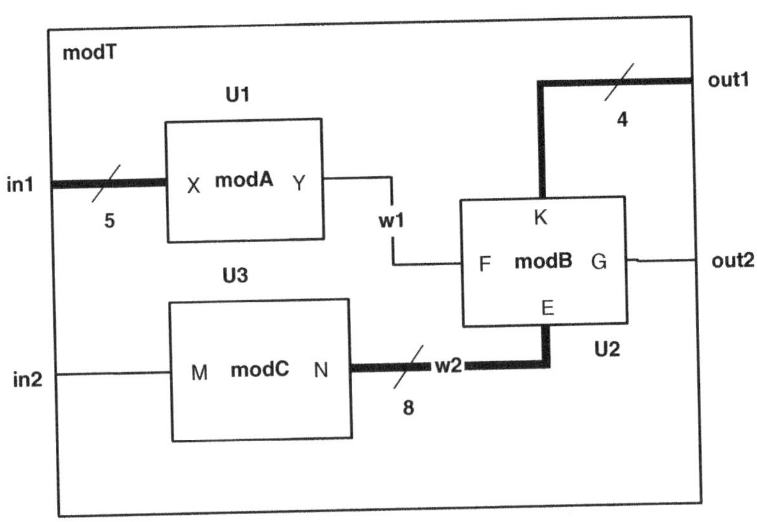

Figure 6.4 Block diagram of a module containing instances of other modules.

instances of other previously defined modules having names `modA`, `modB` and `modC`; the latter are sometimes referred to as *child* modules. The labels `U1–U3` represent the unique instance name for each instantiation; such labels are mandatory, since any given module may be instantiated more than once.

The names of the ports of each child module shown in Figure 6.4 are enclosed within the module's block; inputs enter on the left or bottom edge and outputs exit on the right or top edge.

Listing 6.2 shows the equivalent Verilog description of the block diagram of Figure 6.4.

```
1 module modT(input [4:0] in1,
2          input in2,
3          output [3:0] out1,
4          output out2);
5 wire [7:0] w2;
6 wire w1;

7 modA U1(.X(in1),.Y(w1));
8 modB U2(.F(w1),.E(w2),.K(out1),.G(out2));
9 modC U3(.M(in2),.N(w2));

10 endmodule
```

Listing 6.2 Verilog description for `modT`.

As was the case previously, the line numbers along the left-hand column are included for reference purposes; they do not form part of the module text. Lines 1 to 4 define the module header for `modT`: input `in1` is a 5-bit port and output `out1` is a 4-bit port; all remaining ports are single bit. Lines 5 and 6 declare two internal **`wire`**s used to link modules `modA` and `modC` to `modB`.

The block structure shown in Figure 6.4 is effectively created by the *module instantiation statements* on lines 7, 8, and 9. Each line begins with the name of the module being instantiated; this is followed by a space and then the unique instance name (`U1`, `U2`, ...).

In Verilog, there are two alternative ways of specifying module connectivity: the preferred method, known as *explicit association*, is used in Listing 6.2.

In this notation the ports of the child module are explicitly associated with particular signals by means of the 'dot' (.) notation, whereby the name of the signal being connected to the port is given in parentheses immediately after the selected port name, as shown below:

```
module-name instance-name(.port-name(net-name),...);
```

Explicit association has two important advantages over the second method that is sometimes used to define connectivity (discussed below):

- the connections may be listed in any order;
- the presence of both the port name and the name of the signal to which it is attached minimizes the possibility of errors.

The second method of defining module connectivity is known as *positional association*. In this notation, each port of the instantiated module is connected to the net occupying the corresponding position in the port list of the child module. For example, to instantiate module `modA` using positional association:

```
modA U1(in1, w1); //positional association
```

Clearly, positional association is less robust than explicit association due to the possibility of listing the connected signals in the wrong order. The Verilog compiler may not always report errors such as mismatches in the bit width or port direction caused by the wrong ports being connected to the wrong signals.

Occasionally, it is necessary to leave certain ports of a module unconnected. This can apply to both inputs and outputs. Regardless of whether explicit or positional association is used, *unconnected* ports are indicated by simply leaving blank the space where the connected signal name would normally appear. The two lines shown below illustrate the appearance of unconnected ports using the two alternative formats:

```
//output port 'K' is open circuit
modB U2(.F(w1),.E(w2),.K(),.G(out2));
//input port 'E' is unconnected
modB U4(w1,, out1, out2);
```

When an input is left unconnected, the Verilog simulator will force the corresponding port to take on the high-impedance value `z`.

As mentioned previously, Verilog uses two types of object to model signals in digital hardware:

- net or **`wire`** – must be continuously driven. The primary use is to model connections between continuous assignments and instantiations.
- **`reg`** – retains the last value that was assigned to it. Often (but not exclusively) used to represent storage elements.

Verilog imposes a set of rules regarding the nature of module ports and the type of object they can be connected to in a hierarchical design. Within the confines of a module, ports of direction

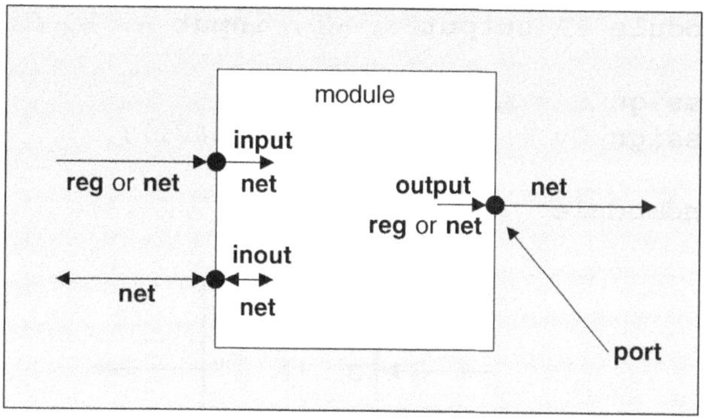

Figure 6.5 Illustration of Verilog port connectivity rules.

input or **inout** are implicitly of type net (defaulting to **wire**). Module output ports can be of either the net or **reg** type.

The **output** and **inout** ports of a module must be connected to nets at the next level up in the hierarchy. However, an input port may be driven by either a net- or a **reg**-type signal.

The above rules are summarized in Figure 6.5.

6.4 VERILOG HDL SIMULATION: A COMPLETE EXAMPLE

In this section, a complete example of a Verilog-HDL design, including a simulation *test-fixture* is presented. One of the key advantages of using an HDL, such as Verilog, is the ability to use the powerful features of the language to create the simulation environment for the design, as well as the design itself. This is the idea behind the so-called *test-fixture* (sometimes referred to as *test-bench* or *test-module*).

The main purpose of the test-fixture is to verify correct operation of the design; this can involve simply generating an input stimulus in order that the output responses may be observed, or more sophisticated techniques may be used to detect subtle design errors in more complex designs.

The principal advantage of the test-fixture results from the fact that it is written in the same standard language as the design and, therefore, provides the flexibility of simulation tool independence, being capable of running on any system that supports the IEEE standard Verilog.

Figure 6.6 shows the Verilog description and symbol for a single-bit binary adder [1]. The module `FA` uses the *dataflow* style of description to capture the behaviour of the logic; continuous assignments on lines 2 and 3 contain Boolean equations for the sum and carry outputs of the adder respectively. In terms of propagation delays, the adder module is ideal. Changes in any of the module inputs `A`, `B` and `Ci` will trigger execution of the two continuous assignments, causing the `S` and `Cy` outputs to be updated after one simulation cycle (delta).

The full adder module shown in Figure 6.6 could be described in a variety of alternative ways, ranging from primitive MOS switch circuitry at the lowest level, to high-level

```
module FA(output S, Cy, input A, B, Ci);

assign S = A^B^Ci;
assign Cy = (A&B)|(A&Ci)|(B&Ci);

endmodule
```

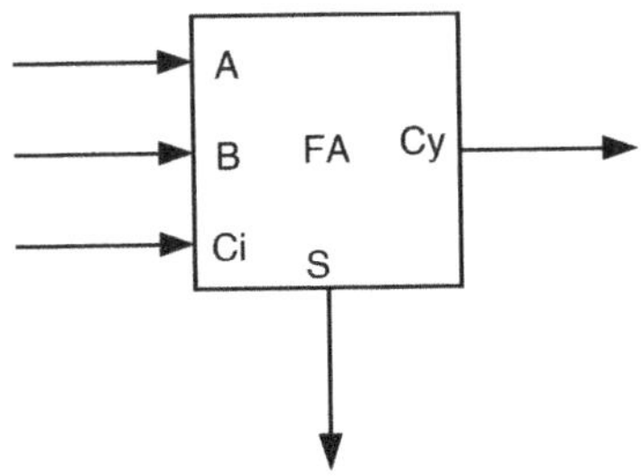

Figure 6.6 Verilog module and symbol for a binary full adder.

behavioural style. In this manner, Verilog supports the idea of *top-down design*, whereby a design is initially captured in an abstract manner to enable rapid verification of the design concept. The design can then be refined by changing its representation into a more detailed form, becoming closer to the eventual hardware technology being targeted.

Having defined the single-bit adder module, Figure 6.7 illustrates how four full adders can be cascaded to form the so-called 4-bit *ripple carry* adder [1].

The module header (lines 1 and 2) for `Add4` now defines outputs and inputs having a range of `3:0`, i.e. 4 bits. The carry-input to the least significant bit and the carry-output from the most significant bit are the only single-bit ports.

The 4-bit adder is constructed using four module instantiations of the full adder module, having instance names `fa0-fa3;` these are situated on lines 4–11. The full adders are interconnected by the carry vector `Cy` (declared on line 3), as shown in the circuit below the listing, along with the external carry input `Cin` and the carry output `Co`, forming the *ripple carry chain*.

Notice the use in the listing in Figure 6.7 of explicit association and *bit selection* in defining the connectivity of the instantiated full adder modules. For example, individual bits of the `A` and `B` input vectors are connected to the corresponding full adder stage input ports by including the relevant bit number in square brackets after the name of the port, as shown below:

```
.A(Ain[1]) //FA port 'A' connects to bit-1 of input vector 'Ain'
```

Although slightly longer, this technique is far clearer and leads to fewer errors being incorporated into the design.

Having constructed the 4-bit adder module, a test-fixture is used to verify the correctness of the design. Listing 6.3 and Figure 6.8 respectively show the Verilog listing and block diagram of a suitable test-fixture for the `Add4` module.

```
module Add4(output [3:0] Sum, output Co,
       input [3:0] Ain, Bin, input Cin);

wire [2:0] Cy;

FA fa0(.S(Sum[0]), .Cy(Cy[0]), .A(Ain[0]),
   .B(Bin[0]), .Ci(Cin));
FA fa1(.S(Sum[1]), .Cy(Cy[1]), .A(Ain[1]),
   .B(Bin[1]), .Ci(Cy[0]));
FA fa2(.S(Sum[2]), .Cy(Cy[2]), .A(Ain[2]),
   .B(Bin[2]), .Ci(Cy[1]));
FA fa3(.S(Sum[3]), .Cy(Co),     .A(Ain[3]),
   .B(Bin[3]), .Ci(Cy[2]));

endmodule
```

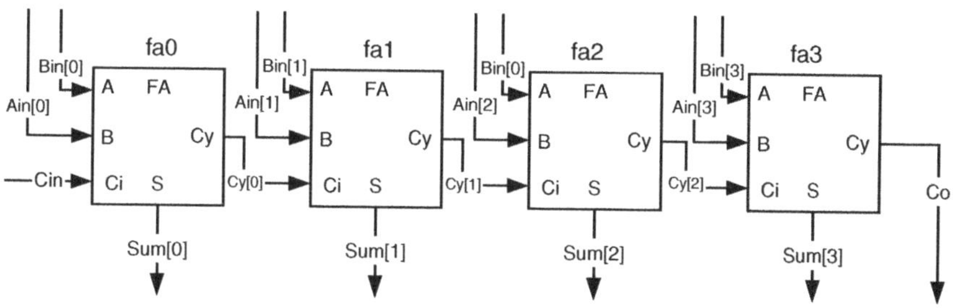

Figure 6.7 Verilog module and circuit for a 4-bit adder.

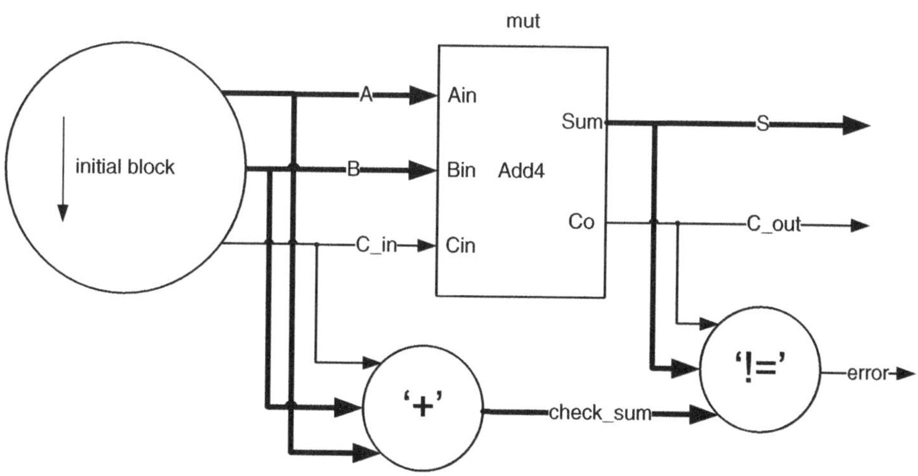

Figure 6.8 Block diagram of 4-bit adder test-module.

```
`timescale 1 ns/ 1 ns
module Test_Add4 (); //test module - no ports needed

//input stimulus
reg[3:0] A, B;
reg C_in;

//wire to hold check sum and error flag
wire[4:0] check_sum;
wire error;

//output responses
wire[3:0] S;
wire C_out;

integer test;

initial //only allowed in test module - runs once only
begin
    {A, B, C_in} = 9'b000000000;
    #100; //wait for 100 time units
    for (test = 0; test < 512; test = test + 1)
    begin //apply all input values
      {A, B, C_in} = test;
      #100;
    end
    $stop; //system command - stops the simulation
end

//instantiate the module-under-test
Add4 mut (.Sum(S), .Co(C_out), .Ain(A), .Bin(B),
       .Cin(C_in));

//add inputs using built-in '+' operator
assign check_sum = A + B + C_in;
//compare with mut output
assign error = (check_sum != {C_out, S} );

endmodule
```

Listing 6.3 Verilog test-module for 4-bit adder.

The block diagram of Figure 6.8 shows the structure of the test fixture. The conventional name given to the module being tested within the test fixture is *module-under-test* or `mut`, as shown above the symbol of the `Add4` module in Figure 6.8. The Verilog test-fixture generates a set of test input stimuli for the adder inputs `A`, `B` and `Cin` by means of a behavioural construct known as an **`initial`** *sequential block*; this is represented by the circle to the left of the adder in Figure 6.8.

In order to perform a basic check that the 4-bit adder performs the correct operation, the built-in Verilog + operator is used to produce a 5-bit result named `check_sum` from the initial block outputs.

The `check_sum` value is compared with the outputs of the 4-bit adder using the built-in Verilog *not-equal-to* operator (!=). A diagnostic output named `error` indicates when there is a mismatch between the outputs of the adder and the result of performing the summation of the stimulus. In this manner, the test-fixture provides a simple single-bit indication of the validity of the `Add4` module outputs.

The test-fixture Verilog description is given in Listing 6.3. This module makes use of several language constructs that have yet to be described. These new elements will be discussed briefly here and covered in more detail in Chapters 7 and 8.

The test-fixture module begins on line 1 with a so-called *compiler directive*. These special directives serve a similar purpose to the *pre-processor* directives found in the C/C++ languages; however, rather than beginning with the hash (#) symbol, Verilog uses the grave accent (`) character to indicate such a directive. The `timescale` directive on line 1 of Listing 6.3 defines a time scale and a time precision, the latter appearing after the '/' character. In this example, both the time scale and precision are specified as 1 ns; this means that any time delay values appearing within the body of the module are interpreted by the simulator as representing a whole number of nanoseconds. The time precision can be set to as small a unit as the femtosecond (10^{-15} s), thus allowing extremely precise timing simulation to be performed. In this example, there is no need for such precision.

The module header on line 2 indicates that the `Test_Add4` module is a test-fixture module rather than a design module by virtue of the fact that there are no inputs and outputs. Note that the empty parentheses after the module name are optional and, therefore, can be omitted without incurring a syntax error; the terminating semicolon is always required, however.

Lines 4 and 5 declare the input stimulus signals that are connected to the inputs of the adder module. The keyword **reg** indicates that these signals must retain their value in between being updated by assignments within the sequential **initial** block starting on line 13.

The outputs of the `module-under-test` (lines 25 and 26) and the continuous assignments on lines 28 and 30 are continuously driven by these statements; therefore, they are declared as **wire**s on lines 10, 11, 7 and 8 respectively.

The main part of the test-fixture is contained within the sequential **initial** block covering lines 13 to 23 in Listing 6.3. As stated previously, all signals that are assigned values by this block must be declared as type **reg**, in order that they retain the value last assigned to them during execution of the block.

The statements enclosed within the **initial** block execute sequentially and once only. This means that this type of block is only suitable for use in a test-fixture; it has no direct equivalent in terms of hardware.

Execution of the **initial** block starts at line 15, at a simulation time of 0 ns. The inputs are initialized to logic 0 using the following sequential statement:

```
{A, B, C_in} = 9'b000000000;
```

The inputs `A`, `B` and `C_in` are collectively assigned zeros by grouping them together using the concatenation operator { }.

Line 16 suspends execution of the sequential block for 100 ns; this allows the `module-under-test` toproduce a response to the input stimulus. The hash (#) symbol represents a time delay in this particular context.

Following on from the initial time delay, lines 17 to 21 contain a **for** loop that iterates through the values 0 to 511_{10} using an **integer** variable `test`, the latter being declared on line 12. Note that **integer** is a reserved word that declares a signed whole number (usually 32 bits in length) that behaves in a similar manner to a **reg**, in that it, too, retains its value in between being updated by assignments within a sequential block.

```
for(test = 0; test < 512; test = test + 1)
begin //apply all input values
    {A, B, C_in} = test;
    #100;
end
```

The body of the **for** loop is a block enclosed between the keywords **begin** and **end**. The first statement (line 19) assigns the least significant 9 bits of the integer variable `test` to the aggregate of the inputs, this apparent mixing of different types either side of an assignment is permitted in Verilog.

The second statement within the **for** loop introduces a 100 ns delay before execution continues with the next iteration of the loop. In this manner, an exhaustive set of input combinations are applied to the adder inputs starting at '000000000_2' and ending at '111111111_2' (511_{10}), each combination being applied for 100 ns.

The value of the loop variable `test` is incremented at the end of the loop and tested at the start of the loop; therefore, when it reaches 512_{10}, the condition `test < 512` becomes false and the loop terminates. An important point to note here is the possibility of a **for** loop being infinite, i.e. never terminating. This would occur if the loop variable `test` had been declared as a **reg** having a length of 9 bits rather than as a 32-bit **integer**. Since the range of values used within the loop and the number of inputs both correspond to a vector of length 9-bits, this may have seemed a logical course of action.

However, a problem occurs when the value of 'test' reaches '111111111_2'.

Incrementing this value by one results in '000000000_2', due to the way in which a 9-bit unsigned binary number overflows. The terminating condition `test < 512` can never be satisfied, since the 9-bit test value can never exceed 511_{10}. Therefore, if a 9-bit **reg** had been used as the loop counter rather than an **integer**, the simulator would carry on applying the same sequence indefinitely while using up increasing amounts disk space to store the results!

One possible solution would be to declare `test` as a 10-bit **reg**; the spare bit allows the loop variable to reach the terminating value of '1000000000_2'.

Having applied an exhaustive set of input values, the simulation is automatically stopped by means of a very common Verilog *system task* on line 22, repeated below:

```
$stop;
```

System tasks are always preceded by the dollar ($) symbol and perform a wide variety of useful functions, ranging from performing detailed timing checks (`$setup()`, `$hold()`, etc.) to outputting simulation data to a file (`$dumpvars`, `$dumpfile("filename")`). The `$stop` system task is often used within a test-fixture to end the simulation run forcibly; examples showing the use of other system tasks will be covered in Chapters 7 and 8.

The description of the test-fixture shown in Listing 6.3 concludes with the continuous assignments on lines 28 and 30 and repeated below:

```
assign check_sum = A + B + C_in;
assign error = (check_sum != {C_out, S} );
```

Table 6.1 Details of a few of the most popular Verilog simulator tools.

Name	Vendor	Web site
Active-HDL® Student Edition	Aldec Incorporated	http://www.aldec.com/education/students/
Modelsim-PE Student**®	Mentor Graphics	http://www.model.com/resources/student_edition/download.asp
Verilogger®	Synapticad	http://www.syncad.com/syn_down.htm
Xilinx® ISE Simulator*	Xilinx	http://www.xilinx.com/ise/logic_design_prod/webpack.htm

*The ISE Simulator is part of the free 'WebPACK' programmable logic design suite available from Xilinx®.
**A vendor-specific version of Modelsim® is also available from Xilinx® (ModelsimXE®).

The above assignments generate a single-bit diagnostic signal named `error` which goes high if there is a discrepancy between the actual output of the 4-bit adder module and the theoretical value predicted by the built-in addition operator. Although this may seem a little unnecessary, given the simplicity of the design, it hopefully illustrates the potential advantage of using the expressive power of the Verilog language to aid in the verification of a more complex design.

Simulation of the 4-bit adder module and associated test-fixture requires a Verilog-2001 [9] compatible simulation tool. There are several excellent Verilog simulators available from a variety of vendors. Table 6.1 contains details of a few of the most popular tools.

Regardless of which of the simulators in Table 6.1 is used, the process of simulation starts with the creation of the Verilog sources. It is normal practice to store each individual module's textual description in a unique ASCII text file (usually named 'module-name.v'). Most simulation tools include a context-sensitive text editor to aid in the creation of the source files; such an editor will include colour-coded keyword highlighting, line numbering and automatic indentation and formatting of the language statements. All the previously mentioned features help the designer to understand and maintain complex designs.

Most simulators make use of the concept of a *project*. This is essentially a repository for all of the Verilog files used in the design. Once written, the source files are added to the project prior to the next step, i.e. compilation. The process of compilation is similar in many ways to that used in other high-level language development systems: the objective is to

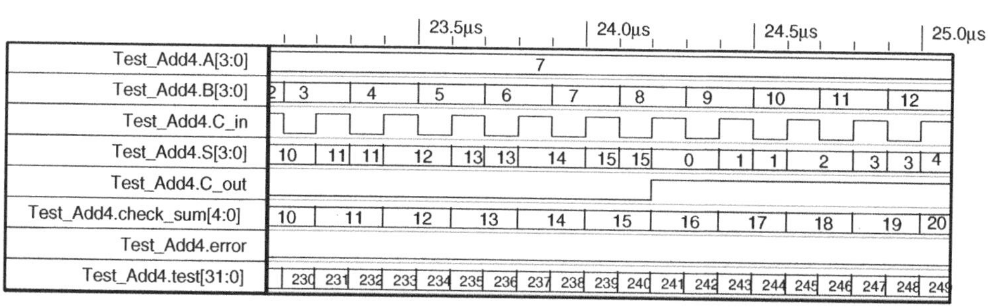

Figure 6.9 Partial simulation result for 4-bit adder test-module.

build an executable model suitable for loading into the simulation kernel, once any syntax errors have been corrected.

All Verilog simulators provide a graphical output in the form of timing waveforms. Figure 6.9 shows the partial result from running the simulation of the `Test_Add4` test-fixture module given in Listing 6.3.

REFERENCES

1. Wakerly J.F. Digital Design: Principles and Practices, 4th edition. New Jersey: Pearson Education, 2006.
2. www.systemc.org [2007 October].
3. www.systemverilog.org [2007 October].
4. www.verilog.com [2007 October]. (Links to IEEE Standards site and other Verilog information.)
5. www.accellera.org/home [2007 October].
6. www.adacore.com/home/ada_answers/ada_overview [2007 October].
7. www.synopsys.com/products/logic/design_compiler.html [2007 October].
8. Weste N.H.E., Eshraghian K. Principles of CMOS VLSI Design. Addison Wesley, 1993; Section 6.2.
9. Ciletti M.D. Advanced Digital Design with the Verilog HDL. New Jersey: Pearson Education, 2003; Appendix I – Verilog-2001.

In addition to the references listed, there are a considerable number of excellent internet sites containing tutorial and reference material on Verilog; many of these can be located using a standard internet search engine by entering the keyword 'Verilog'

7

Elements of Verilog HDL

This chapter introduces the basic lexical elements of the Verilog HDL. In common with other high-level languages, Verilog defines a set of types, operators and constructs that make up the vocabulary of the language. Emphasis is placed on those aspects of the language that support the description of synthesizable combinatorial and sequential logic.

7.1 BUILT-IN PRIMITIVES AND TYPES

7.1.1 Verilog Types

As mentioned in Chapter 6, Verilog makes use of two basic types: nets and registers. Generally, nets are used to establish connectivity and registers are used to store information, although the latter does not always imply the presence of sequential logic.

Within each category there exist several variants; these are listed in Table 7.1. All of the type names, listed in Table 7.1, are Verilog reserved words; the most commonly used types are shown in bold.

Along with the basic interconnection net type **wire**, two additional predefined nets are provided to model power supply connections: **supply0** and **supply1**.

These special nets possess the so-called 'supply' drive strength (the strongest; it cannot be overridden by another value) and are used whenever it is necessary to tie input ports to logic 0 or logic 1. The following snippet of Verilog shows how to declare and use power supply nets:

```
module ...

supply0 gnd;
supply1 vdd;
  nand g1(y, a, b, vdd); //tie one input of nand gate high

endmodule
```

The power supply nets are also useful when using Verilog to describe switch-level MOS circuits. However, the Verilog switch-level primitives [1] (**nmos**, **pmos**, **cmos**, etc.) are not

Table 7.1 Verilog types.

Nets (connections)	Registers (storage)
wire	
tri	
supply0	**reg**
supply1	**integer**
wand	real
wor	time
tri0	realtime
tri1	
triand	
trireg	
trior	

generally supported by synthesis tools; therefore, we will not pursue this area any further here.

Of the remaining types of net shown in the left-hand column of Table 7.1, most are used to model advanced net types that are not supported by synthesis tools; the one exception is the net type **tri**. This net is exactly equivalent to the **wire** type of net and is included mainly to improve clarity. Both **wire** and **tri** nets can be driven by multiple sources (continuous assignments, primitives or module instantiations) and can, therefore, be in the high-impedance state (z) when none of the drivers are forcing a valid logic level. The net type **tri** can be used instead of **wire** to indicate that the net spends a significant amount of the time in the high-impedance state.

Nets such as **wire** and **tri** cannot be assigned an initial value as part of their declaration; the default value of these nets at the start of a simulation is high impedance (z).

The handling of multiple drivers and high-impedance states is built in to the Verilog HDL, unlike some other HDLs, where additional IEEE-defined packages are required to define types and supporting functions for this purpose.

The right-hand column of Table 7.1 lists the register types provided by Verilog; these have the ability to retain a value in-between being updated by a sequential assignment and, therefore, are used exclusively inside sequential blocks. The two most commonly used register variables are **reg** and **integer**; the remaining types are generally not supported by synthesis tools and so will not be discussed further.

There are some important differences between the **reg** and **integer** types that result in the **reg** variable being the preferred type in many situations.

A **reg** can be declared as a 1-bit object (i.e. no size range is specified) or as a vector, as shown by the following examples:

```
reg a, b; //single-bit register variables
reg [7:0] busa; //an 8-bit register variable
```

As shown above, a **reg** can be declared to be of any required size; it is not limited by the word size of the host processor.

An **integer**, on the other hand, cannot normally be declared to be of a specified size; it takes on the default size of the host machine, usually 32 or 64 bits.

The other difference between the **integer** and **reg** types relates to the way they are handled in arithmetic expressions. An **integer** is stored as a two's complement signed number and is handled in arithmetic expressions in the same way, i.e. as a *signed* quantity (provided that all operands in the expression are also signed). In contrast, a **reg** variable is by default an *unsigned* quantity.

If it is necessary to perform signed two's complement arithmetic on **reg**s or **wire**s, then they can be qualified as being **signed** when they are declared. This removes the host-dependent word length limit imposed by the use of the **integer** type:

```
reg signed [63:0] sig1; //a 64-bit signed reg
wire signed [15:0] sig2; //a 16-bit signed wire
...
```

The use of the keyword **signed** to qualify a signal as being both positive and negative also applies to module port declarations, as shown in the module header below:

```
module mod1(output reg signed [11:0] dataout,
            input signed [7:0] datain,
            output signed [31:0] dataout2);
...
```

Finally, both the **integer** and **reg** types can be assigned initial values as part of their declarations, and in the case of the **reg** this can form part of the module port declaration, as shown below:

```
module mod1(output reg clock = 0,
            input [7:0] datain = 8'hFF,
            output [31:0] dataout2 = 0);

integer i = 3;
...
```

The differences discussed above mean that the **reg** and **integer** variables have different scopes of application in Verilog descriptions. Generally, **reg** variables are used to model actual hardware registers, such as counters, state registers and data-path registers, whereas **integer** variables are used for the computational aspects of a description, such as loop counting. The example in Listing 7.1 shows the use of the two types of register variable.

The Verilog code shown describes a 16-bit synchronous binary up-counter. The module makes use of two **always** *sequential* blocks – a detailed description of sequential blocks is given in the Chapter 8.

The first sequential block, spanning lines 5 to 11 of Listing 7.1, describes a set of flip flops that are triggered by the positive edges (logic 0 to logic 1) of the 'clock' input.

The state of the flip flops is collectively stored in the 16-bit **reg**-type output signal named q, declared within the module header in line 2. Another 16-bit **reg**-type signal, named t, is declared in line 3. This vector is the output of a combinational circuit described by the sequential block spanning lines 12 to 20. This illustrates the point that a **reg**-type signal does not always represent sequential logic, being necessary wherever a signal must retain the value last assigned to it by statements within a sequential block. The **always** block starting on line 12 responds to changes in the outputs of the flip flops q and updates the values of t accordingly. The updated values of the t vector then determine the next values of the q outputs at the subsequent positive edge of the 'clock' input.

```
1 module longcnt (input clock, reset, output reg [15:0] q);
2
3 reg [15:0] t; //flip-flop outputs and inputs
4 //sequential logic
5 always @ (posedge clock)
6 begin
7         if (reset)
8             q <= 16'b0;
9         else
10            q <= q ^ t;
11 end

12 always @(q)   //combinational logic
13 begin: t_block
14        integer i;    //integer used as loop-counter
15        for (i = 0; i < 16; i = i + 1)
16            if (i == 0)
17                t[i] = 1'b1;
18            else
19                t[i] = q[i-1] & t[i-1] ;
20 end
21 endmodule
```

Listing 7.1 Use of Verilog types **reg** and **integer**.

The second sequential block (lines 12–20) is referred to as a *named block*, due to the presence of the label t_block after the colon on line 13. Naming a block in this manner allows the use of local declarations of both **reg**s and **integer**s for use inside the confines of the block (between **begin** and **end**). In this example, the **integer** i is used by the **for** loop spanning lines 15–19, to process each bit of the 16-bit **reg** t, such that apart from t[0], which is always assigned a logic 1, the *i*th bit of t (t[i]) is assigned the logical AND (&) of the $(i-1)$th bits of q and t. The sequential **always** block starting on line 12 describes the iterative logic required to implement a synchronous binary ascending counter.

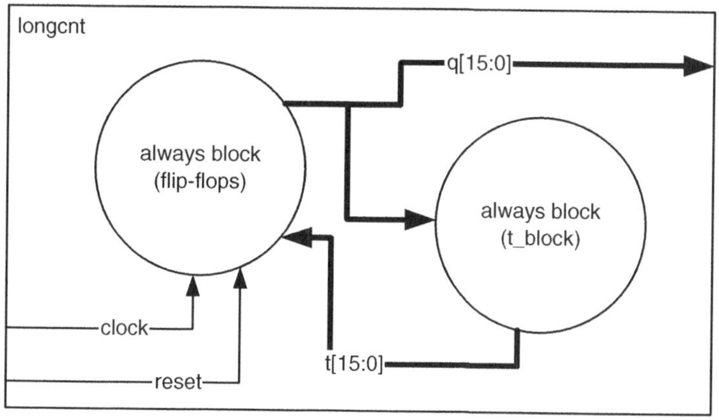

Figure 7.1 Block diagram of module `longcnt`.

Figure 7.1 shows the structure of the `longcnt` module given in Listing 7.1.

Simulation of a 4-bit version of the binary counter module `longcnt` results in the waveforms shown in Figure 7.2. The waveforms in Figure 7.2 clearly show the `q` outputs counting up in ascending binary, along with the corresponding `t` vector pulses causing the `q` output bits to 'toggle' state at the appropriate times. For example, when the `q` output is '0111_2', the `t` vector is '1111_2' and so all of the output bits toggle (change state) on the next positive edge of 'clock'.

7.1.2 Verilog Logic and Numeric Values

Each individual bit of a Verilog HDL **`reg`** or **`wire`** can take on any one of the four values listed in Table 7.2. Verilog also provides built-in modelling of signal strength; however, this feature is generally not applicable to synthesis and, therefore, we will not cover it here.

Of the four values listed in Table 7.2, logic 0 and logic 1 correspond to Boolean false and true respectively. In fact, any nonzero value is effectively true in Verilog, as it is in the C/C++

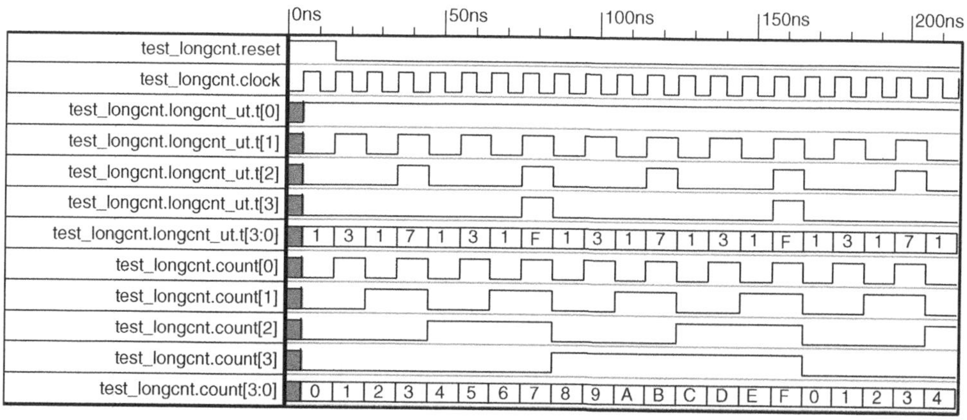

Figure 7.2 Simulation results for module `longcnt` (4-bit version)

Table 7.2 Four-valued logic.

Logic value	Interpretation
0	Logic 0 or false
1	Logic 1 or true
x	Unknown (or don't care)
z	High impedance

programming languages. Relational operators all result in a 1-bit result indicating whether the comparison is true (1) or false (0).

Two meta-logical values are also defined. These model unknown states (x) and high impedance (z); the x is also used to represent 'don't care' conditions in certain circumstances. At the start of a simulation, at time zero, all **reg**s are initialized to the unknown state x, unless they have been explicitly given an initial value at the point of declaration. On the other hand, **wire**s are always initialized to the *undriven* state z.

Once a simulation has commenced, all **reg**s and **wire**s should normally take on meaningful numeric values or high impedance; the presence of x usually indicates a problem with the behaviour or structure of the design.

Occasionally, the unknown value x is deliberately assigned to a signal as part of the description of the module. In this case, the x indicates a so-called *don't care* condition, which is used during the logic minimization process underlying logic synthesis.

Verilog provides a set of built-in pre-defined logic gates, these primitive elements respond to unknown and high-impedance inputs in a sensible manner. Figure 7.3 shows the simulation results for a very simple two-input AND gate module using the built-in **and** primitive. The simulation waveforms show how the output of the module below responds when its inputs are driven by x and z states.

```
module valuedemo(output y, input a, b);
and g1(y, a, b);
endmodule
```

Referring to the waveforms in Figure 7.3, during the period 0–250 ns the **and** gate output y responds as expected to each combination of inputs a and b. At time 250 ns, the a input is driven to the z state (indicated by the dotted line) and the gate outputs an x (shaded regions) between 300 and 350 ns, since the logical AND of logic 1 and z is undefined. Similarly, during the interval 400–450 ns, the x on the b input also causes the output y to be an x.

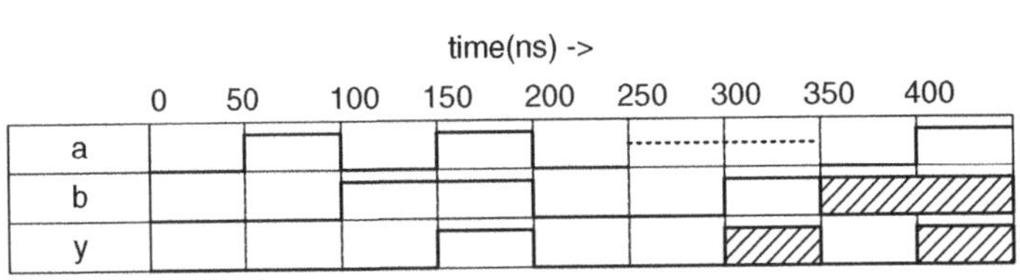

Figure 7.3 The **and** gate response to x and z.

However, during the intervals 350–400 ns and 250–300 ns, one of the inputs is low, thus causing y to go low. This is due to the fact that anything logically ANDed with a logic 0 results in logic 0.

7.1.3 Specifying Values

There are two types of number used in Verilog HDL: *sized* and *unsized*. The format of a sized number is

```
<size>'<base><number>.
```

Both the `<size>` and `<base>` fields are optional; if left out, the number is taken to be in decimal format and the size is implied from the variable to which the number is being assigned.

The `<size>` is a decimal number specifying the length of the number in terms of binary bits; `<base>` can be any one of the following:

binary	b or B
hexadecimal	h or H
decimal (default)	d or D
octal	o or O

The actual value of the number is specified using combinations of the digits from the set ⟨0, 1, 2, 3, 4, 5, 6, 7, 8, 9, A, B, C, D, E, F⟩. Hexadecimal numbers may use all of these digits; however, binary, octal and decimal are restricted to the subsets ⟨0, 1⟩, ⟨0–7⟩ and ⟨0–9⟩ respectively. Below, are some examples of literal values written using the format discussed above:

```
4'b0101 //4-bit binary number
12'hefd //12-bit hex number
16'd245 //16-bit decimal number
1'b0, 1'b1 //logic-0 and logic-1
```

Generally, it is not necessary to specify the size of a number being assigned to an **`integer`**-type variable, since such objects are unsized (commonly occupying 32 bits or 64 bits, depending upon the platform).

The literals x and z (unknown and high impedance) may be used in binary-, hexadecimal- and octal-based literal values. An x or z sets 4 bits in a hex number, 3 bits in an octal number and 1 bit in a binary number.

Furthermore, if the most-significant digit of a value is 0, z or x, then the number is automatically extended using the same digit so that the upper bits are identical.

Below, are some examples of literal values containing the meta-logical values x and z:

```
12'h13x //12-bit hex number '00010011xxxx' in binary
8'hx //8-bit hex number 'xxxxxxxx' in binary
16'bz //16-bit binary number 'zzzzzzzzzzzzzzzz'
11'b0 //11-bit binary number '00000000000'
```

As shown above, the single-bit binary states of logic 0 and logic 1 are usually written in the following manner:

```
1'b0 //logic-0
1'b1 //logic-1
```

Of course, the single decimal digits 0 and 1 can also be used in place of the above. As mentioned previously, in Verilog, the values `1'b0` and `1'b1` correspond to the Boolean values 'false' and 'true' respectively.

7.1.4 Verilog HDL Primitive Gates

The Verilog HDL provides a comprehensive set of built-in primitive logic and three-state gates for use in creating gate-level descriptions of digital circuits. These elements are all synthesizable; however, they are more often used in the output gate-level Verilog net-list produced by a synthesis tool.

Figures 7.4 and 7.5 show the symbolic representation and Verilog format for each of the primitives [1]. The use of the primitive gates is fairly self-explanatory; the basic logic gates, such as AND, OR, etc., all have single-bit outputs but allow any number of inputs (Figure 7.4 shows two-input gates only). The Buffer and NOT gates allow multiple outputs and have a single input.

The three-state gates all have three terminals: output, input and control. The state of the control terminal determines whether or not the buffer is outputting a high-impedance state or not.

For all gate primitives, the output port must be connected to a net, usually a **wire**, but the inputs may be connected to nets or register-type variables.

An optional delay may be specified in between the gate primitive name and the instance label; these can take the form of simple propagation delays or contain separate values for *rise-time*, *fall-time* and *turnoff-time* delays [1], as shown by the examples below:

```
//AND gate with output rise-time of 10 time units
//and fall-time of 20 time units
and #(10, 20) g1 (t3, t1, a);

//three-state buffer with output rise-time of 15 time unit,
//fall-time of 25 time units
//and turn-off delay time of 20 time units
bufif1 #(15, 25, 20) b1 (dout, din, c1);
```

Figure 7.6 shows a simple example of a gate-level Verilog description making use of the built-in primitives; each primitive gate instance occupies a single line from numbers 4 to 8 inclusive. A single propagation delay value of 10 ns precedes the gate instance name; this means that changes at the input of a gate are reflected at the output after this delay, regardless of whether the output is rising or falling. The actual units of time to be used during the simulation are defined using the `timescale` compiler directive; this immediately precedes the module to which it applies, as shown in line 1 in Figure 7.6.

Gate Symbol	Verilog Instantiation
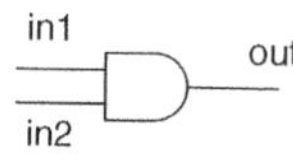	**and** a1(out, in1, in2);
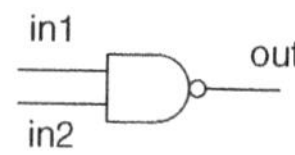	**nand** na1(out, in1, in2);
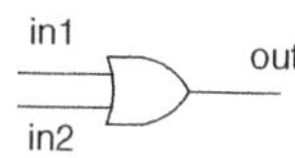	**or** o1(out, in1, in2);
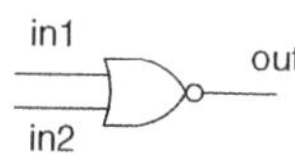	**nor** no1(out, in1, in2);
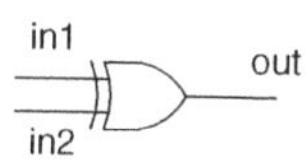	**xor** xo1(out, in1, in2);
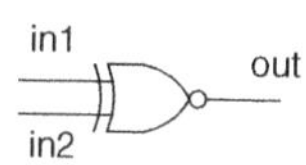	**xnor** xno1(out, in1, in2);
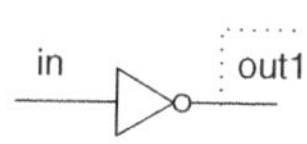	**not** nt1(out1, in);
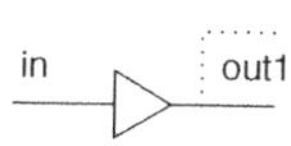	**buf** b1(out1, in);

Figure 7.4 Verilog primitive logic gates.

Gate Symbol	Verilog Instantiation
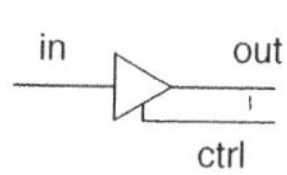	**bufif1** g1(out, in, ctrl);
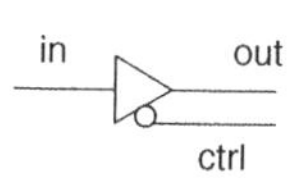	**bufif0** g1(out, in, ctrl);
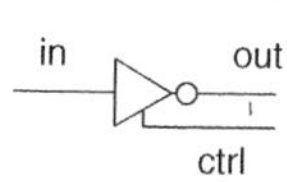	**notif1** g1(out, in, ctrl);
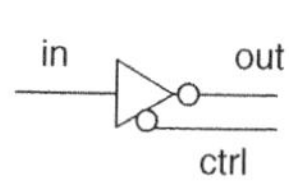	**notif0** g1(out, in, ctrl);

Figure 7.5 Verilog primitive three-state gates

```
`timescale  1 ns /1 ns

module x_or_s(output y, input a, b);

wire t1, t2, t3, t4;

and #10 g1(t3, t1, a);
and #10 g2(t4, t2, b);
not #10 g3(t1, b);
not #10 g4(t2, a);
or  #10 g5(y, t3, t4);

endmodule
```

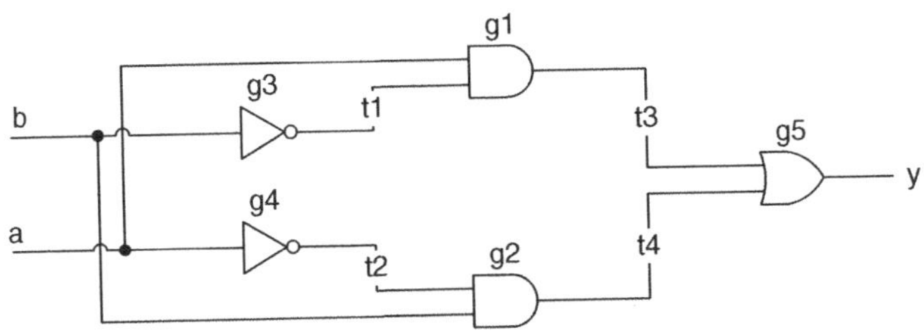

Figure 7.6 Gate-level logic circuit and Verilog description.

Gate delays are *inertial*, meaning that input pulses which have a duration of less than or equal to the gate delay do not produce a response at the output of the gate, i.e. the gate's inertia is not overcome by the input change. This behaviour mirrors that of real logic gates.

Gate delays such as those used in Figure 7.6 may be useful in estimating the performance of logic circuits where the propagation delays are well established, e.g. in the model of a TTL discrete logic device. However, Verilog HDL descriptions intended to be used as the input to logic synthesis software tools generally do not contain any propagation delay values, since these are ignored by such tools.

7.2 OPERATORS AND EXPRESSIONS

The Verilog HDL provides a powerful set of operators for use in digital hardware modelling. The full set of Verilog operators is shown in Table 7.3. The table is split into four columns, containing (from left to right) the category of the operator, the symbol used in the language for the operator, the description of the operator and the number of operands used by the operator.

Inspection of Table 7.3 reveals the similarity between the Verilog operators and those of the C/C++ languages. There are, however, one or two important differences and enhancements provided by Verilog in comparison with C/C++. The main differences between the C-based languages and Verilog, in terms of operators, are summarized overleaf:

Table 7.3 Verilog operators.

Operator type	Symbol	Operation	Operands
Arithmetic	*	Multiply	2
	/	Divide	2
	+	Add	2
	−	Subtract	2
	%	Modulus	2
	**	Raise to power	2
Logical	!	Logical negation	1
	&&	Logical AND	2
	||	Logical OR	2
Relational	>	Greater than	2
	<	Less than	2
	>=	Greater than or equal	2
	<=	Less than or equal	2
Equality	==	Equality	2
	!=	Inequality	2
	===	Case equality	2
	!==	Case inequality	2
Bitwise	~	Bitwise NOT	1
	&	Bitwise AND	2
	|	Bitwise OR	2
	^	Bitwise exclusive OR	2
	^~ or ~^	Bitwise exclusive NOR	2
Reduction	&	Reduction AND	1
	~&	Reduction NAND	1
	|	Reduction OR	1
	~|	Reduction NOR	1
	^	Reduction EXOR	1
	^~ or ~^	Reduction EXNOR	1
Shift	>>	Shift right	2
	<<	Shift left	2
	>>>	Shift right signed	2
	<<<	Shift left signed	2
Concatenation	{ }	Concatenate	Any number
Replication	{ { } }	Replicate	Any number
Conditional	? :	Conditional	3

- Verilog provides a powerful set of *unary* logical operators (so-called *reduction* operators) that operate on all of the bits within a single word.
- Additional 'case' equality/inequality operators are provided to handle high-impedance (z) and unknown (x) values.
- The curly braces '{' and '}' are used in the *concatenation* and *replication* operators instead of block delimiters (Verilog uses **`begin`** ... **`end`** for this).

The operators listed in Table 7.3 are combined with operands to form an expression that can appear on the right hand side of a continuous assignment statement or within a sequential block.

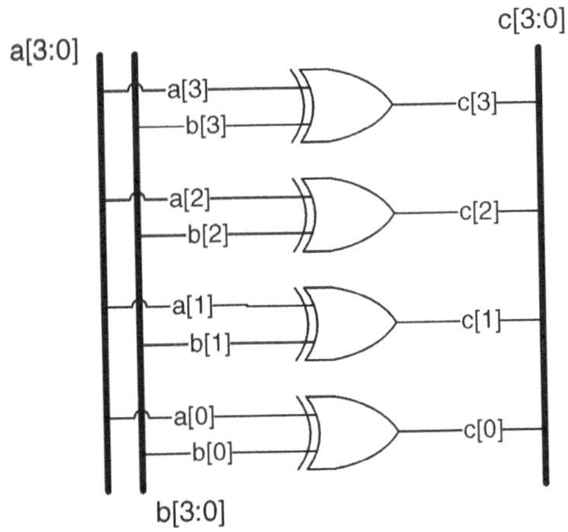


```
// 4 2-input Exor gates
assign c = a[3:0] ^ b[3:0];
```

Figure 7.7 Exclusive OR of part-selects.

The operands used to form an expression can be any combination of **wire**s, **reg**s and **integer**s; but, depending on whether the expression is being used by a continuous assignment or a sequential block, the target must be either a **wire** or a **reg** respectively.

In the case of multi-bit objects (buses), an operand can be the whole object (referenced by the name of the object), *part-select* (a subset of the bits within a multi-bit bus) or individual bit, as illustrated by the following examples.

Given the following declarations, Figures 7.7–7.10 show a selection of example continuous assignments and the corresponding logic circuits.

```
wire  [7:0] a, b; //8-bit wire
wire  [3:0] c;    //4-bit wire
wire  d;    //1-bit wire
```

Figure 7.7 illustrates the use of the bitwise exclusive OR operator on two 4-bit operands. As shown by the logic circuit in Figure 7.7, part-selects `[3:0]` of the two 8-bit wires, `a` and `b`, are processed bit-by-bit to produce the output `c`.

Individual bits of a **wire** or **reg** are accessed by means of the bit-select (square brackets [ ]) operator. Figure 7.8 shows the continuous assignment statement and logic for an AND gate with an inverted input.

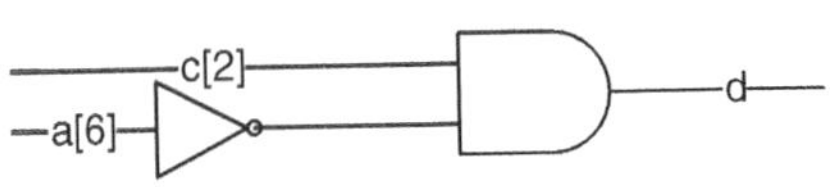


```
//2-input And with inverted I/P
assign d = c[2] & ~ a[6];
```

Figure 7.8 Logical AND of bit-selects.

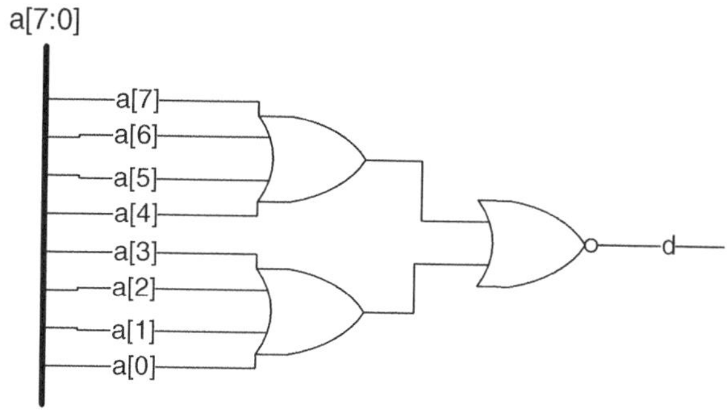


```
//8-input Nor gate using reduction NOR
assign d = ~|a;
```

Figure 7.9 Reduction NOR operator.

The bit-wise reduction operators, shown in Table 7.3, are unique to Verilog HDL. They provide a convenient method for processing all of the bits within a single multi-bit operand. Figure 7.9 shows the use of the reduction-NOR operator. This operator collapses all of the bits within the operand a down to a single bit by ORing them together; the result of the reduction is then inverted, as shown by the equivalent expression below:

```
assign d=~(a[7] |a[6] |a[5] |a[4] |a[3] |a[2] |a[1] |a[0] );
```

The target d of the continuous assignment statement in Figure 7.9 will be logic 1 if all 8 bits of a are logic 0, otherwise d is assigned logic 0. All of the reduction operators act on one operand positioned to the right of the operator. Those that are inverting, such as reduction NOR (~|) and reduction NAND (~&), combine all of the bits of the operand using bitwise OR or bitwise AND prior to inverting the single-bit result. The bitwise reduction operators provide a convenient means of performing multi-bit logical operations without the need to instantiate a primitive gate. Finally, if any bit or bits within the operand are high impedance (z), the result is generated as if the corresponding bits were unknown (x).

The last example, shown in Figure 7.10, illustrates the use of the conditional operator (? :) to describe a set of four 2-to-1 multiplexers, the *true expression* a[3:0] is assigned to c when the *control expression*, in this case d, is equal to 1'b1. When the single-bit signal d is equal to logic 0 (1'b0), the target c is assigned the *false expression* a[7:4] .

The unsigned shift operators (≪, ≫ in Table 7.3) shuffle the bits within a multi-bit **wire** or **reg** by a number of bit positions specified by the second operand. These operators shift logic 0s into the vacant bit positions; because of this, care must be taken when shifting two's complement (signed) numbers. If a negative two's complement number is shifted right using the '≫' operator, then the sign bit is changed from a 1 to a 0, changing the polarity of the number from negative to positive.

Right-shifting of two's complement numbers can be achieved by means of the 'shift right signed' (⋙) operator (provided the **wire** or **reg** is declared as **signed**) or by using the replication/concatenation operators (see later).

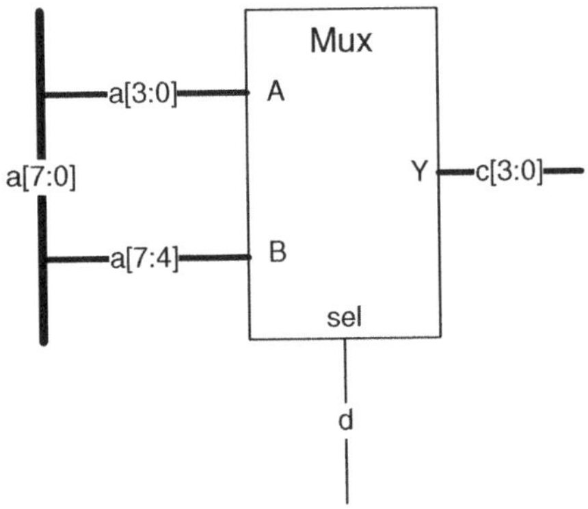


```
//8-to-4 multiplexer using conditional operator
assign c = d ? a[3:0] : a[7:4];
```

Figure 7.10 An 8-to-4 multiplexer.

The following assignments illustrate the use of the unsigned shift operators:

```
//declare and initialize X
reg [3:0] X=4'b1100;
Y=X>>1; //Result is 4'b0110
Y=X<<1; //Result is 4'b1000
Y=X>>2; //Result is 4'b0011
```

Verilog supports the use of arithmetic operations on multi-bit **reg** and **wire** objects as well as **integer**s; this is very useful when using the language to describe hardware such as counters (+, − and %) and digital signal processing systems (* and /).

With the exception of type **integer**, the arithmetic and comparison operators treat objects of these types as *unsigned* by default. However, as discussed in Section 7.1, **reg**s and **wire**s (as well as module ports) can be qualified as being **signed**. In general, Verilog performs signed arithmetic only if all of the operands in an expression are signed; if an operand involved in a particular expression is unsigned, then Verilog provides the system function `$signed()` to perform the conversion if required (an additional system function named `$unsigned()` performs the reverse conversion).

Listing 7.2 illustrates the use of the multiply, divide and shifting operators on signed and unsigned values. As always, the presence of line numbers along the left-hand column is for reference purposes only.

```
//test module to demonstrate signed/unsigned arithmetic
module test_v2001_ops();
reg [7:0] a=8'b01101111; //unsigned value (111_10)
reg signed [3:0] d=4'b0011; //signed value (+3_10)
reg signed [7:0] b=8'b10010110; //signed value (-106_10)
```

```
5 reg signed [15:0] c; //signed value
6 initial
7 begin
8     c = a * b; // unsigned value * signed value
9     #100;
10    c = $signed (a) * b; // signed value * signed value
11    #100;
12    c = b / d; // signed value ÷ signed value
13    #100;
14    b = b >>> 4; //arithmetic shift right
15    #100;
16    d = d << 2; // shift left logically
17    #100;
18    c = b * d; // signed value * signed value
19    #100;
20    $stop;
21 end
22 endmodule
```

Time	a	d	b	c	Line
0	111	3	−106	16 650	8
100	111	3	−106	−11 766	10
200	111	3	−106	−35	12
300	111	3	−7	−35	14
400	111	−4	−7	−35	16
500	111	−4	−7	28	18

Listing 7.2 Signed and unsigned arithmetic.

The table shown below the Verilog source listing in Listing 7.2 shows the results of simulating the module `test_v2001_ops();` the values of `a`, `b`, `c` and `d` are listed in decimal.

The statements contained within the **`initial`** sequential block starting on line 6 execute from top to bottom in the order that they are written; the final `$stop` statement, at line 20, causes the simulation to terminate. The result of each statement is given along with the corresponding line number in the table in Listing 7.2.

The statement on line 8 assigns the product of an unsigned and a signed value to a signed value. The unsigned result of $16\,650_{10}$ is due to the fact that one of the operands is unsigned and, therefore, the other operand is also handled as if it were unsigned, i.e. 150_{10} rather than -106_{10}.

The statement on line 10 converts the unsigned operand `a` to a signed value before multiplying it by another signed value; hence, all of the operands are signed and the result, therefore, is signed ($-11\,766_{10}$).

The statement on line 12 divides a signed value (-106_{10}) by another signed value ($+3_{10}$), giving a signed result (-35_{10}). The result is truncated due to integer division.

Line 14 is a signed right-shift, or arithmetic right-shift (⋙ operator). In this case, the sign-bit (most significant bit (MSB)) is replicated four times and occupies the leftmost bits, effectively dividing the number by 16_{10} while maintaining the correct sign. In binary, the result is '11111001_2', which is -7_{10}.

A logical shift-left (≪) is performed on a signed number on line 16. Logical shifts always insert zeros in the vacant spaces and so the result is '1100_2', or -4_{10}.

Finally, on line 18, two negative signed numbers are multiplied to produce a positive result.

The use of the keyword **signed** and the system functions `$signed` and `$unsigned` are only appropriate if the numbers being processed are two's complement values that represent bipolar quantities. The emphasis in this book is on the design of FSMs where the signals are generally either single-bit or multi-bit values used to represent a machine state. For this reason, the discussion presented above on signed arithmetic will not be developed further.

The presence of the meta-logical values `z` or `x` in a **reg** or **wire** being used in an arithmetic expression results in the whole expression being unknown, as illustrated by the following example:

```
//assigning values to two 4-bit objects
in1 = 4'b101x;
in2 = 4'b0110;
sum = in1 + in2; //sum = 4'bxxxx due to 'x' in in1
```

Figure 7.11 shows a further example of the use of the Verilog bitwise logical operators. The continuous assignment statement on lines 4 to 7 makes use of the AND (&), NOT (~) and OR (|) operators; note that there is no need to include parentheses around the inverted inputs, since the NOT operator (~) has a higher precedence than the AND (&) operator. However, the parentheses around the ANDed terms are required, since the '&' and '|' operators have the same precedence.

Figure 7.12 shows an alternative way of describing the same logic described by Figure 7.11. Here, a *nested conditional operator* is used to select one of four inputs `i0–i3`, under the control of a 2-bit input `s1, s0` and assign it to the output port named `out`.

There is no limit to the degree of nesting that can be used with the conditional operator, other than that imposed by the requirement to maintain a certain degree of readability.

The listing in Figure 7.13 shows another use of the conditional operator. Here, it is used on line 3 to describe a so-called 'three-state buffer' as an alternative to using eight instantiations of the built-in primitive `bufif1` (see Figure 7.5).

When the `enable` input is at logic 1, the `Dataout` port is driven by the value applied to `Datain`; on the other hand, when the `enable` input is at logic 0, the `Dataout` port is effectively undriven, being assigned the value '`zzzzzzzz`'.

In addition to ports of direction **input** and **output**, Verilog provides for ports that allow two-way communication by means of the **inout** keyword. Figure 7.14 illustrates a simple bidirectional interface making use of an **inout** port. It is necessary to drive bidirectional ports to the high-impedance state when they are acting as an input, hence the inclusion of the 8-bit three-state buffer on line 3. The Verilog simulator makes use of a built-in resolution mechanism to predict correctly the value of a **wire** that is subject to multiple drivers. In the current example, the bidirectional port `Databi` can be driven to a logic 0 or logic 1 by an external signal; hence, it

```
//A 4-to-1 multiplexer described using bitwise operators
module mux4_to_1(output out,
       input i0, i1, i2, i3, s1, s0);

assign out = (~s1 & ~s0 & i0)|
       (~s1 & s0 & i1)|
       (s1 & ~s0 & i2)|
       (s1 & s0 & i3);

endmodule
```

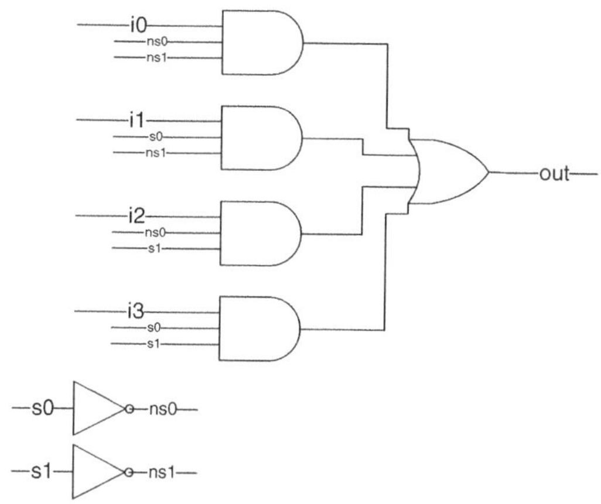

Figure 7.11 A 4-to-1 multiplexer described using bitwise operators.

```
//A 4-to-1 mux described using the conditional operator
module mux4_to_1(output out,
       input i0, i1, i2, i3, s1, s0);

assign out = s1 ? ( s0 ? i3 : i2)
       : (s0 ? i1 : i0);

endmodule
```

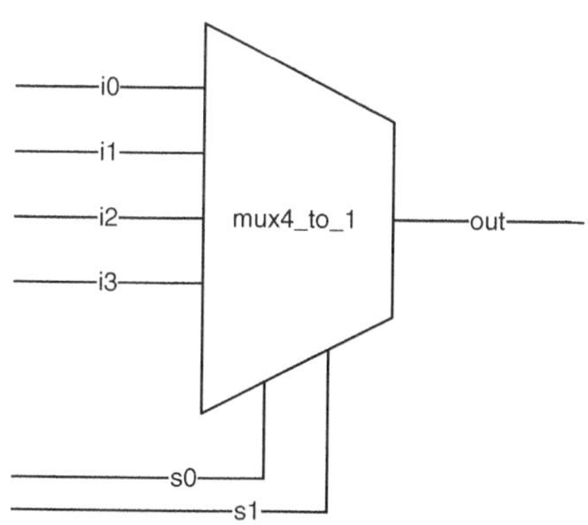

Figure 7.12 A 4-to-1 multiplexer described using nested conditional operators.

```
//An 8-bit three-state buffer
module Tribuff8(input [7:0] Datain, input enable,
        output [7:0] Dataout);

assign Dataout = (enable == 1'b1)? Datain : 8'bz;

endmodule
```

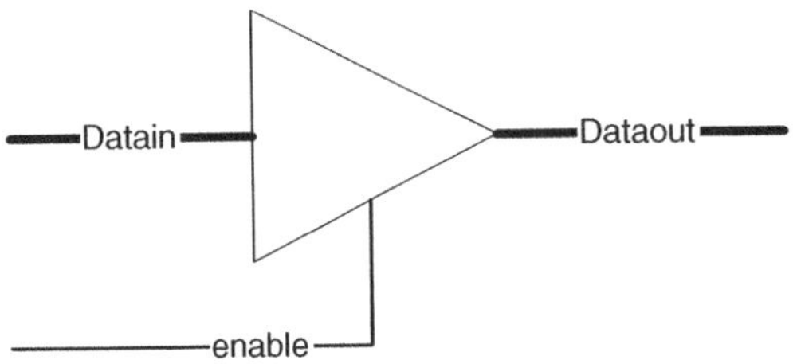

Figure 7.13 An 8-bit three-state buffer.

has two drivers. The presence of a logic level on the `Databi` port will override the high-impedance value being assigned to it by line 3; hence, `Datain` will take on the value of the incoming data applied to `Databi` as a result of the continuous assignment on line 4.

The simple combinational logic example given in Figure 7.15 illustrates the use of the logical OR operator (||) and the equality operator (==). The module describes a three-input majority voter that outputs a logic 1 when two or more of the inputs is at logic 1.

On line 4, the individual input bits are grouped together to form a 3-bit value on **wire** `abc`. The concatenation operator ({ }) is used to join any number of individual 1-bit or multi-bit

```
1  //An 8-bit bi-directional port using a three-state buffer
2  module Bidir(input [7:0] Databuff,
               output [7:0] Datain,
               input enable, inout [7:0] Databi);

3  assign Databi = (enable == 1'b1)?Databuff : 8'bz;
4  assign Datain = Databi;

5  endmodule
```

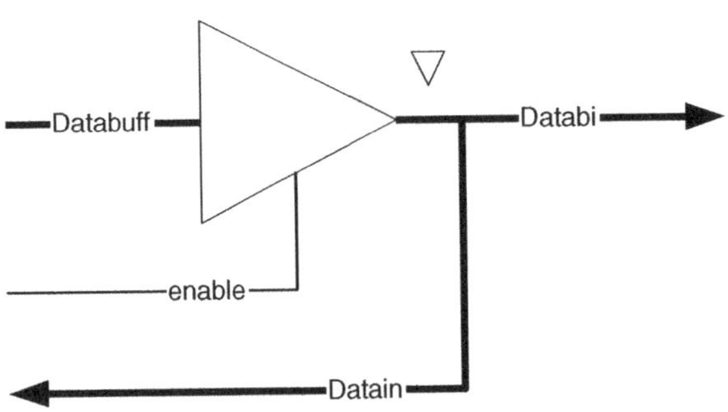

Figure 7.14 A bidirectional bus interface.

```
1 //A 3-input majority voter with 10 ns delay
2 `timescale 1 ns/ 1ns
3 module maj3(input a, b, c, output m);

4 wire [2:0] abc = {a, b, c}; //join inputs together

5 assign  #10 m = (abc == 3'b110)||
6                 (abc == 3'b101)||
7                 (abc == 3'b011)||
8                 (abc == 3'b111);  //  'on' terms

9 endmodule
```

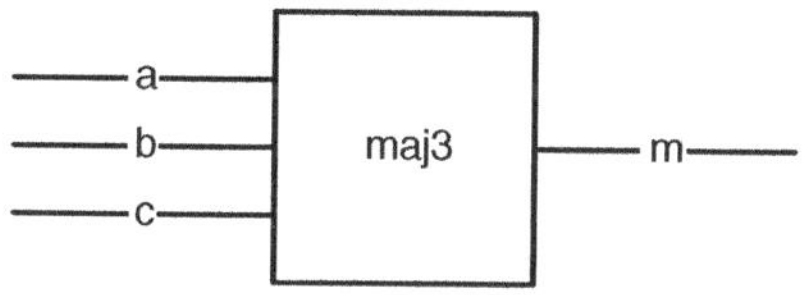

Figure 7.15 A three-input majority voter.

wires or **reg**s together into one bus signal. Line 4 illustrates the use of the combined **wire** declaration and continuous assignment in a single statement.

The continuous assignment on lines 5 to 8 in Figure 7.15 incorporates a delay, such that any input change is reflected at the output after a 10 ns delay. This represents one method of modelling propagation delays in modules that do not instantiate primitive gates.

The expression on the right-hand side of the assignment operator on line 5 makes use of the logical OR operator (||) rather than the bitwise OR operator (|). In this example, it makes no difference which operator is used, but occasionally the choice between bitwise and object-wise is important, since the latter is based on the concept of Boolean true and false. In Verilog, any bit pattern other than all zeros is considered to be true; consider the following example:

```
wire[3:0] a = 4'b1010; //true
wire[3:0] b = 4'b0101; //true
wire[3:0] c = a & b; //bit-wise result is 4'b0000 (false)
wire d = a && b; //logical result is 1'b1 (true)
```

Verilog-HDL provides a full set of relational operators, such as *greater than*, *less than* and *greater than or equal to*, as shown in Table 7.3. The following example, given in Figure 7.16, illustrates the use of these operators in the description of a memory address decoder.

The purpose of the module shown in Figure 7.16 is to activate one of four active-low 'chip select' outputs `Csbar[3:0]`, depending upon what particular range of hexadecimal address values is present on the 16-bit input address. Such a decoder is often used to implement the *memory map* of a microprocessor-based system.

Each of the continuous assignments on lines 4, 6, 8 and 10 responds to changes in the value of the input `Address`. For example, `Csbar[2]` is driven to logic 0 when `Address` changes to a hexadecimal value within the range 1500_{16} to $16FF_{16}$ inclusive.

```
//16-bit Address Decoder
module Addr_dec(input [15:0] Address,
                output [3:0]  Csbar);

assign Csbar[0] = ~((Address >= 0) &&
                    (Address <= 16'h03FF));
assign Csbar[1] = ~((Address >= 16'h0800) &&
                    (Address <= 16'h12FF));
assign Csbar[2] = ~((Address >= 16'h1500) &&
                    (Address <= 16'h16FF));
assign Csbar[3] = ~((Address >= 16'h1700) &&
                    (Address <= 16'h18FF));

endmodule
```

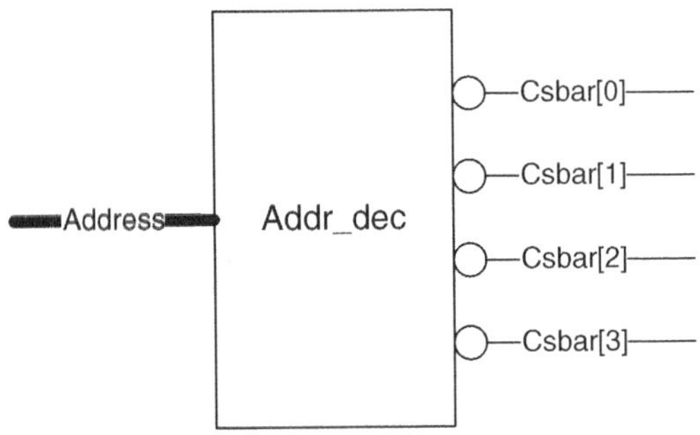

Figure 7.16 Memory address decoder using the relational operators.

Another example of the use of the relational operators is shown in Figure 7.17. This shows the Verilog description of a 4-bit magnitude comparator.

The module given in Figure 7.17 makes use of the relational and equality operators, along with logical operators, to form the logical expressions contained within the conditional operators on the right-hand side of the continuous assignments on lines 5 and 7. Note the careful use of parentheses in the test expression contained within line 7 for example:

```
((a>b) || ( (a==b) && (agtbin==1'b1))) ?
```

The parentheses force the evaluation of the logical AND expression before the logical OR expression, the result of the whole expression is logical 'true' or 'false', i.e. `1'b1` or `1'b0`. The result of the Boolean condition selects between logic 1 and logic 0 and assigns this value to the `agtbout` output. The expression preceding the question mark (?) on lines 7 and 8 could have been used on its own to produce the correct output on port `agtbout`; the conditional operator is used purely to illustrate how a variety of operators can be mixed in one expression.

The expression used on the right-hand side of the continuous assignment on line 9, shown below, makes use of the bitwise logical operators to illustrate that, in this case, the outcome is exactly the same as that which would be produced by the logical operators:

```
altbout= (a<b) | ((a==b) & altbin);
```

```
//4-bit magnitude comparator
module mag4comp(input [3:0] a, b,
       input aeqbin, agtbin, altbin,
       output aeqbout, agtbout, altbout);

assign aeqbout = ((a == b) && (aeqbin == 1'b1))?
                      1'b1 : 1'b0;

assign agtbout = ((a > b) || ((a == b) &&
          (agtbin == 1'b1))) ? 1'b1 : 1'b0;

assign altbout = (a < b) | ((a == b) & altbin);

endmodule
```

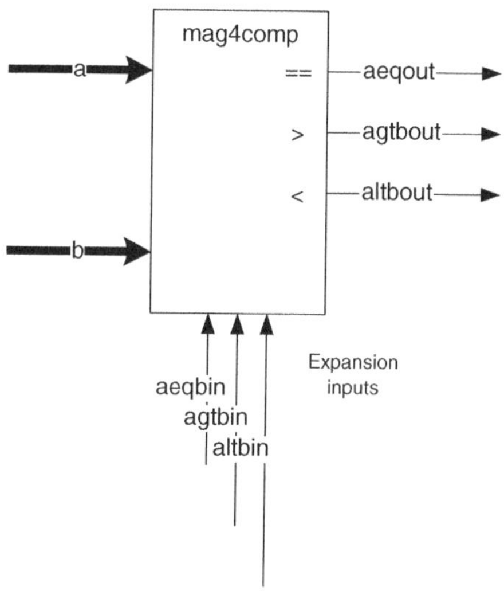

Figure 7.17 A 4-bit magnitude comparator.

The relational and logical operators used in the above examples will always produce a result of either logic 0 or logic 1 provided the operands being compared do not contain unknown (x) or high-impedance (z) values in any bit position. In the event that an operand does contain a meta-logical value, these operators will generate an unknown result (x), as illustrated by the examples below:

```
reg[3:0] A=4'b1010;
reg[3:0] B=4'b1101;
reg[3:0] C=4'b1xxx;

A <= B //Evaluates to logic-1
A>B //Evaluates to logic-0
A && B //Evaluates to logic-1
```

```
C == B //Evaluates to x
A < C //Evaluates to x
C || B //Evaluates to x
```

In certain situations, such as in simulation test-fixtures, it may be necessary to detect when a module outputs an unknown or high-impedance value; hence, there exists a need to be able to compare values that contain x and z.

Verilog HDL provides the so-called *case-equality* operators ('===' and '!==') for this purpose. These comparison operators compare xs and zs, as well as 0s and 1s; the result is always 0 or 1. Consider the following examples:

```
reg[3:0] K = 4'b1xxz;
reg[3:0] M = 4'b1xxz;
reg[3:0] N = 4'b1xxx;

K === M //exact match, evaluates to logic-1
K === N //1-bit mismatch, evaluates to logic-0
M !== N //Evaluates to logic-1
```

Each of the three **reg** signals declared above is initialized to an unknown value; the three comparisons that follow yield 1s and 0s, since all four possible values (0, 1, x, z) are considered when performing the case-equality and case-inequality bit-by-bit comparisons.

The last operators in Table 7.3 to consider are the *replication* and *concatenation* operators; both make use of the curly-brace symbol ({ }) commonly used to denote a **begin**...**end** block in the C-based programming languages.

The concatenation ({ }) operator is used to append, or join together, multiple operands to form longer objects. All of the individual operands being combined must have a defined size, in terms of the number of bits. Any combination of whole objects, part-selects or bit-selects may be concatenated, and the operator may be used on either or both sides of an assignment. The assignments shown below illustrate the use of the concatenation operator:

```
// A = 1'b1, B = 2'b00, C = 3'b110
Y = {A, B} ; //Y is 3'b100
Z = {C, B, 4'b0000} ; //Z is 9'b110000000
W = {A, B[0] , C[1] } ; //W is 3'b101
```

The following Verilog statements demonstrate the use of the concatenation operator on the left-hand side (target) of an assignment. In this case, the 'carry out' **wire** c_out occupies the MSB of the 5-bit result of adding two 4-bit numbers and a 'carry input':

```
wire[3:0] a, b, sum;
wire c_in, c_out;

//target is 5-bits long [4:0]
{c_out, sum} = a + b + c_in;
```

The replication operator can be used on its own or combined with the concatenation operator. The operator uses a *replication constant* to specify how many times to replicate an expression. If `j` is the expression being replicated and `k` is the replication constant, then the format of the replication operator is as follows:

```
{k{j}}
```

The following assignments illustrate the use of the replication operator.

```
//a=1'b1, b=2'b00, c=2'b10

//Replication only
Y={4{a}} //Y is 4'b1111

//Replication and concatenation
Y={4{ a} , 2{b}} //Y is 8'b11110000
Y={3{ c} , 2{1'b1}} //Y is 8'b10101011
```

One possible use of replication is to extend the sign-bit of a two's complement signed number, as shown below:

```
//a two's comp value (-54₁₀)
wire[7:0] data=8'b11001010;

//arithmetic right shift by 2 places
//data is 8'b11110010 (-14₁₀)
assign data={3{ data[7]} , data[6:2]} ;
```

The above operation could have been carried out using the arithmetic shift right operator '>>>'; however, the **`wire`** declaration for `data` would have to include the **`signed`** qualifier.

7.3 EXAMPLE ILLUSTRATING THE USE OF VERILOG HDL OPERATORS: HAMMING CODE ENCODER

This section presents a complete example involving the use of some of the Verilog HDL operators discussed previously. Figure 7.18 shows the block symbol representation and Verilog HDL description of a Hamming code [2] encoder for 8-bit data. The function of the module `Hamenc8` is to generate a set of *parity check* bits from an incoming 8-bit data byte; the check bits are then appended to the data to form a 13-bit Hamming codeword [2]. Such a codeword provides error-correcting and -detecting capabilities, such that any single-bit error (including the check bits) can be corrected and any 2-bit error (double error) can be detected.

The details of how the Hamming code achieves the above error-correcting and -detecting capabilities are left to the interested reader to explore further in Reference [2].

```
1  //Hamming Encoder for an 8-bit Data word
2  module Hamenc8(input [7:0] Data,
3                 output [4:0] Parout);

4  //define masks to select bits to xor for each parity bit
5  localparam MaskP1 = 8'b01011011;
6  localparam MaskP2 = 8'b01101101;
7  localparam MaskP3 = 8'b10001110;
8  localparam MaskP4 = 8'b11110000;

9  assign Parout[4:1] = {^(Data & MaskP4),
                         ^(Data & MaskP3),
                         ^(Data & MaskP2),
                         ^(Data & MaskP1)};

10 assign Parout[0] = ^{Parout[4:1], Data};

11 endmodule
```

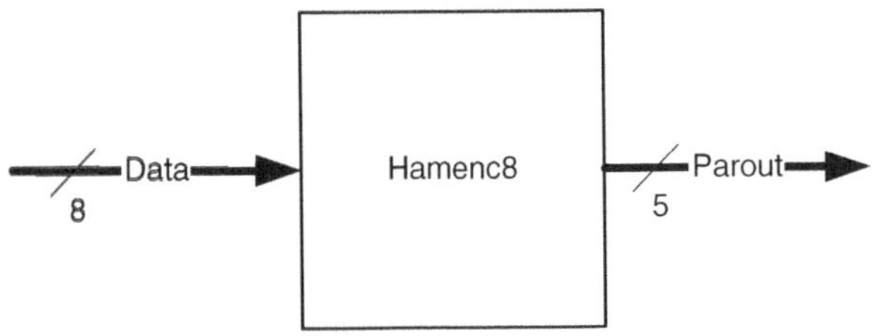

Figure 7.18 An 8-bit Hamming code encoder.

Lines 2 and 3 of Figure 7.18 define the module header for the Hamming encoder, the output `Parout` is a set of five parity check bits generated by performing the exclusive OR operation on subsets of the incoming 8-bit data value appearing on input port `Data`.

The bits of the input data that are to be exclusive ORed together are defined by a set of masks declared as *local parameters* on lines 5 to 8. The **`localparam`** keyword allows the definition of parameters or constant values that are *local* to the enclosing module, i.e. they cannot be overridden by external values. For example, mask `MaskP1` defines the subset of data bits that must be processed to generate `Parout[1]` , as shown below:

```
Bit position -> 76543210
     MaskP1 = 8'b01011011;
Parout[1] = Data[6] ^Data[4] ^Data[3] ^Data[1] ^Data[0] ;
```

The above modulo-2 summation is achieved in module `Hamenc8` by first masking out the bits to be processed using the bitwise AND operation (&), then combining these bit values using the reduction exclusive OR operator. These operations are performed for each parity bit as part of the continuous assignment on line 9:

```
^(Data & MaskP1)
```

The concatenation operator ({ }) is used on the right-hand side of the continuous assignment on line 9 to combine the four most significant parity check bits in order to assign them to `Parout [4:1]`.

The double-error-detecting ability of the Hamming code is provided by the overall parity check bit `Parout[0]`. This output is generated by modulo-2 summing (exclusive OR) all of the data bits along with the aforementioned parity bits, this being achieved on line 10 of Figure 7.18 by a combination of concatenation and reduction, as repeated below:

```
assign Parout[0] = ^{Parout[4:1], Data};
```

In performing the above operation, certain data bits are eliminated from the result due to cancellation, this is caused by the fact that the exclusive OR operation results in logic 0 when the same bit is combined an *even* number of times. The overall parity bit is therefore given by the following expression:

```
Parout[0] = Data[7] ^Data[5] ^Data[4] ^Data[2] ^Data[1] ^Data[0];
```

Data bits `Data[6]` and `Data[3]` are not included in the above equation for the overall parity. This is reflected in the logic diagram of the Hamming encoder, shown in Figure 7.19; this circuit could represent the output produced by a logic synthesis software tool after processing the Verilog description of the Hamming encoder shown in Figure 7.18.

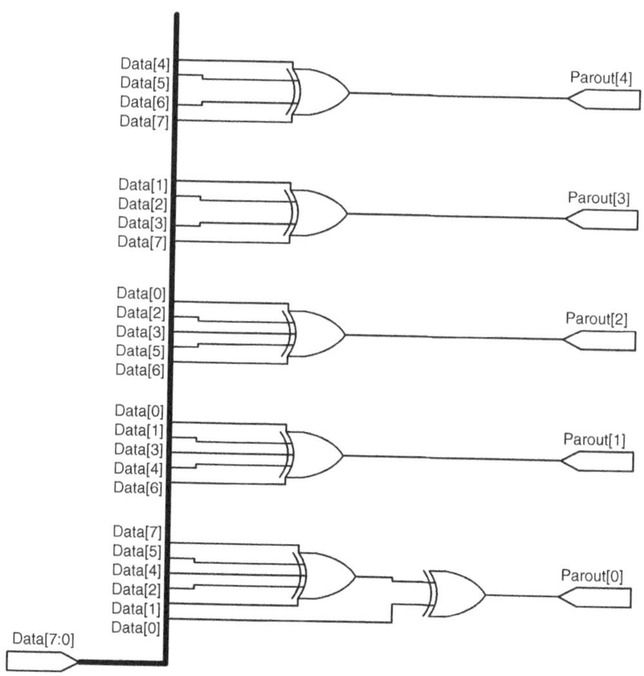

Figure 7.19 Hamming encoder logic diagram.

7.3.1 Simulating the Hamming Encoder

The operation of the Hamming encoder module shown in Figure 7.18 could be verified by simulation in an empirical manner. This would involve applying a set of random input data bytes and comparing the resulting parity check bit outputs against the values predicted from the encoder Boolean equations.

An alternative, and more systematic, approach would be to make use of a Hamming code decoder to decode the encoder output automatically, thus providing a more robust checking mechanism (assuming the Hamming decoder is correct of course!).

The use of a Hamming decoder module also allows the investigation of the error-correcting and -detecting properties of the Hamming code, by virtue of being able to introduce single and double errors into the Hamming code prior to processing by the decoder.

A Verilog HDL description of an 8-bit Hamming code decoder is given in Listing 7.3.

```
//Verilog description of a 13-bit Hamming Code Decoder
module Hamdec8 (input [7:0] Datain,
     input[4:0] Parin,
     output reg[7:0] Dataout,
     output reg[4:0] Parout,
     output reg NE, DED, SEC);

  //define masks to select bits to xor for each parity
localparam MaskP1 = 8'b01011011;
localparam MaskP2 = 8'b01101101;
localparam MaskP4 = 8'b10001110;
localparam MaskP8 = 8'b11110000;

reg [4:1] synd; //error syndrome
reg P0; //regenerated overall parity

always @(Datain or Parin)
begin

  //assign default outputs (assumes no errors)
NE = 1'b1;
DED = 1'b0;
SEC = 1'b0;
Dataout = Datain;
Parout = Parin;
P0 = ^{ Parin, Datain} ; //overall parity

  //generate syndrome bits
synd[4] = (^(Datain & MaskP8))^Parin[4] ;
synd[3] = (^(Datain & MaskP4))^Parin[3] ;
```

```
synd[2] = (^(Datain & MaskP2))^Parin[2] ;
synd[1] = (^(Datain & MaskP1))^Parin[1] ;
if ((synd == 0) && (P0 == 1'b0)) //no errors
  ; //accept default o/p
else if (P0 == 1'b1) //single error (or odd no!)
begin
   NE = 1'b0;
   SEC = 1'b1;
 //correct single error
case   (synd)
   0:  Parout[0] =~Parin[0] ;
   1:  Parout[1] =~Parin[1] ;
   2:  Parout[2] =~Parin[2] ;
   3:  Dataout[0] =~Datain[0] ;
   4:  Parout[3] =~Parin[3] ;
   5:  Dataout[1] =~Datain[1] ;
   6:  Dataout [2] =~Datain[2] ;
   7:  Dataout[3] =~Datain[3] ;
   8:  Parout[4] =~Parin[4] ;
   9:  Dataout[4] =~Datain[4] ;
   10: Dataout[5] =~Datain[5] ;
   11: Dataout[6] =~Datain[6] ;
   12: Dataout[7] =~Datain[7] ;
default:
    begin
       Dataout = 8'b00000000;
       Parout = 5'b00000;
    end
  endcase
end
else if ((P0 == 0) && (synd != 4'b0000))
begin //double error
   NE = 1'b0;
   DED = 1'b1;
   Dataout = 8'b00000000;
   Parout = 5'b00000;
end
end //always

endmodule
```

Listing 7.3 An 8-bit Hamming code decoder.

The module header on line 1 of Listing 7.3 defines the interface of the decoder, the 13-bit Hamming code input is made up from `Datain` and `Parin` and the corrected outputs are on ports `Dataout` and `Parout`. Three diagnostic outputs are provided to indicate the status of the incoming code:

`NE`: no errors (`Datain` and `Parin` are passed through unchanged)
`DED`: double error detected (`Dataout` and `Parout` are set to all zeros)
`SEC`: single error corrected (a single-bit error has been corrected[1]).

Note that all of the Hamming decoder outputs are qualified as being of type **`reg`**; this is due to the behavioural nature of the Verilog description, i.e. the outputs are assigned values from within a sequential **`always`** block (starting on line 8).

Lines 2 to 5 define the same set of 8-bit masks as those declared within the encoder module; they are used in a similar manner within the decoder to generate the 4-bit code named `synd`, declared in line 6. This 4-bit code is known as the *error syndrome*; it is used in combination with the regenerated overall parity bit `P0` (line 7) to establish the extent and location of any errors within the incoming Hamming codeword.

The main part of the Hamming decoder is contained within the **`always`** sequential block starting on line 8 of Listing 7.3. The statements enclosed between the **`begin`** and **`end`** keywords, situated on lines 9 and 55 respectively, execute sequentially whenever there is a change in either (or both) of the `Datain` and `Parin` ports. The **`always`** block represents a behavioural description of a combinational logic system that decodes the Hamming codeword.

At the start of the sequential block, the module outputs are all assigned default values corresponding to the 'no errors' condition (lines 10 to 14); this ensures that the logic described by the block is combinatorial. Following this, on lines 15 to 20 inclusive, the overall parity and syndrome bits are generated using expressions similar to those employed within the encoder description.

Starting on line 21, a sequence of conditions involving the overall parity and syndrome bits is tested, in order to establish whether or not any errors are present within the incoming codeword.

If the overall parity is zero and the syndrome is all zeros, then the input codeword is free of errors and the presence of the null statement (`;`) on line 22 allows the default output values to persist.

If the first condition tested by the **`if...else`** statement is false, then the next condition (line 23) is tested to establish whether a single error has occurred. If the overall parity is a logic 1, then the decoder assumes that a single error has occurred; the statements between lines 24 and 47 flag this fact by first setting the 'single error' (`SEC`) output high before going on to correct the error. The latter is achieved by the use of the **`case`** statement on lines 27 to 46; the value of the 4-bit syndrome is used to locate and invert the erroneous bit.

Finally, the **`if...else`** statement tests for the condition of unchanged overall parity combined with a nonzero syndrome; this indicates a double error. Under these circumstances, the Hamming decoder cannot correct the error and, therefore, it simply asserts the 'double error detected' output and sets the `Dataout` and `Parout` port signals to zero.

[1]An odd number of erroneous bits greater than one would be handled as a single error, usually resulting in an incorrect output.).

The Hamming code encoder and decoder are combined in a Verilog test module named `TestHammingCcts`, shown in Listing 7.5, and in block diagram form in Figure 7.20. An additional module, named `InjectError`, is required to inject errors into the valid Hamming code produced by the Hamming encoder prior to being decoded by the Hamming decoder.

The `InjectError` module is given in Listing 7.4.

```
//Module to inject errors into Hamming Code
1 module InjectError(input [7:0] Din,
        input [4:0] Pin,
        output [7:0] Dout,
        output [4:0] Pout,
        input [12:0] Ein);
2 assign { Dout, Pout} ={ Din, Pin} ^ Ein;

3 endmodule
```

Listing 7.4 The 13-bit error injector module.

This uses a single continuous assignment to invert selectively one or more of the 13 bits of the incoming Hamming codeword by exclusive ORing it with a 13-bit *error mask* named `Ein`, in line 2.

As shown in the block diagram of Figure 7.20 and Listing 7.5, the test module comprises instantiations of the encoder, decoder and error injector (lines 33 to 46) in addition to two **`initial`** sequential blocks named `gen_data` and `gen_error`, these being situated on lines 13 and 20 respectively of Listing 7.5.

```
// Verilog test fixture for Hamming Encoder and Decoder
1 `timescale 1ns / 1ns
2 module TestHammingCcts();

  //Hamming encoder data input
3   reg [7:0] Data;
  //Error mask pattern
4   reg [12:0] Error;
  //Hamming encoder output
5   wire [4:0] Par;
  //Hamming code with error
6   wire [7:0] EData;
7   wire [4:0] EPar;
// Hamming decoder outputs
8   wire DED;
9   wire NE;
```

```
wire SEC;
wire [7:0] Dataout;
wire [4:0] Parout;

initial //generate exhaustive test data
begin : gen_data
   Data = 0;
   repeat (256)
     #100 Data = Data + 1;
   $stop;
end

initial //generate error patterns
begin : gen_error
   Error = 13'b0000000000000;
   #1600;
   Error = 13'b0000000000001;
     #100;
   repeat (100) //rotate single error
     #100 Error = {Error[ 11:0] , Error[ 12]} ;
   Error = 13'b0000000000011;
     #100;
   repeat (100) //rotate double error
     #100 Error = {Error[ 11:0] , Error[ 12]} ;
end

//instantiate modules

Hamenc8 U1 (.Data(Data),
            .Parout(Par));

Hamdec8 U2 (.Datain(EData),
            .Parin (EPar),
            .Dataout(Dataout),
            .DED(DED),
            .NE(NE),
            .Parout(Parout),
            .SEC(SEC));
InjectError U3 (.Din(Data),
            .Ein(Error),
            .Pin(Par),
            .Dout(EData),
            .Pout(EPar));
endmodule
```

Listing 7.5 Hamming encoder/decoder test-fixture module.

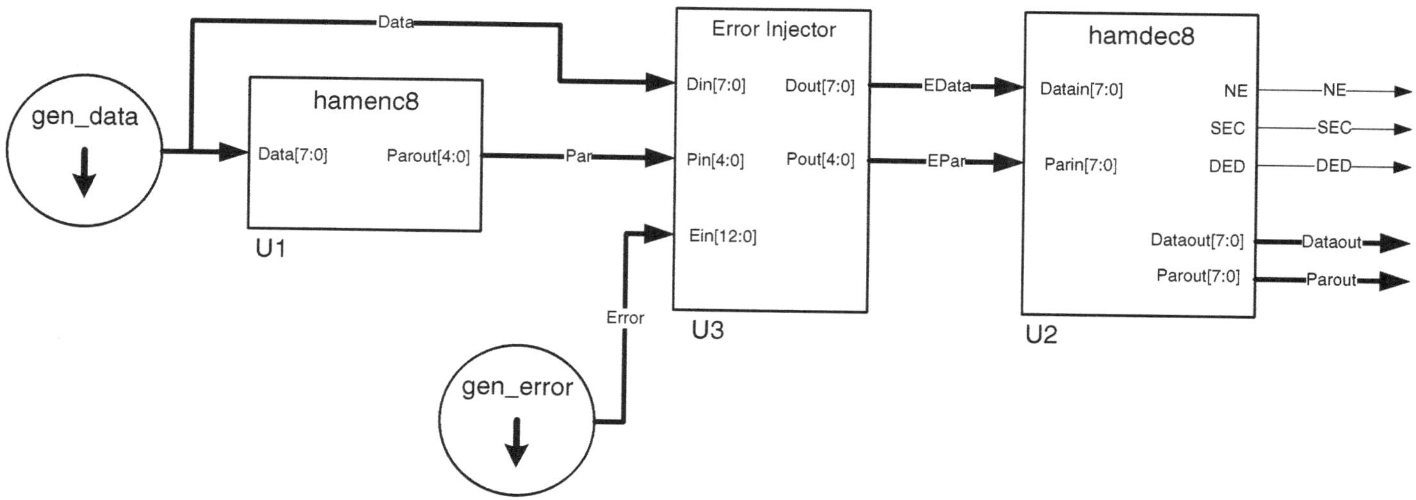

Figure 7.20 Hamming encoder/decoder test-fixture block diagram.

The `gen_data` **`initial`** block uses the **`repeat`** loop to generate an exhaustive set of 8-bit input data values ascending from 0_{10} to 255_{10} at intervals of 100 ns, stopping the simulation on line 18 by means of the `$stop` command. The `gen_error` block, on lines 20 to 32, starts by initializing the error mask `Error` to all zeros and allows it to remain in this state for 1600 ns in order to verify the 'no error' condition.

On line 24 of Listing 7.5, the error mask is set to `13'b0000000000001`, thereby introducing a single-bit error into the least significant bit of the Hamming codeword. After applying this pattern for 100 ns, a **`repeat`** loop (lines 26 and 27) is used to rotate the single-bit error through all 13 bits of the error mask at intervals of 100 ns for 100 iterations, this sequence

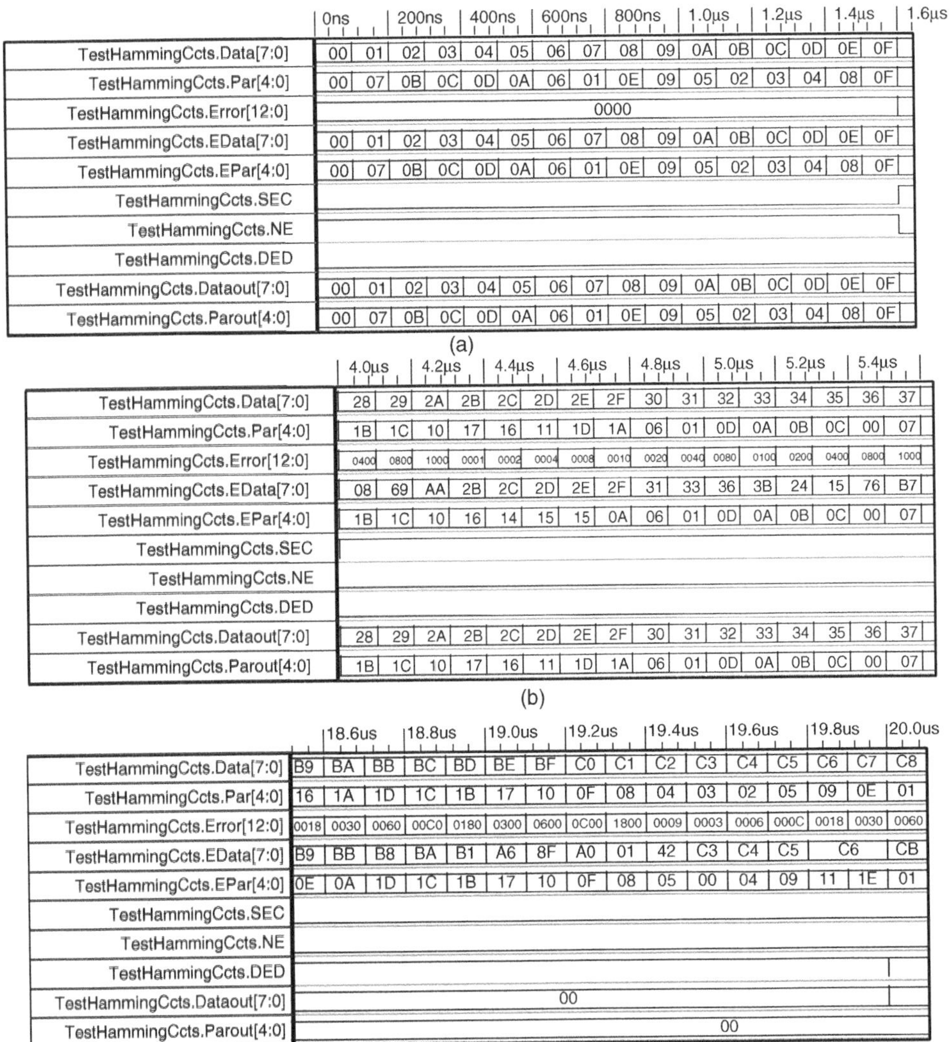

Figure 7.21 `TestHammingCcts` simulation results showing: (a) no errors; (b) single-error correction; (c) double-error detection.

will demonstrate the Hamming decoder's ability to correct a single-bit error in any bit position for a variety of test data values.

Finally, on lines 28 to 31, the error mask is reinitialized to a value of `13'b0000000000011`. This has the effect of introducing a *double-error* into the Hamming codeword. As above, this pattern is rotated through all 13-bits of the Hamming code 100 times, at intervals of 100 ns, in order to verify the decoder's ability to detect double errors for a variety of test data. Simulation results for the `TestHammingCcts` test module are shown in Figure 7.21a–c.

Figure 7.21a shows the first 16 test pattern results corresponding to an error mask value of all zeros (third waveform from the top), i.e. no errors. The top two waveforms are the 8-bit data (`Data`) and 5-bit parity (`Par`) values representing the 13-bit Hamming code output of the Hamming encoder module; all waveforms are displayed in hexadecimal format.

The outputs of the error injector module (`EData` and `EPar`), shown on the fourth and fifth waveforms, are identical to the top two waveforms due to the absence of errors. The diagnostic outputs, `SEC`, `NE` and `DED`, correctly show the 'no errors' output asserted, while the bottom two waveforms show the Hamming code being passed through the decoder unchanged.

Figure 7.21b shows a selection of test pattern results corresponding to an error mask containing a single logic 1 (third waveform from the top), i.e. a single error.

The outputs of the error injector module (`EData` and `EPar`), shown on the fourth and fifth waveforms, differ when compared with the top two waveforms by a single bit (e.g. $2A_{16}$, 10_{16} becomes AA_{16}, 10_{16} at time 4.2 μs). The diagnostic outputs, `SEC`, `NE` and `DED`, correctly show the 'single error corrected' output asserted, while the bottom two waveforms confirm that the single error introduced into the original Hamming code (top two waveforms) has been corrected after passing through the decoder.

Figure 7.21c shows a selection of test pattern results corresponding to an error mask containing two logic 1s (third waveform from the top), i.e. a double error.

The outputs of the error injector module (`EData` and `EPar`), shown on the fourth and fifth waveforms, differ when compared with the top two waveforms by 2 bits (e.g. BC_{16}, $1C_{16}$ becomes BA_{16}, $1C_{16}$ at time 18.8 μs). The diagnostic outputs, `SEC`, `NE` and `DED`, correctly show the 'double error detected' output asserted, while the bottom two waveforms confirm that the double error introduced into the original Hamming code (top two waveforms) has been detected and the output codeword is set to all zeros.

In summary, this section has presented a realistic example of the use of the Verilog HDL operators and types to describe a Hamming code encoder, decoder and test module. The behavioural style of description has been used to illustrate the power of the Verilog language in describing a relatively complex combinatorial logic system in a high-level manner.

Chapter 8 covers those aspects of the Verilog language concerned with the description of sequential logic systems, in particular the FSM.

REFERENCES

1. Ciletti M.D. Modeling, Synthesis and Rapid Prototyping with the Verilog HDL. New Jersey: Prentice Hall, 1999.
2. Wakerly J.F. Digital Design: Principles and Practices, 4th Edition. New Jersey: Pearson Education, 2006 (Hamming codes: p. 61, section 2.15.3).

8

Describing Combinational and Sequential Logic using Verilog HDL

8.1 THE DATA-FLOW STYLE OF DESCRIPTION: REVIEW OF THE CONTINUOUS ASSIGNMENT

We have already come across numerous examples in the previous chapters of Verilog designs written in the so-called data-flow style. This style of description makes use of the parallel statement known as a *continuous assignment.* Predominantly used to describe combinational logic, the flow of execution of continuous assignment statements is dictated by events on signals (usually **wire**s) appearing within the expressions on the left- and right-hand sides of the continuous assignments. Such statements are identified by the keyword **assign**. The keyword is followed by one or more assignments terminated by a semicolon.

All of the following examples describe combinational logic, this being the most common use of the continuous assignment statement:

```
//some continuous assignment statements
   assign A = q[0], B = q[1], C = q [2];

   assign out = (~s1 & ~s0 & i0) |
     (~s1 & s0 & i1) |
     (s1 & ~s0 & i2) |
     (s1 & s0 & i3);

   assign #15 { c_out, sum} = a + b + c_in;
```

The continuous assignment statement forms a static binding between the **wire** being assigned on the left-hand side of the = operator and the expression on the right-hand side of the assignment operator. This means that the assignment is continuously active and ready to respond to any

```
1 module latch (output q, input data, en);
2 assign q = en ? data : q;
3 endmodule
```

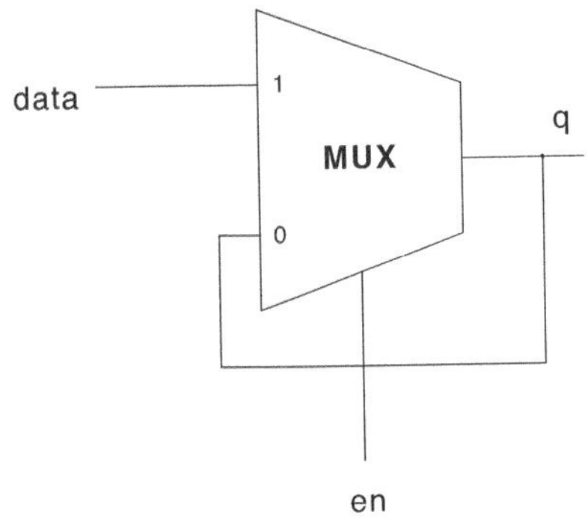

Figure 8.1 Describing a level-sensitive latch using a continuous assignment.

changes to variables appearing in the right-hand side expression (the inputs). Such changes result in the evaluation of the expression and updating of the target **wire** (output). In this manner, a continuous assignment is almost exclusively used to describe combinatorial logic.

As mentioned previously, a Verilog module may contain any number of continuous assignment statements; they can be inserted anywhere between the module header and internal **wire**/**reg** declarations and the **endmodule** keyword.

The expression appearing on the right-hand side of the assignment operator may contain both **reg**- and **wire**-type variables and make use of any of the Verilog operators mentioned in Chapter 7.

The so-called target of the assignment (left-hand side) must be a **wire**, since it is *continuously driven*. Both single-bit and multi-bit wires may be the targets of continuous assignment statements.

It is possible, although not common practice, to use the continuous assignment statement to describe sequential logic, in the form of a level-sensitive latch.

The conditional operator (? :) is used on the right-hand side of the assignment on line 2 of the listing shown in Figure 8.1. When `en` is true (logic 1) the output `q` is assigned the value of the input `data` *continuously*. When `en` goes to logic 0, the output `q` is assigned itself, i.e. feedback maintains the value of `q`, as shown in the logic diagram below the Verilog listing.

It should be noted that the use of a continuous assignment to create a level-sensitive latch, as shown in Figure 8.1, is relatively uncommon. Most logic synthesis software tools will issue a warning message on encountering such a construct.

8.2 THE BEHAVIOURAL STYLE OF DESCRIPTION: THE SEQUENTIAL BLOCK

The Verilog HDL *sequential block* defines a region within the hardware description containing *sequential statements*; these statements execute in the order they are written, in just the same way as a conventional programming language. In this manner, the sequential block provides a mechanism for creating hardware descriptions that are *behavioural* or *algorithmic*. Such a style lends itself ideally to the description of synchronous sequential logic, such as counters and FSMs; however, sequential blocks can also be used to describe combinational functions.

A discussion of some of the more commonly used Verilog sequential statements will reveal their similarity to the statements used in the C language. In addition to the two types of sequential block described below, Verilog HDL makes use of sequential execution in the so-called **task** and **function** elements of the language. These elements are beyond the scope of this book; the interested reader is referred to Reference [1].

Verilog HDL provides the following two types of sequential block:

- The **always** block. This contains sequential statements that execute repetitively, usually in response to some sort of trigger mechanism. An **always** block acts rather like a continuous loop that never terminates. This type of block can be used to describe any type of digital hardware.
- The **initial** block. This contains sequential statements that execute from beginning to end *once only*, commencing at the start of a simulation run at time zero. Verilog **initial** blocks are used almost exclusively in simulation *test fixtures*, usually to create test input stimuli and control the duration of a simulation run. This type of block is not generally used to describe synthesizable digital hardware, although a simulation model may contain an **initial** statement to perform an initialization of memory or to load delay data.

The two types of sequential block described above are, in fact, *parallel statements*; therefore, a module can contain any number of them. The order in which the **always** and **initial** blocks appear within the module does not affect the way in which they execute. In this sense, a sequential block is similar to a continuous assignment: the latter uses a single expression to assign a value to a target whenever a signal on the right-hand side undergoes a change, whereas the former executes a sequence of statements in response to some sort of triggering event.

Figure 8.2 shows the syntax of the **initial** sequential block, along with an example showing how the construct can be used to generate a clock signal.

As can be seen in lines 3 to 8, an **initial** block contains a sequence of one or more statements enclosed within a **begin**...**end** block. Occasionally, there is only a single statement enclosed within the initial block; in this case, it is permissible to omit the **begin**...**end** bracketing, as shown in lines 12 and 13. It is recommended, however, that the bracketing is included, regardless of the number of sequential statements, in order to minimize the possibility of syntax errors.

Figure 8.2 also includes an example **initial** block (lines 14 to 21), the purpose of which is to generate a repetitive clock signal. A local parameter named `PERIOD` is defined in line 14; this sets the time period of the clock waveform to 100 time-units. The execution of the **initial** block starts at time zero at line 18, where the `CLK` signal is initialized to logic 0; note that the signal `CLK` must be declared as a **reg**, since it must be capable of retaining the value last assigned to it by statements within the sequential block. Also note that the initialization of `CLK`

```
1  //general syntax of the initial sequential block
2  //containing more than one statement
3  initial
4  begin
5     //sequential statement 1
6     //sequential statement 2
7     ...
8  end
9
10 //general syntax of the initial sequential block
11 //containing one statement (no need for begin...end)
12 initial
13    //sequential statement


14 localparam PERIOD = 100;  //clock period

15 reg CLK;

16 initial
17 begin
18    CLK = 1'b0;
19    forever  //an endless loop!
20      #(PERIOD/2) CLK = ~CLK;
21 end
```

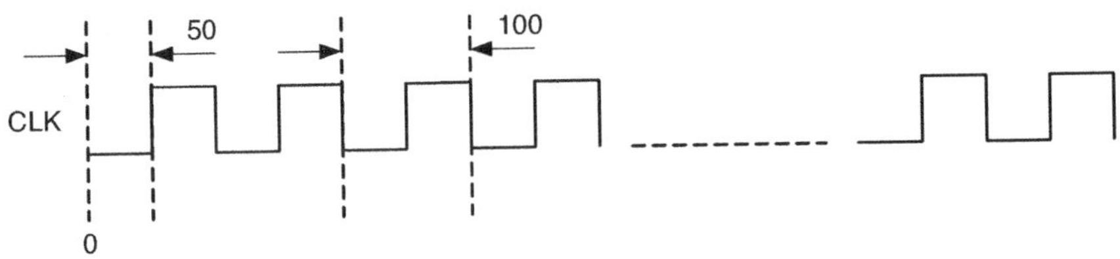

Figure 8.2 Syntax of the **`initial`** block and an example.

could have been included as part of its declaration in line 15, as shown below:

```
15 reg CLK = 1'b0;
```

Following initialization of CLK to logic 0, the next statements to execute within the initial block are lines 19 and 20 of the listing in Figure 8.2. These contain an *endless loop* statement known as a **`forever`** loop, having the general syntax shown below:

```
forever
begin
  //sequential statement 1
```

```
    //sequential statement 2
  ...
  end
```

In common with the **initial** block itself, the **forever** loop may contain a single statement or a number of statements that are required to repeat indefinitely; in the latter case, it must include the **begin**...**end** bracketing shown above. The example shown in Figure 8.2 contains a single *delayed sequential assignment* statement in line 20 (the use of the hash symbol # within a sequential block indicates a time delay). The effect of this statement is to invert the `CLK` signal every 50 time-units repetitively; this results in the `CLK` signal having the waveform shown at the bottom of Figure 8.2.

As it stands, the Verilog description contained in lines 14–21 of Figure 8.2 could present a potential problem to a simulator, in that most such tools have a command to allow the simulator to effectively *run forever* (e.g. 'run –all' in Modelsim®). The **forever** loop in lines 19 and 20 would cause a simulator to run indefinitely, or at least until the host computer ran out of memory to store the huge amount of simulation data generated.

There are two methods by which the above problem can be solved:

1. Include an additional **initial** block containing a `$stop` system command.
2. Replace the **forever** loop with a **repeat** loop.

The first solution involves adding the following statement:

```
   //n is the no. of clock pulses required
  initial #(PERIOD*n) $stop;
```

The above statement can be inserted anywhere after line 14 within the module containing the statements shown in Figure 8.2. The execution of the initial block in line 16 commences at the same time as the statement shown above (0 s); therefore, the delayed `$stop` command will execute at an absolute time equal to `n* PERIOD` seconds. The result is a simulation run lasting exactly `n` clock periods. It should be noted that, in order for the above statement to compile correctly, the variable `n` would have to be replaced by an actual positive number or would have to have been previously declared as a local parameter.

The second solution involves modifying the **initial** block in lines 16–21 of the listing given in Figure 8.2 to that shown below:

```
initial
begin
   CLK = 1'b0;
    repeat (n) //an finite loop
begin
   #(PERIOD/2) CLK = 1'b1;
   #(PERIOD/2) CLK = 1'b0;
end
```

```
9  $stop;
10 end
```

The **repeat** loop is a sequential statement that causes one or more statements to be repeated a fixed number of times. In the above case, the variable n defines the number of whole clock periods required during the simulation run. In this example, the *loop body* contains two delayed assignments to the **reg** named CLK; consequently, the **begin**. . .**end** bracketing is required.

Each repetition of the **repeat** loop lasts for 100 time-units, i.e. one clock period. Once all of the clock pulses have been applied, the **repeat** loop terminates and the simulation is stopped by the system command in line 9 above.

An important point to note regarding the **repeat** and **forever** loops is that neither can be synthesized into a hardware circuit; consequently, these statements are exclusively used in Verilog test-fixtures or within simulation models.

Listing 8.1a–e shows the various formats of the Verilog HDL sequential block known as the **always** block. The most general form is shown in Listing 8.1a: the keyword **always** is followed by the so-called *event expression*; this determines when the sequential statements in the block (between **begin** and **end**) execute. The @(event expression) is required for both combinational and sequential logic descriptions.

In common with the **initial** block, the **begin**. . .**end** block delimiters can be omitted if there is only one sequential statement subject to the **always** @ condition. An example of this is shown in Listing 8.1e.

(a)
```
always @(event_expression)
begin
  //sequential statement 1
  //sequential statement 2
  ...
end
```

(b)
```
always @(input1 or input2 or input3...)
begin
  //sequential statement 1
  //sequential statement 2
  ...
end
```

(c)
```
always @(input1, input2, input3...)
begin
  //sequential statement 1
  //sequential statement 2
  ...
end
```

(d)
```
always @( * )
begin
```

```
    //sequential statement 1
    //sequential statement 2
  ...
end
```

(e)

```
always @(a)
    y = a * a;
```

Listing 8.1 Alternative formats for the **always** sequential block: (a) General form of the always sequential block; (b) **always** sequential block with **or**-separated list; (c) **always** sequential block with comma-separated list; (d) **always** sequential block with wildcard event expression; (e) **always** sequential block containing a single sequential statement.

Unlike the **initial** block, the sequential statements enclosed within an **always** block execute repetitively, in response to the *event expression*. After each execution of the sequential statements, the **always** block usually suspends at the beginning of the block of statements, ready to execute the first statement in the sequence. When the *event expression* next becomes true, the sequential statements are then executed again. The exact nature of the *event expression* determines the nature of the logic being described; as a general guideline, any of the forms shown in Listing 8.1 can be used to describe combinational logic. However, the format shown in Listing 8.1b is most commonly used to describe sequential logic, with some modification (see later).

Also in common with the **initial** block, signals that are assigned from within an **always** block must be **reg**-type objects, since they must be capable of retaining the last value assigned to them during suspension of execution.

It should be noted that the **always** block could be used in place of an **initial** block, where the latter contains a **forever** loop statement. For example, the following **always** block could be used within a test module to generate the clock waveform shown in Figure 8.2:

```
1 localparam PERIOD = 100; //clock period

2 reg CLK = 1'b0;

3 always
4 begin
5    #(PERIOD/2) CLK = 1'b1;
6    #(PERIOD/2) CLK = 1'b0;
7 end
```

The **always** sequential block, shown in lines 3 to 7 above, does not require an *event expression* since the body of the block contains sequential statements that cause execution to be suspended for a fixed period of time.

This example highlights an important aspect of the **always** sequential block: it must contain either at least one sequential statement that causes suspension of execution or the keyword

always must be followed by an *event expression* (the presence of both is ambiguous and, therefore, is not allowed).

The absence of any mechanism to suspend execution in an **always** block will cause a simulation tool to issue an error message to the effect that the description contains a zero-delay infinite loop, and the result is that the simulator will 'hang', being unable to proceed beyond time zero.

In summary, the use of an **always** block in a test module, as shown above, is not recommended owing to the need to distinguish clearly between modules that are intended for synthesis and implementation and those that are used during simulation only.

8.3 ASSIGNMENTS WITHIN SEQUENTIAL BLOCKS: BLOCKING AND NONBLOCKING

An **always** sequential block will execute whenever a signal change results in the *event expression* becoming true. In between executions, the block is in a state of suspension; therefore, any signal objects being assigned to within the block must be capable of remembering the value that was last assigned to them. In other words, signal objects that are assigned values within sequential blocks are not *continuously driven*. This leads to the previously stated fact that only **reg**-type objects are allowed on the left-hand side of a sequential assignment statement.

The above restriction regarding objects that can be assigned a value from within a sequential block does not apply to those that appear in the *event expression*, however. A sequential block can be triggered into action by changes in both **reg**s and/or **wire**s; this means that module input ports, as well as gate outputs and continuous assignments, can cause the execution of a sequential block and, therefore, behavioural and data-flow elements can be mixed freely within a hardware description.

8.3.1 Sequential Statements

Table 8.1 contains a list of the most commonly used sequential statements that may appear within the confines of a sequential block (**initial** or **always**); some are very similar to those used in the C language, while others are unique to the Verilog HDL.

A detailed description of the semantics of each sequential statement is not included in this section; instead, each statement will be explained in the context of the examples that follow. It should also be noted that Table 8.1 is not exhaustive; there are several less commonly used constructs, such as *parallel blocks* (**fork...join**) and *procedural continuous assignments*, that the interested reader can explore further in Reference [1].

With reference to Table 8.1, items enclosed within square brackets ([ ]) are optional, curly braces ({ }) enclose repeatable items, and all bold keywords must be lower case.

Table 8.1 The most commonly used Verilog HDL sequential statements.

Sequential statement	Description
`=`	Blocking sequential assignment
`<=`	Nonblocking sequential assignment
`;`	Null statement. Also required at the end of each statement

begin
`{ seq_statements}`
end
	Block or compound statement. Always required if there is more than one sequential statement

if `(expr)`
`seq_statement`
`[`**else**
`seq_statement ]`
	Conditional statement, expression (`expr`) must be in parentheses. The **else** part is optional and the statement may be nested. Multiple statements require **begin**. . .**end** bracketing

case `(expr)`
`{ { value,} : seq_statement}`
`[`**default**`: seq_statement ]`
endcase
	Multi-way decision, the expression (`expr`) must be in parentheses. Multiple `values` are allowed in each limb, but no overlapping values are allowed between limbs. Default limb is required if previous values do not cover all possible values of expression. Multiple statements require **begin**. . .**end** bracketing

forever `seq_statement`	Unconditional loop. Multiple statements require **begin**. . .**end** bracketing
repeat `(expr)` `seq_statement`	Fixed repetition of `seq_statement` a number of times equal to `expr`. Multiple statements require **begin**. . .**end** bracketing
while `(expr)` `seq_statement`	Entry test loop (same as C) repeats as long as `expr` is nonzero. Multiple statements require **begin**. . .**end** bracketing
for `(exp1; exp2; exp3)` `seq_statement`	Universal loop construct (same as C). Multiple statements require **begin**. . .**end** bracketing
`#(time_value) seq_statement`	Suspends a block for `time_value` time-units
`@(event_expr) seq_statement`	Suspends a block until `event_expr` triggers

The *continuous assignment* parallel statement makes use of the = assignment operator exclusively. As shown in Table 8.1, sequential assignments can make use of two different types of assignment:

- blocking assignment – uses the = operator;
- nonblocking assignment – uses the <= operator.

The difference between the above assignments is quite subtle and can result in simulation and/or synthesis problems if not fully understood.

The *blocking* assignment is the most commonly used type of sequential assignment when describing combinational logic. As the name suggests, the target of the assignment is updated before the next sequential statement in the sequential block is executed, in much the same way as in a conventional programming language. In other words, a blocking assignment 'blocks' the execution of the subsequent statements until it has completed. Another aspect of blocking sequential assignments is that they effectively overwrite each other when assignments are made to the same signal. An example of this is seen in the Hamming code decoder example at the end of Chapter 7 (see Listing 7.3), where the decoder outputs are initialized to a set of default values prior to being conditionally updated by subsequent statements.

On encountering a *nonblocking* assignment, the simulator schedules the assignment to take place at the beginning of the next simulation cycle, this normally occurs at the end of the sequential block (or at the point when the sequential block is next suspended). In this manner, subsequent statements are not blocked by the assignment, and all assignments are scheduled to take place at the same point in time.

Nonblocking assignments can be used to assign several **reg**-type objects synchronously, under control of a common clock. This is illustrated by the example shown in Figure 8.3.

The three nonblocking assignments on lines 17, 18 and 19 of the listing shown in Figure 8.3 are all scheduled to occur at the positive edge of the signal named 'CLK'. This is achieved by means of the event expression on line 15 making use the event qualifier **posedge** (derived from **pos**itive-**edge**), i.e. the execution of the **always** sequential block is triggered by the logic 0 to logic 1 transition of the signal named `CLK`. This particular form of triggering is commonly used to describe *synchronous sequential logic* and will be discussed in detail later in this chapter.

The nonblocking nature of the assignments enclosed within the sequential block means that the value being assigned to `R2` at the first positive edge of the clock, for example, is the current value of `R1`, i.e. 'unknown' (`1'bx`). The same is true for the value being assigned to `R3` at the second positive edge of `CLK;` that is, the current value of `R2,` which is also `1'bx`. Hence, the initial unknown states of `R1`, `R2` and `R3` are successively changed to logic 0 after three clock pulses; in this manner, the nonblocking assignments describe what is, in effect, a 3-bit shift register, as shown in Figure 8.4.

Figure 8.5 shows an almost identical listing to Figure 8.3, apart from the three assignments in lines 17, 18 and 19, which in this case are of the blocking variety. The initial value of **regs** `R1`, `R2` and `R3` is unknown as before, and the **reg** `R0` is initialized at time zero to logic 0.

The effect of the blocking assignments is apparent in the resulting simulation result shown in Figure 8.5: all three signals change to logic 0 at the first positive edge of the `CLK`. This is due to

```
`timescale 1 ns/ 1 ns
module non_blocking_assignmnts();

reg R1, R2, R3, R0, CLK;

initial
begin
    R0 = 1'b0;
    CLK = 1'b0;
    repeat(3)
    begin
        #50 CLK = 1'b1;
        #50 CLK = 1'b0;
    end
    $stop;
end

always @(posedge CLK)
begin //a sequence of non-blocking assignments
    R1 <= R0;
    R2 <= R1;
    R3 <= R2;
end

endmodule
```

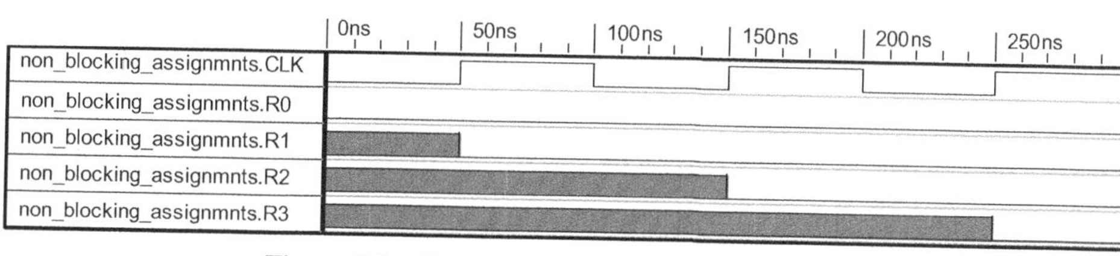

Figure 8.3 Illustration of nonblocking assignments.

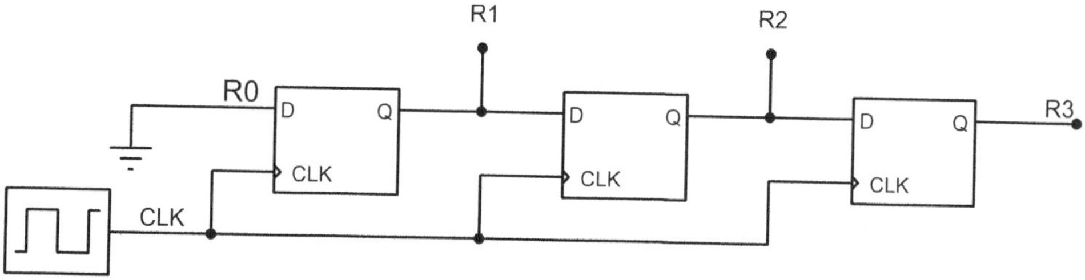

Figure 8.4 Nonblocking assignment equivalent circuit.

```
`timescale 1 ns/ 1 ns
module  blocking_assignmnts();

reg R1, R2, R3, R0, CLK;

initial
begin
    R0 = 1'b0;
    CLK = 1'b0;
    repeat (3)
    begin
        #50 CLK = 1'b1;
        #50 CLK = 1'b0;
    end
    $stop;
end

always @( posedge CLK )
begin //a sequence of blocking assignments
    R1 = R0;
    R2 = R1;
    R3 = R2;
end

endmodule
```

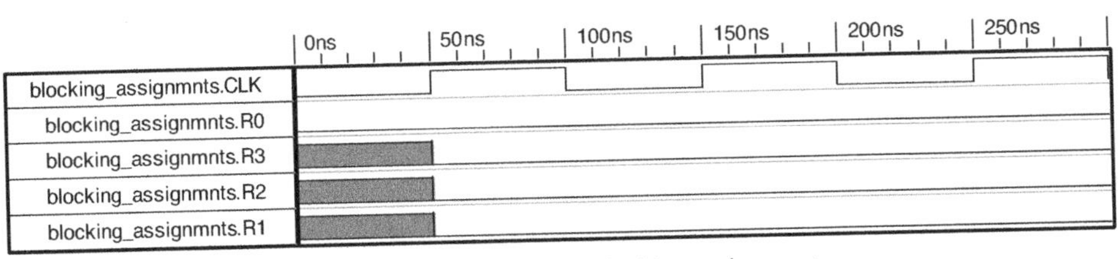

Figure 8.5 Illustration of blocking assignments.

the fact that the blocking assignment updates the signal being assigned prior to the next statement in the sequential block. The result is that the three assignments become what is, in effect, one assignment of the value of `R0` to `R3`. The equivalent circuit of the **`always`** block listed in Figure 8.5 is shown in Figure 8.6.

The choice of whether to use blocking or nonblocking assignments within a sequential block depends on the nature of the digital logic being described. Generally, it is recommended that nonblocking assignments are used when describing synchronous sequential logic, whereas blocking assignments are used for combinational logic.

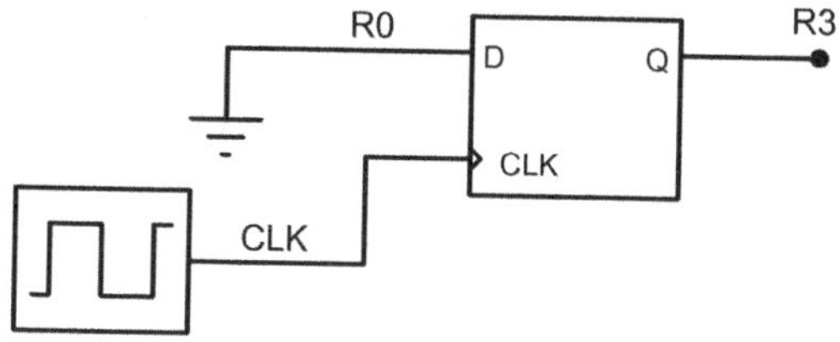

Figure 8.6 Blocking assignment equivalent circuit.

Sequential blocks intended for use within test modules are usually of the **`initial`** type; therefore, blocking assignments are the most appropriate choice.

A related point regarding the above guidelines is that blocking and nonblocking assignments should not be mixed within a sequential block.

8.4 DESCRIBING COMBINATIONAL LOGIC USING A SEQUENTIAL BLOCK

The rich variety of sequential statements that can be included within a sequential block means that the construct can be used to describe virtually any type of digital logic. Figure 8.7 shows the Verilog HDL description of a multiplexer making use of an **`always`** sequential block.

The module header in line 1 declares the output port `out` as a **`reg`**, since it appears on the left-hand side of an assignment within the sequential block. This example illustrates that despite the keyword **`reg`** being short for *register*, it is often necessary to make use of the **`reg`** object when describing purely combinational logic.

```
1 module mux(output reg out, input a, b, sel);

2 always @(a or b or sel)
3 begin
4    if (sel)
5       out = a;
6    else
7       out = b;
8 end
9 endmodule
```

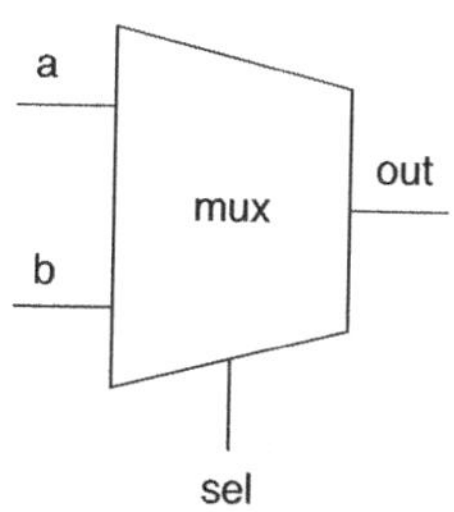

Figure 8.7 A two-input multiplexer described using an **`always`** block.

The event expression in line 2 of the listing in Figure 8.7 includes all of the inputs to the block in parentheses and separated by the keyword **`or`**. This format follows the original Verilog-1995 style; the more recent versions of the language allow either a comma-separated list or the use of the wildcard '`*`' to mean any **`reg`** or **`wire`** referenced on the right-hand side of an assignment within the sequential block.

Regardless of the event expression format used, the meaning is the same, in that any input change will trigger execution of the statements within the block.

The sequential assignments in lines 5 and 7 are of the nonblocking variety, as recommended previously. The value assigned to `out` is either the `a` input or the `b` input, depending on the state of the select input `sel`.

One particular aspect of using an **`always`** sequential block to describe combinational logic is the possibility of creating an *incomplete assignment*. This occurs when, for example, an **`if`**...**`else`** statement omits a final **`else`** part, resulting in the **`reg`** target signal retaining the value that was last assigned to it.

In terms of hardware synthesis, such an incomplete assignment will result in a latch being created. Occasionally, this may have been the exact intention of the designer; however, it is a more common situation that the designer has inadvertently omitted a final **`else`** or forgotten to assign a default value to the output. In either case, most logic synthesis software tools will issue warning messages if they encounter such a situation.

The following guidelines should be observed when describing purely combinational logic using an **`always`** sequential block:

Include *all* of the inputs to the combinatorial function in the *event expression* using one of the formats shown in Listing 8.1b–d.

To avoid the creation of unwanted latches, ensure either of the following is applicable:

- assign a default value to all outputs at the top of the **`always`** block, prior to any sequential statement such as **`if`**, **`case`**, etc.;
- in the absence of default assignments, ensure that all possible combinations of input conditions result in a value being assigned to the outputs.

The example in Figure 8.8 illustrates the points discussed above regarding incomplete assignments.

The designer of the module `latch_implied` listed in Figure 8.8 has used an **`always`** block to describe the behaviour of a selector circuit. The 2-bit input `sel[1:0]` selects one of three inputs `a`, `b` or `c` and feeds it through to the output `y`.

The assumption has been made that `y` will be driven to logic 0 if `sel` is equal to `2'b11`. This is, of course, incorrect: the omission of a final **`else`** clause results in `y` retaining its current value (since it is a **`reg`**), hence the presence of the feedback connection between the `y` output and the lower input of the left-hand multiplexer of the circuit shown in Figure 8.8. The synthesis tool has correctly *inferred* a latch from the semantics of the **`if`**...**`else`** statement and the **`reg`** object.

There are two alternative ways in which the listing in Figure 8.8 may be modified in order to remove the presence of the inferred latch in the synthesized circuit. These are shown in Figure 8.9a and b, with the corresponding latch-free circuit shown in Figure 8.9c.

```
module latch_implied(input a, b, c,
                     input [1:0] sel,
                     output reg y);
always @(*)//wildcard triggering
begin
  if (sel == 2'b00)
    y = a;
  else if (sel == 2'b01)
    y = b;
  else if (sel == 2'b10)
    y = c;
end
endmodule
```

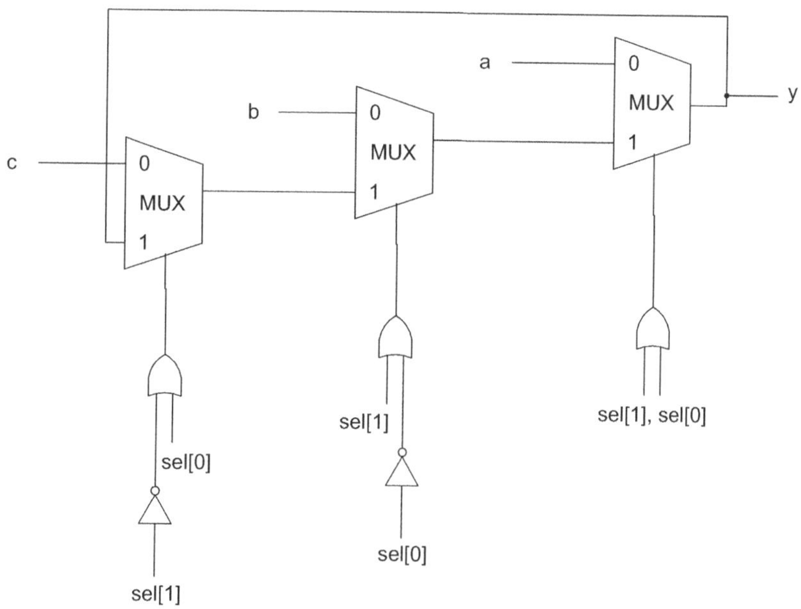

Figure 8.8 Example showing latch inference.

The listing shown in Figure 8.9a adds a final **else** part in lines 12 and 13; this has the effect of always guaranteeing the output y is assigned a value under all input conditions. Figure 8.9b achieves the same result by assigning a default value of logic 0 to output y in line 6.

Of the alternative strategies for latch removal exemplified above, the use of default assignments at the beginning of the sequential block is the more straightforward of the two to apply; therefore, this is the recommended approach to eliminating this particular problem.

The following examples further illustrate how the Verilog HDL can be used to describe a combinational logic function using an always sequential block. The first example, shown in Figure 8.10, describes a three-input to eight-output decoder (similar to the TTL device known as the 74LS138).

The function of the `ttl138` module is to decode a 3-bit input ⟨`A`, `B`, `C`⟩, and assert one of eight active-low outputs. The decoding is enabled by the three `G` inputs ⟨`G1`, `G2A`, `G2B`⟩, which must be set to the value ⟨1, 0, 0⟩. If the enable inputs are not equal to ⟨1, 0, 0⟩, then all of the `Y` outputs are set high.

This behaviour is described using an **`always`** sequential block that responds to changes on all inputs, starting in line 3 of the listing shown in Figure 8.10. The `Y` outputs are set to a default value of all ones in line 5 and this is followed by an **`if`** statement that conditionally asserts one of the

(a)

```
module data_selector(input a, b, c,
                     input  [1:0] sel,
                     output reg    y);
 always @(a, b, c, sel) //same as '*'
 begin
   if (sel == 2'b00)
     y = a;
   else if (sel == 2'b01)
     y = b;
   else if (sel == 2'b10)
     y = c;
   else //final else removes latch
     y = 1'b0;
 end
 endmodule
```

(b)

```
 module data_selector(input a, b, c,
                      input [1:0] sel,
                      output reg y);

 always @(a or b or c or sel)
 begin
   y = 1'b0; //default assignment
   if (sel == 2'b00)
     y = a;
   else if (sel == 2'b01)
     y = b;
   else if (sel == 2'b10)
     y = c;
 end

 endmodule
```

Figure 8.9 Removal of unwanted latching feedback: (a) removal of latch using final **`else`** part; (b) removal of latch using assignment of default output value; (c) synthesized circuit for (a) and (b).

(c)

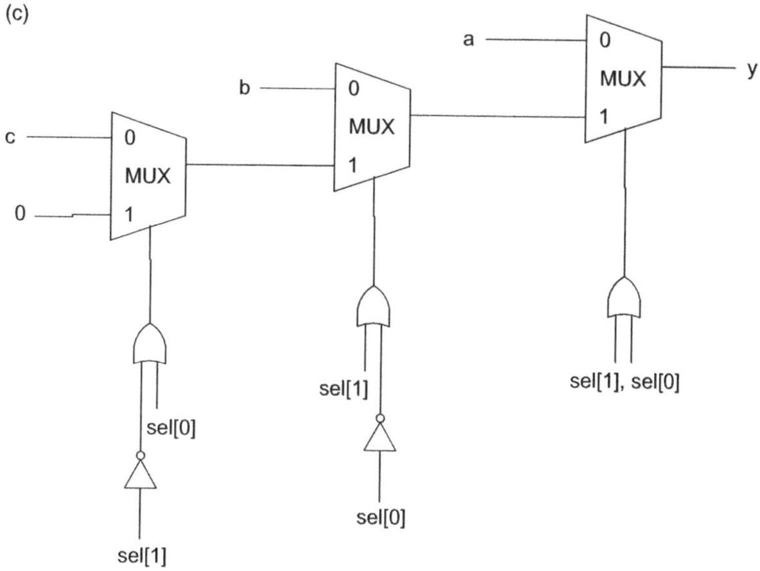

Figure 8.9 (*Continued*).

`Y` outputs to logic 0, depending on the decimal equivalent (0–7) of ⟨`A`, `B`, `C`⟩, in lines 6 and 7 respectively.

Simulation of the `ttl138` module is achieved using the Verilog test-fixture shown in Figure 8.11. The test-fixture module shown in Figure 8.11 makes use of a so-called *named sequential block* starting in line 6. The name of the block, `gen_tests`, is an optional label that

```
1 module ttl138(input A, B, C, G1, G2A, G2B,
2                   output reg [7:0] Y);

3 always @(A, B, C, G1, G2A, G2B)
4 begin
5    Y = 8'hFF; //set default output
6    if (G1 & ~G2A & ~G2B)
7          Y[{A, B, C}] = 1'b0;
8 end

9 endmodule
```

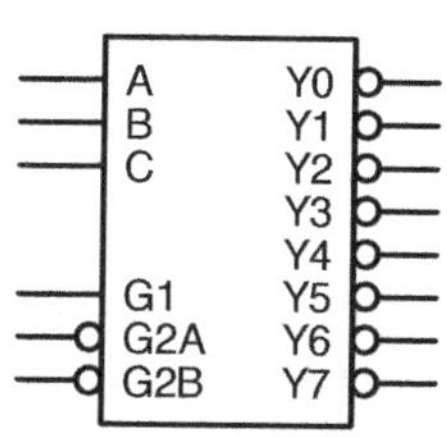

Figure 8.10 Three-to-eight decoder Verilog description and symbol.

must be placed after a colon following the keyword **begin**. Naming a sequential block in this manner (both **always** and **initial** blocks may be named) allows items, such as **reg**s and **integer**s, to be declared and made use of within the confines of the block. These locally declared objects may only be referenced from outside the block in which they are declared by preceding the object name with the block name; for example, the integer t in the listing of Figure 8.11 could be referenced outside of the **initial** block as follows:

```
gen_tests.t
```

The use of locally declared objects, as described above, allows the creation of a more structured description. However, it should be noted that, at the time of writing, not all logic synthesis tools recognize this aspect of the Verilog language.

The integer t is used within the **initial** block to control the iteration of the **for** loop situated between lines 9 and 12 inclusive. The purpose of the loop is to apply an exhaustive set of input states to the ⟨A, B, C⟩ inputs of the decoder. The syntax and semantics of the Verilog **for** loop is very similar to that of its C-language equivalent, as shown below:

```
for (initialization; condition; increment) begin
        sequential statements
end
```

The above is equivalent to the following:

```
initialization;
while (condition) begin
sequential statements
...
increment;
end
```

In line 10 it can be seen how Verilog allows the 32-bit integer to be assigned directly to 3-bit concatenation of the input signals without the need for conversion.

The timing simulation results are also included in Figure 8.11; these clearly show the decoding of the 3-bit input into a one-out-of-eight output during the first 800 ns. During the last 200 ns of the simulation, the enable inputs are set to 3'b000 and then 3'b011 in order to show all of the Y outputs going to logic 1 as a result of the decoder being disabled.

Finally, it should be noted that the very simple description of the decoder given in Figure 8.10 is not intended to be an accurate model of the actual TTL device; rather, it is a simple behavioural model intended for fast simulation and synthesis.

A second example is shown in Figure 8.12. This shows the Verilog source description and symbolic representation of a *majority voter* capable of accepting an *n*-bit input word. The function of this module is to drive a single-bit output named maj to either a logic 1 or logic 0 corresponding to the majority value of the input bits. Clearly, such a module requires an odd number of input bits greater than or equal to 3 in order to produce a meaningful output.

The module header (lines 2 and 3 of the listing in Figure 8.12) includes a **parameter** named n to set the number of input bits, having a default value of 5. The use of a parameter

```
`timescale 1 ns/ 1 ns
module test_ttl138;

reg A, B, C, G1, G2A, G2B;
wire [7:0] Y;

initial
begin : gen_tests
     integer t;
     {G1, G2A, G2B} = 3'b100;
     for (t = 0; t <= 7; t = t + 1) begin
          {A, B, C} = t;
          #100;
     end
     //disable the decoder
     {G1, G2A, G2B} = 3'b000;
     #100;
     {G1, G2A, G2B} = 3'b011;
     #100;
     $stop;
end

ttl138 uut(.A(A),
           .B(B),
           .C(C),
           .G1(G1),
           .G2A(G2A),
           .G2B(G2B),
           .Y(Y));

endmodule
```

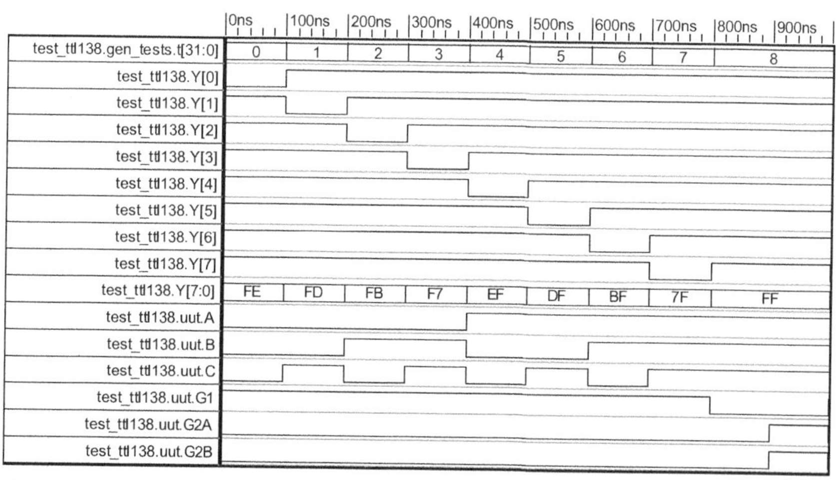

Figure 8.11 Test fixture and simulation results for the three-to-eight decoder module.

```
1   // n-bit majority voter, (n must be odd and >= 3)
2   module majn #(parameter n = 5)
3               (input [n-1:0] A, output maj);

4   integer num_ones, bit;

5   reg is_x;

6   always @(A)
7   begin
8     is_x = 1'b0;
9     num_ones = 0;
10    for (bit = 0; bit < n; bit = bit + 1) begin
11      if ((A[bit] === 1'bx)||(A[bit] === 1'bz))
12        is_x = 1'b1;
13      else if (A[bit] == 1'b1)
14        num_ones = num_ones + 1;
15    end
16  end

17  assign maj = (is_x == 1'b1)? 1'bx :
18               (n - num_ones) < num_ones;

19  endmodule
```

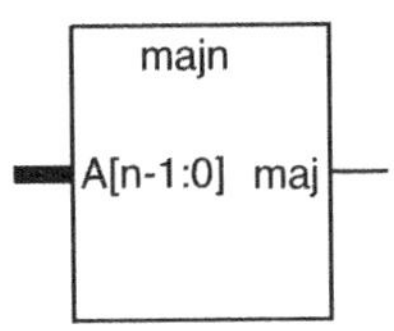

Figure 8.12 Verilog description and symbol for an *n*-bit majority voter.

makes the majority voter module potentially more useful due to it being *scalable*, i.e. the user simply sets the parameter to the desired value as part of the module instantiation.

Two *register*-type objects, in the form of **`integer`**s are declared in line 4. The first, `num_ones`, is used to keep track of the number of logic 1s contained in the input `A`, and the second, named `bit`, is used as a loop counter within the **`for`** loop situated in lines 10–15. A single-bit **`reg`** named `is_x` is declared in line 5 to act as a flag to record the presence of any unknown or high-impedance input bits.

The behaviour of the majority voter is described using an **`always`** sequential block commencing in line 6 of the listing show in Figure 8.12. The block is triggered by changes in the input word `A`, and starts by initializing `is_x` and `num_ones` to their default values of zero. The **`for`** loop then scans through each bit of the input word, first checking for the presence of an unknown or high-impedance state and then incrementing `num_ones` each time a logic 1 is detected. Note the use of the *case-equality* operator (===) in line 11 to compare each input bit of `A` explicitly with the meta-logical values `1'bx` and `1'bz`:

```
(A[bit] === 1'bx)||(A[bit] === 1'bz)
```

On completion of the **`for`** loop in line 15, the sequential block suspends until subsequent events on the input `A`.

The output maj is continuously assigned a value based on the outcome of the **always** block. The expression in lines 17 and 18 assigns 1'bx to the output subject to the conditional expression being true, thereby indicating the presence of an unknown or high impedance among the input bits. In the absence of any unknown input bits, the output is determined by comparing the number of logic 1s within A (num_ones) with the total number of bits in A (n):

```
(n - num_ones) < num_ones
```

It is left to the reader to verify that the above expression is true (false), i.e. yields a logic 1 (logic 0) if num_ones is greater (less) than the number of logic 0s in the *n*-bit input A.

The simulation of a 7-bit majority voter module is carried out using the test module shown in Figure 8.13. This test module instantiates a 7-bit ($n = 7$) majority voter in line 5. The **initial** block starting in line 6 sets the input to all zeros in line 8 and then applies an exhaustive set of input values by means of a **repeat** loop in lines 9–12 inclusive. The expression 1 << 7, used to set the number of times to execute the **repeat** loop, effectively raises the number 2 to the power 7, by shifting a single logic 1 to the left seven times. This represents an alternative to using the 'raise-to-the-power' operator '**', which is not supported by all simulation and synthesis tools.

After applying all known values to the A input of the majority voter module, the test module then applies two values containing the meta-logical states (lines 14–17) in order to verify that the module correctly detects an unknown input.

Figure 8.13 also shows a sample of the simulation results produced by running the test module. Inspection of the results reveals that the module correctly outputs a logic 1 when four or more, i.e. the majority of the inputs, are at logic 1. The behaviour of the internal objects num_ones and is_x can also be seen to be correct.

8.5 DESCRIBING SEQUENTIAL LOGIC USING A SEQUENTIAL BLOCK

With the exception of the simple level-sensitive latch given in Figure 8.1, Verilog HDL descriptions of sequential logic are exclusively constructed using the **always** sequential block. The reserved words **posedge** (positive edge) and **negedge** (negative edge) are used within the event expression to define the sensitivity of the sequential block to changes in the clocking signal. Figure 8.14 shows the general forms of the **always** block that are applicable to purely synchronous sequential logic, i.e. logic systems where *all* signal changes occur either on the rising (a) or falling (b) edges of the global clock signal.

The use of both **posedge** and **negedge** triggering is permitted within the same event expression at the beginning of an **always** block; however, this does not usually imply *dual-edge* clocking. The use of both of the aforementioned event qualifiers is used to describe synchronous sequential logic that includes an *asynchronous* initialization mechanism, as will be seen later in this section.

Figure 8.15 shows the symbol and Verilog description of what is perhaps the simplest of all synchronous sequential logic devices: the positive-edge-triggered *D*-type flip flop.

The module header, in line 1 of the listing in Figure 8.15, declares the output Q to be a **reg**-type signal, owing to the fact that it must retain a value in between active clock edges. The use of the keyword **reg** is not only compulsory, but also highly appropriate in this case, since Q represents the state of a single-bit register.

```
`timescale 1 ns/ 1 ns
module test_majn;

reg [6:0] Ain;
wire M;

majn #(.n(7)) maj7(.A(Ain), .maj(M));

initial
begin
   Ain = 0;
   repeat (1 << 7) begin
      #100;
      Ain = Ain + 1;
   end
   #100;
   Ain = 7'b1001x01;
   #100;
   Ain = 7'b000zz11;
   #100;
   $stop;
end
endmodule
```

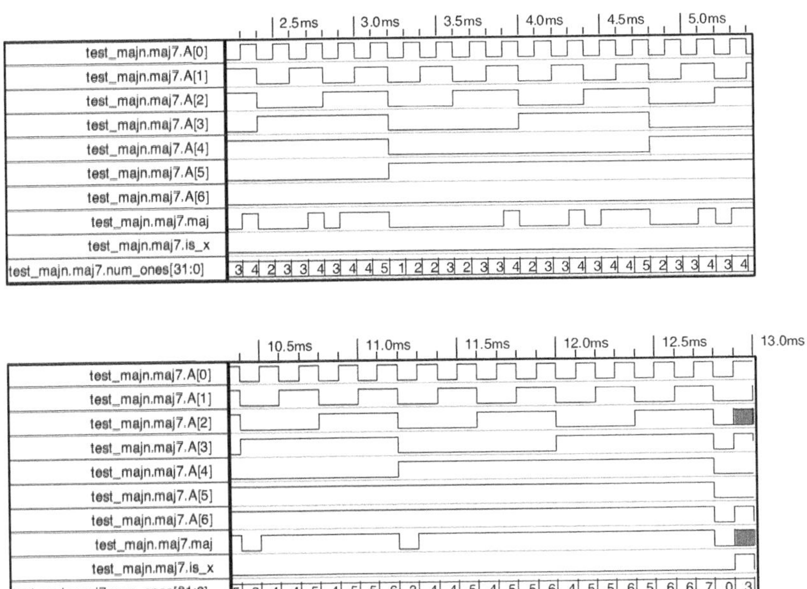

Figure 8.13 Test fixture and simulation results for the n-bit majority voter.

(a)

```
always @ (posedge clock)
begin
  //sequential statement 1
  //sequential statement 2
  ...
end
```

(b)

```
always @ (negedge clock)
begin
  //sequential statement 1
  //sequential statement 2
  ...
end
```

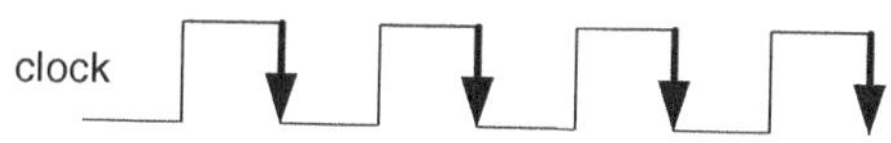

Figure 8.14 General forms of the **always** block when describing synchronous sequential logic: (a) positive-edge-triggered sequential logic; (b) negative-edge-triggered sequential logic.

The **always** sequential block in lines 2 and 3 contains a single sequential statement (hence the absence of the **begin**...**end** bracketing) that performs a nonblocking assignment of the input value D to the stored output Q on each and every positive edge of the input named CLK. In this manner, the listing given in Figure 8.15 describes an ideal functional model of a flip flop: unlike a real device, it does not exhibit propagation delays, nor are there any data *set-up* and *hold* times that must be observed. To include such detailed timing aspects would result in a far more complicated model, and this is not required for the purposes of logic synthesis.

As mentioned previously, it is conventional to use the *nonblocking* assignment operator when describing sequential logic. However, it is worth noting that the above flip-flop description would perform identically if the assignment in line 3 was of the blocking variety. This is due to the fact that there is only one signal being assigned a value from within the **always** block.

```
1 module dff(output reg Q, input D, CLK);

2 always @(posedge CLK)
3    Q <= D;

4 endmodule
```

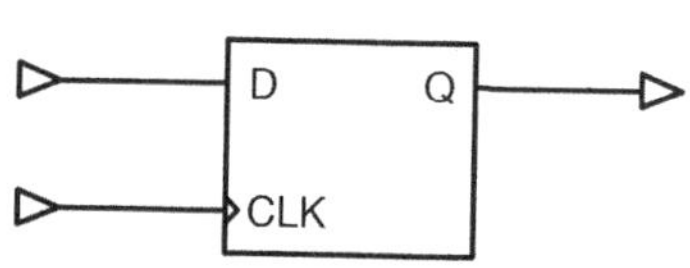

Figure 8.15 A positive-edge-triggered *D*-type flip-flop.

```
1  `timescale 1 ns/ 1 ns
2  module test_dff();

3  reg CLK, D;

4  wire Q;

5  initial
6  begin
7     D = 1'b0;
8     repeat (3) @(negedge CLK);
9     D = 1'b1;
10 end

11 initial
12 begin
13    CLK = 1'b0;
14    #100;
15    repeat(4) begin
16          #50 CLK = 1'b1;
17          #50 CLK = 1'b0;
18    end
19    $stop;
20 end

21 dff dut(.Q(Q), .D(D), .CLK(CLK));

22 endmodule
```

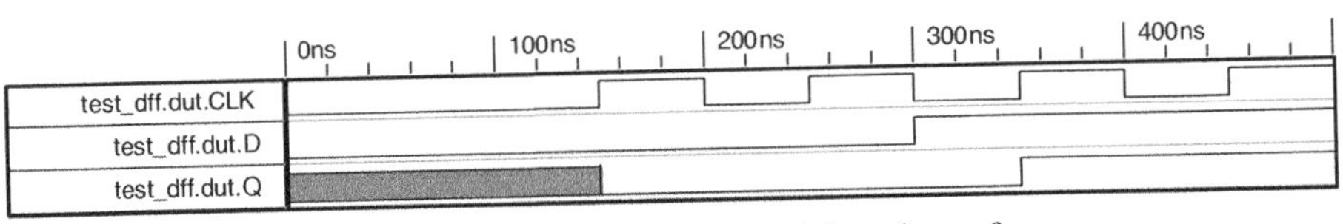

Figure 8.16 *D*-type flip-flop test module and waveforms.

Figure 8.16 shows a Verilog test-module and corresponding simulation waveform results for the *D*-type flip flop. This test module makes use of two initial sequential blocks to produce the `D` and `CLK` inputs of the flip flop. Line 8 illustrates the use of the `@(event_expression)` statement within a test module; in this case, the **`repeat`** loop waits for three consecutive negative-edge transitions to occur on the `CLK` before setting the data input `D` to a logic 1.

Inspection of the timing waveforms below the listing in Figure 8.16 shows that the `Q` output of the flip flop remains in an unknown state (shaded) until the first 0-to-1 transition of the clock; in other words, the flip-flop is initialized *synchronously*. In addition, the change in the data input `D` appears to occur at the *second* falling-edge of the clock, despite the fact that the **`repeat`** loop specifies *three* iterations; this apparent discrepancy is due to the change from the **`initial`** state of `CLK`, i.e. `1'bx`, to `1'b0` at time zero, being equivalent to a negative edge at the very start of

```
1 // A 4-bit UP Counter with asynchronous reset
2 module cntr4(input clock, reset,
3              output reg [3:0] count);

4 always @(posedge reset or posedge clock)
5 begin
6   if (reset == 1'b1)
7     count <= 4'b0000;
8   else //synchronous part
9     count <= count + 1;
10 end

11 endmodule
```

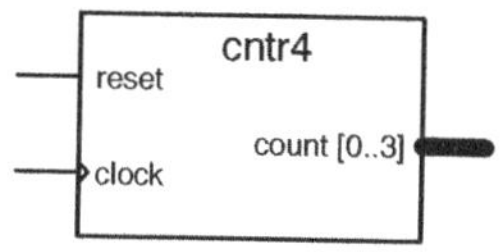

Figure 8.17 Verilog description of a 4-bit counter.

the simulation run. Finally, it can be seen that the `Q` output of the flip-flop changes state coincident with the rising edge of the clock, in response to the change from logic 0 to logic 1 on the data input at the preceding clock falling edge.

The following examples illustrate how the **always** sequential block is used to describe a number of common sequential logic building blocks.

Figure 8.17 shows the symbol and Verilog description for a 4-bit binary counter having an active-high *asynchronous* reset input. The input named `reset` takes priority over the synchronous `clock` input and, when asserted, forces the counter output to zero immediately. This aspect of the behaviour is achieved by means of the reference to **posedge** `reset` in the event expression in line 4 along with the use of the **if...else** statement in lines 6–9 of the listing in Figure 8.17.

The presence of the event qualifier **posedge** before the input `reset` might imply that the module has two clocking mechanisms. However, when this is combined with the test for `reset==1'b1` in line 6, the overall effect is to make `reset` act as an asynchronous input that overrides the clock.

When the `reset` input is at logic 0, a rising edge on the `clock` input triggers the **always** block to execute, resulting in the count being incremented by the sequential assignment statement located within the **else** part of the **if** statement (see line 9).

Consistent with previous sequential logic modules, the 4-bit counter makes use of nonblocking assignments directly to the 4-bit output signal, this having been declared within the module header as being of type **reg**, in line 3. Note that Verilog allows an output port such as `count` to appear on either side of the assignment operator, allowing the value to be either written to or read from. This is evident in line 9 of the listing in Figure 8.17, where the current value of `count` is incremented and the result assigned back to `count`.

Figure 8.18 shows a test module and the corresponding simulation results for the 4-bit counter. The waveforms clearly show the count incrementing on each positive edge of the clock input, until the asynchronous reset input `RST` is asserted during the middle of the `count` = 8 state, immediately forcing the count back to zero.

```
`timescale 1 ns/ 1 ns
module test_cntr4();

reg CLK, RST;
wire [3:0] Q;

initial
begin
      RST = 1'b1;
      repeat (3) @(negedge CLK);
      RST = 1'b0;
      repeat (8) @(negedge CLK);
      RST = 1'b1;
      @(negedge CLK);
      RST = 1'b0;
end

initial
begin
      CLK = 1'b0;
      #100;
      repeat(30) begin
            #50 CLK = 1'b1;
            #50 CLK = 1'b0;
      end
      $stop;
end

cntr4 dut(.clock(CLK), .reset(RST), .count(Q));

endmodule
```

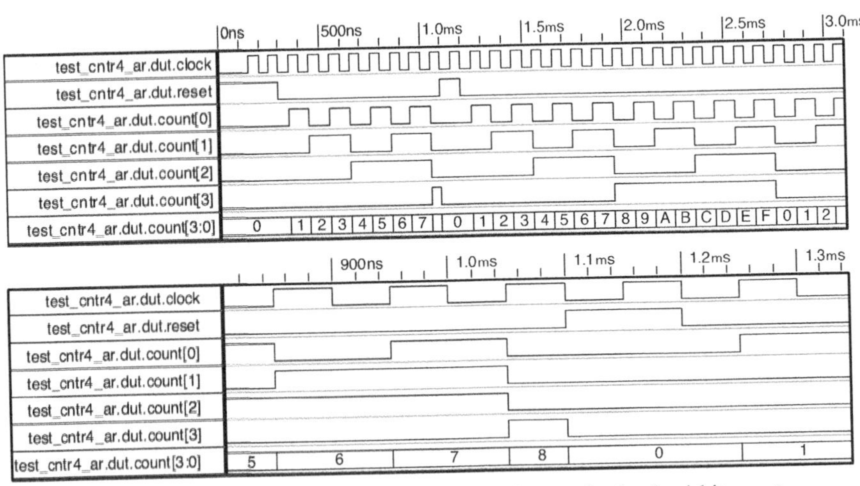

Figure 8.18 Verilog test-module and simulation results for the 4-bit counter.

```
1 //A 4-bit shift register with
2 //asynch active-low reset and shift enable
3 module shift4(input clock, clrbar, shift, serial,
4               output reg [3:0] q);

5 always @(negedge clrbar or posedge clock)
6 begin
7   if (clrbar == 1'b0)
8     q <= 4'b0;
9   else if (shift == 1'b1) //synchronous part
10    q <= {q[2:0], serial};
11 end

12 endmodule
```

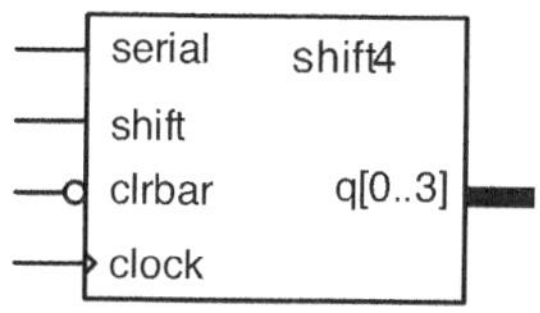

Figure 8.19 Verilog description of a 4-bit shift register.

As expected, the 4-bit count value automatically *wraps around* to zero on the next positive edge of the clock when the count of all-ones (`4'b1111`) is reached.

The next example of a common sequential logic module is given in Figure 8.19, showing the Verilog description and symbol for a 4-bit shift register. The module header declares an active-low asynchronous clear input named `clrbar` and a synchronous control input named `shift`, the latter enables the contents of the shift register (4-bit output **`reg`** `q`) to shift left on the active clock edge.

The sequential **`always`** block is triggered by the following event expression in line 5 of the listing shown in Figure 8.19:

```
always @(negedge clrbar or posedge clock)
```

The presence of the qualifier **`negedge`** indicates that it is the logic 1 to logic 0 transition (negative edge) of the input `clrbar` that triggers execution of the sequential block. This, in conjunction with the test for `clrbar` being equal to logic 0, at the start of the **`if...else`** statement in line 7, implements the asynchronous active-low initialization.

In line 9, the input `shift` is compared with logic 1 at each positive edge of the clock input. If this is true, then the following statement updates the output `q`:

```
q <= {q[2:0], serial};
```

The above sequential assignment shuffles the least significant three bits of q into the three most significant bit positions while simultaneously clocking the serial data input (`serial`) into the least significant bit position. In other words, a single-bit, left-shift operation is performed for each clock cycle that `shift` is asserted.

The corresponding test module for the shift register is provided in Figure 8.20. The module `test_shift4` is very similar to the test module shown in Figure 8.18 for the 4-bit counter. Two

```
`timescale 1 ns/ 1 ns
module test_shift4();

reg CLK, CLRB, SFT, SER;

wire [3:0] Q;

initial
begin
      CLRB = 1'b0;
      SFT = 1'b0;
      SER = 1'b1;
      repeat (2) @(negedge CLK);
      CLRB = 1'b1;
      repeat (3) @(negedge CLK);
      SFT = 1'b1;
      repeat (6) @(negedge CLK);
      CLRB = 1'b0;
      @(negedge CLK);
      CLRB = 1'b1;
      repeat (6) begin
            @(negedge CLK);
            SER = ~SER;
      end
end

initial
begin
      CLK = 1'b0;
      #100;
      repeat(30) begin
            #50 CLK = 1'b1;
            #50 CLK = 1'b0;
      end
      $stop;
end

shift4 dut(.clock(CLK), .clrbar(CLRB),
            .shift(SFT), .serial(SER), .q(Q));

endmodule
```

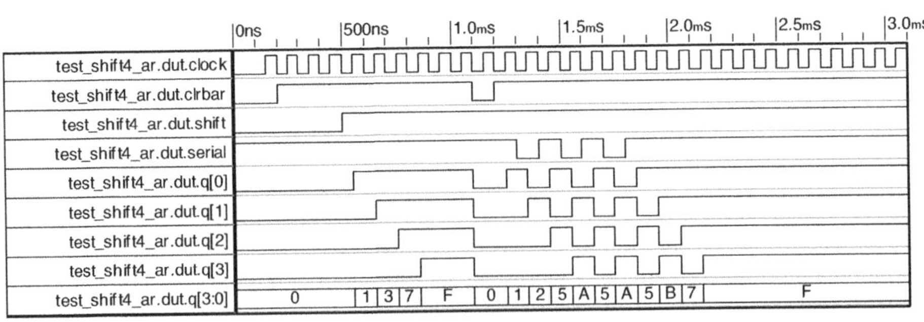

Figure 8.20 Verilog test-module and simulation results for the 4-bit shift register.

```
1 //D-Type FF with asynch. Set and Reset
2   module dff_asr(output reg q, qb,
3                  input d, clk, set, reset);

4   always @(posedge clk or posedge set
5             or posedge reset)
6   begin
7     if (reset) begin //reset has highest priority
8       q <= 0;
9       qb <= 1;
10    end else if (set) begin //set has second highest
11      q <= 1;
12      qb <= 0;
13    end else begin //clock when set and reset are low
14      q <= d;
15      qb <= ~d;
16    end
17  end
18  endmodule
```

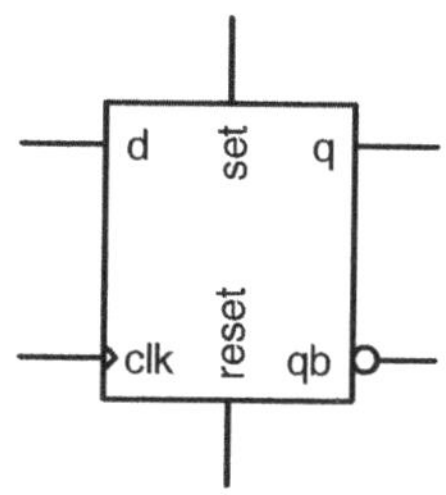

Figure 8.21 *D*-type flip-flop with asynchronous set and reset.

initial sequential blocks are used, one to provide an input stimulus and the other a set of clock pulses; the resulting simulation waveforms are also shown in Figure 8.20.

The previous two examples have shown how a sequential logic module can be described having either a single active-high or active-low asynchronous reset. The following example shows how both asynchronous reset and set inputs can be accommodated, if required.

Figure 8.21 shows the Verilog module and symbol for a *D*-type flip-flop having true and complementary outputs along with both a `set` input and a `reset` input for asynchronous initialization to either logic 1 or logic 0 respectively. Note that, in general, although this example makes use of only active-high control inputs, any combination of active-high and active-low control can be described by use of the **posedge** and **negedge** event qualifiers.

Lines 4 and 5 of the listing given in Figure 8.21 **or** together three inputs to form the event expression, one of which (`clk`) is the synchronous clock. This event expression, combined with the nested **if...else...if...else** statement, implements the hierarchical reset and set operations in conjunction with synchronous clocking. Notice the use of the **begin...end** bracketing to enclose the two assignments that make up each part of the **if...else** statement.

```
1  //An 8-bit register with synchronous reset
2  module REG8SR(output reg [7:0] Dataout,
3           input [7:0] Datain,
4           input Rst, Clk);

5  always @(posedge Clk) //triggers on 'Clk' only
6  begin
7    if (Rst)
8      Dataout <= 0;
9    else
10     Dataout <= Datain;
11 end
12 endmodule
```

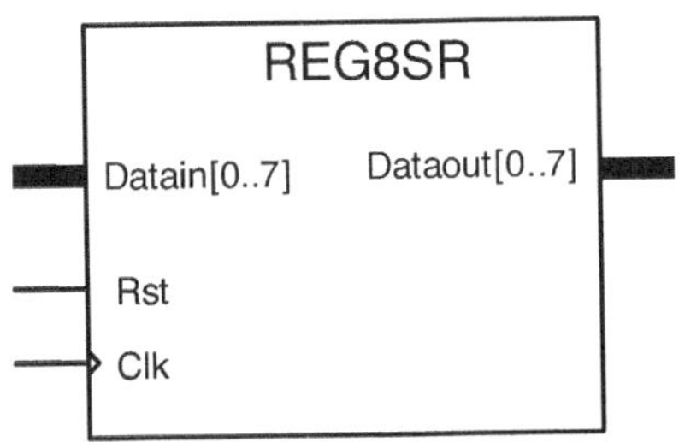

Figure 8.22 Example of a module using synchronous reset.

In certain situations it may be necessary, or indeed desirable, to perform all initialization *synchronously*. In this case, all assignments to the **reg**-type outputs of a sequential logic module are synchronized to the positive or negative edges of the master clock input.

The example shown in Figure 8.22 illustrates how the above can be implemented. The figure shows a Verilog module and symbol for a fully synchronous 8-bit data register. The event expression in line 5 of the listing shown in Figure 8.22 refers only to the positive edge of the `Clk` input. Therefore, all assignments to `Dataout` are subject to this condition, including the reset operation that occurs when `Rst` is at logic 1.

The last example in this section is a Verilog design that makes use of various aspects from previous examples, such as scalability, synchronous clocking and behavioural modelling.

Figure 8.23 shows the listing and symbolic representation for a so-called *universal register/counter* capable of performing a number of useful operations, in addition to having scalable input and output data ports. The latter is achieved by means of a **parameter** named `size` declared in the module header.

The module `unireg`, as well as being a parallel data register, is capable of performing the function of an up/down counter as well as providing left and right shifting. The number of bits that make up the register is defined by a parameter in line 2 of the listing, and, as shown, it is set to a default value of 8.

```
//Scalable Universal Register/Counter
module unireg #(parameter size = 8)
         (input clock, serinl, serinr,
         input [2:0] mode,
         input [size-1:0] datain,
         output reg [size-1:0] dataout,
         output termcnt);
always @(posedge clock) //synchronous counter
begin
  case (mode)
    0 : dataout <= 0;            //clear
    1 : dataout <= datain;       //parallel load
    2 : dataout <= dataout + 1;      //increment
    3 : dataout <= dataout - 1;      //decrement
    4 : begin  //shift left using  '<<' operator
      dataout <= dataout << 1;
      dataout[0] <= serinl;
    end
    //shift right using concatenation
    5 : dataout <= {serinr, dataout[size-1:1]};
    default : dataout <= dataout;  //refresh
  endcase
end

//continuous assignment to detect zero
assign termcnt = (mode == 3) ? ~|dataout :
                  ((mode == 2) ? &dataout : 0);

endmodule
```

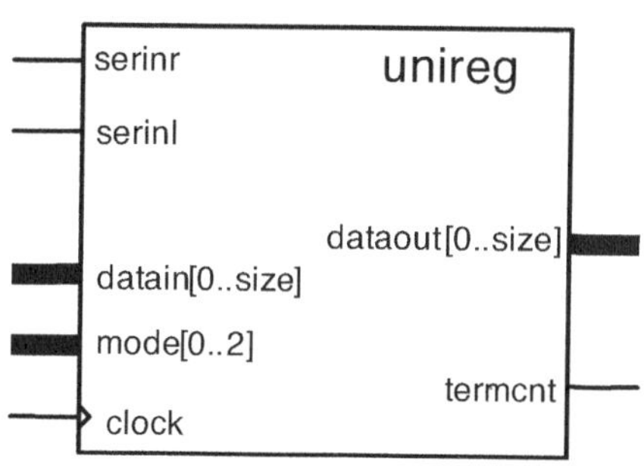

Figure 8.23 A universal counter/register module.

The `dataout` port of the `unireg` module constitutes the register itself; this is declared in line 6 of the module header. Each operation that the register performs is synchronized with the positive edges of the `clock` input; the nature of the operation is determined by a 3-bit control input named `mode` declared in line 4. The function selection nature of the `mode` input is implemented using a **`case...endcase`** statement between lines 10 and 22; each possible value of `mode` corresponds to one of the unique branches situated in lines 11–21. There are a total of seven operating modes, the last (`mode` = `6` or `7`) being covered by the final **`default`** branch in line 21.

Serial data inputs are provided for left and right shifting, via input ports `serinl` and `serinr` respectively. With reference to the listing in Figure 8.23, lines 15–18 correspond to the shift left operation (`mode` = `4`), where the register bits are shifted to the left by one position and the serial data present on input port `serinl` is loaded into bit 0 of the register. This synchronous data movement is achieved through the use of two nonblocking assignments in lines 16 and 17.

A mode value of 5 corresponds to a right shift. This corresponds to line 20 of the listing, where the *concatenation* operator is used to move the most significant `size-1` bits into the least significant `size-1` bit positions. The leftmost bit (MSB) of the register is loaded with the serial data applied to the `serinr` input port.

Operating modes 0 to 3 are self-explanatory; these correspond to the sequential assignments situated in lines 11–14 of the listing in Figure 8.23.

The remaining mode of operation is covered by the **`default`** branch of the **`case`** statement; this is the *refresh* mode, corresponding to a `mode` value of 6 or 7. The default sequential assignment simply assigns the register with the current value of `dataout`, i.e. itself. This could have been achieved in an alternative manner, as shown below:

```
default: ; // refresh using null statement
```

The *null statement* (`;`) is a 'do nothing' statement; in the above context it indicates that the `dataout` register is to retain its current value by virtue of not being updated. The choice of whether to use this method of retaining or refreshing the value stored in a **`reg`**-type signal, as opposed to the method shown in line 21, is a matter of personal preference.

The last output port of the `unireg` module is a **`wire`**-type signal named `termcnt`, which is a shortened form of 'terminal count'. The purpose of this output is to indicate when the register has reached the maximum or minimum value when operating in count-up or count-down mode respectively.

The flexible nature of the `dataout` register length makes it difficult to compare it with a fixed maximum value such as `8'hFF`; this problem is overcome by the use of the conditional operator and the bitwise reduction operators, as shown in the continuous assignment in lines 25 and 26 of the listing of Figure 8.23, and repeated below:

```
assign termcnt = (mode == 3) ? ~|dataout: ((mode == 2) ? &dataout: 0);
```

The above expression detects when the operating mode is either 'count-up' (2) or 'count-down' (3) and respectively assigns the reduction AND or the reduction NOR of `dataout` to the `termcnt` port. It is straightforward to appreciate that the expression will result in a logic 1 if `mode` is equal to 2 (3) and all of the register bits are logic 1 (logic 0), otherwise the above expression will be a logic 0.

Figure 8.24 includes a listing of a test module named `Test_unireg`, the purpose of which is to allow simulation of the universal register/counter described above. The module contains a declaration of a local parameter (`test_size`) in line 3 that is effectively a constant value for use within the enclosing module. In this case, the local parameter `test_size` is assigned the value 4. This corresponds to the number of bits contained in the parallel data input **`reg`**, and data output **`wire`**, connected to the register (see lines 7 and 9), as well as being used to override the value of the **`parameter`** that sets the width of the instantiated universal register/counter (`size`). This latter use of a local parameter, to determine the value of a **`parameter`** used in a scalable module, is implemented in line 12 of the test module shown in Figure 8.24.

The test module shown in Figure 8.24 includes two **`initial`** sequential blocks, the first of which generates a repetitive clock signal in lines 20–25 inclusive. The second **`initial`** block, spanning lines 26–49, generates a sequence of stimulus signals to exercise the various operating modes of the universal register/counter. The results of running the simulation are shown below the listing in Figure 8.24.

After clearing the register to zero by forcing the `mode` input to zero, the register is then set to counting-up mode (2) for 30 clock cycles. Inspection of the simulation waveforms clearly shows the data output bits counting up in binary, during which the terminal count (`termcnt`) output goes high coincident with a data output value of all ones.

The test module then sets the mode control to count-down mode (3) for a further 30 clock cycles. The data output bits follow a descending sequence and, as expected, the terminal count output is asserted when the state of all zeros is reached. The other operating modes of the universal register/counter are activated by subsequent statements in the initial block, shifting left (`mode = 4`) and shifting right (`mode = 5`), parallel load (`mode = 1`) and refresh (`mode = 7`) between lines 38 and 47; the simulation is stopped by the system command in line 48.

8.6 DESCRIBING MEMORIES

This section presents some very simple modules that can be used as rudimentary simulation models of RAM and ROM. These modules lack the timing accuracy and sophistication of the Verilog simulation models that are occasionally provided by commercial memory-device manufacturers. However, they can nevertheless be used effectively whenever a fast, functional model is required as part of a larger system simulation.

The Verilog descriptions discussed in this section serve to further reinforce some of the aspects that have already been covered, such as scalability and the use of parameters, as well as behavioural modelling with sequential blocks. In addition to these important elements of Verilog, the memory models presented here make use of other features not yet covered in previous chapters; these are as follows:

- arrays – the principle mechanism used to model a memory;
- bidirectional ports – the ability to use a single port as an input or output;
- memory initialization – loading a memory array with values from a file.

```
`timescale 1 ns/1 ns

module Test_unireg();

localparam test_size = 4; //size of the unireg

//inputs
reg clock, serinl, serinr;
reg [2:0] mode;
reg [test_size-1:0] datain;

//outputs
wire [test_size-1:0] dataout;
wire termcnt;

//instantiate the unireg module, 4-bits in size
unireg #(.size(test_size))
          mut(.clock(clock),
          .serinl(serinl),
          .serinr(serinr),
          .mode(mode),
          .datain(datain),
          .dataout(dataout),
          .termcnt(termcnt));

initial //generate a 100 ns clock
begin
     clock = 0;
     forever
          #50 clock = ~clock;
end

initial  //apply test inputs
begin
     serinl = 0;
     serinr = 1;
     mode = 0;
     datain = 'h9;
     #200 mode = 2;
     repeat (30)  //wait for 30 clock edges
          @(posedge clock);
     mode = 3;
     repeat (30)
```

Figure 8.24 Test module and simulation results for universal register/counter.

```
            @(posedge clock);
        mode = 4;
        repeat (8)
            @(posedge clock);
        mode = 5;
        repeat (8)
            @(posedge clock);
        mode = 1;
        #400 mode = 2;
        #800 mode = 7;
      #1000;
      $stop;
  end
endmodule
```

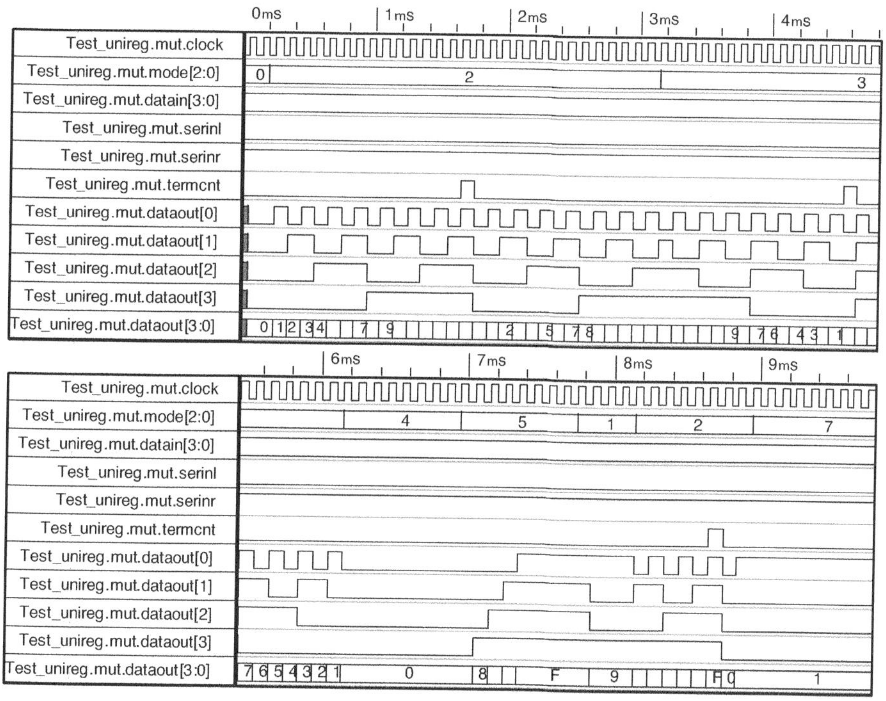

Figure 8.24 *(Continued).*

The Verilog language does not support the creation of a new and distinct composite type such as an array or record; instead, an array of **reg**s can be declared using the following syntax (an array of **wire**s can be declared in a similar manner):

```
//An array of m, n-bit regs
reg [ n-1:0] mem[0:m-1] ;
```

The above line declares an array having *m* elements, each one comprising an *n*-bit **reg**. In this manner, the object named `mem` can be viewed as a two-dimensional array of bits, i.e. a memory.

The capabilities of the Verilog language in terms of array handling were considerably enhanced with the release of the Verilog-2001 standard, with multidimensional arrays and the ability to reference an individual bit directly being two of the key improvements. The aforementioned new features provided by the update are not required by the simple memory models presented here, however; for further information, see Reference [2].

The other feature commonly made use of in memory models is *bidirectional data communication*. Most RAMs make use of a bidirectional three-state data bus to allow both read and write accesses using a single set of bus wires. The Verilog language provides for this by means of the **inout** port mode, along with the built-in simulation support for the high-impedance state in conjunction with the resolution of multiple signal drivers. It should be noted that the **inout** port is modelled as a **wire** having one or more *drivers*. During a read operation, for example, the **inout** port is driven by the value being accessed from the memory array; otherwise it is driven to the high-impedance state. During a write operation to a RAM, the port is driven by an external source which, combined with the high-impedance value being driven onto the data bus by the memory module itself, automatically resolves to a value to be written into the memory array.

Figure 8.25 shows the symbol and Verilog description of a simple and flexible RAM module. The model is general purpose insofar as it provides scalable address and data buses, allowing different-sized memories to be instantiated.

Line 4 of the listing of the module named `ram` declares the parameters `Awidth` and `Dwidth`. These define the width of the `address` and `data` ports subsequently declared in lines 6 and 7 of the module header. Three active-low control signals are declared in line 5, having the following functionality:

- `web` – write-enable, writes data into the memory array when low;
- `ceb` – chip-enable, enables the memory for reading or writing;
- `oeb` – output-enable, drives the data from the memory array onto the data port during a read operation.

The length of the memory array is equal to the number 2 raised to the power of the number of address input bits, i.e. $2^{\texttt{Awidth}}$. The local parameter declared on line 8 computes this value by means of the shift-left operator (since, as mentioned previously, not all simulators support the '`**`' operator).

The **localparam** `Length` is then used in the declaration of the memory array in line 9 of the listing in Figure 8.25.

Lines 11 and 12 describe the logic for a memory read operation using a continuous assignment, as repeated below:

```
assign data = (~ceb & ~oeb & web) ?
        mem[address] : 'bz;
```

The above statement is executed whenever a change occurs in any of the signals on the right-hand side of the assignment operator (=); this includes all of the memory control inputs, as well as the address value.

```
1  //A generic static random access memory
2  //Awidth is no. of address lines
3  //Dwidth is no. of data lines

4  module ram #(parameter Awidth = 8, Dwidth = 8)
5          (input web, oeb, ceb,
6          inout [Dwidth-1:0] data,
7          input [Awidth-1:0] address);

8  localparam Length = (1 << Awidth);

9  reg [Dwidth-1:0] mem[0:Length-1]; //memory array

10 //memory read
11 assign data = (~ceb & ~oeb & web) ?
12         mem[address] : 'bz;

13 //memory write
14 always @(posedge web) //occurs on 0-1 transition on web
15   if ((ceb == 1'b0) && (oeb == 1'b1))
16     mem[address] = data;

17 endmodule
```

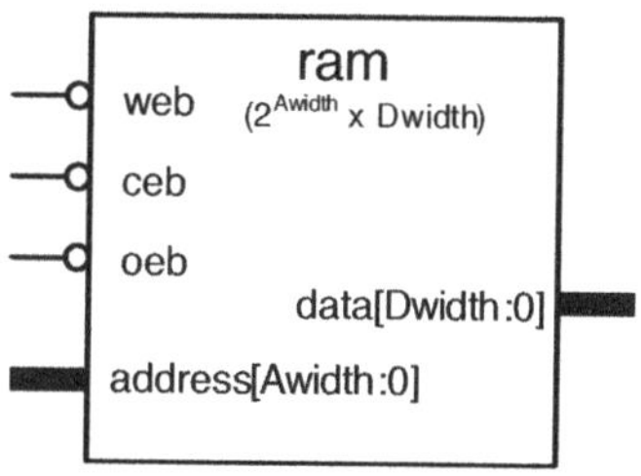

Figure 8.25 Verilog description and symbol for a simple RAM.

The inclusion of the condition that 'write-enable' must be a logic 1 during a read limits the possibility of a so-called *bus contention*, the result of trying to perform a read and a write simultaneously.

The memory word being read is accessed using the familiar array indexing notation ([ ]) found in the C language and also when accessing individual bits or bit ranges of a multi-bit `reg` or `wire`.

It should be pointed out that the Verilog-1995 language does not allow part- or bit-selects to be used in conjunction with an array access, this being one of the enhancements introduced with the update resulting in Verilog-2001. This limitation does not affect the simple memory models discussed here, since all accesses to memory arrays are to whole words only.

The use of a continuous assignment in lines 11 and 12 of the listing in Figure 8.25 is consistent with the definition of the `data` port as mode `inout`, effectively making it behave as a `wire`. The continuous assignment will drive the bidirectional data ports of the memory module with the high-impedance state if the condition preceding the '?' is false.

The memory write operation is implemented by the sequential **always** block in lines 14–16 of the listing in Figure 8.25. The incoming data value is latched into the memory array at the rising edge of the active-low 'write-enable' control input, providing the memory is enabled and not attempting to perform a read. In this case, the bidirectional data ports of the memory are being used as input **wire**s; the Verilog simulator automatically resolves the value on the data port from the combination of the high-impedance state being assigned by the continuous assignment in lines 11 and 12 and the value being driven onto the port from the external source.

Figure 8.26 shows the Verilog source description of a test module for the `ram` model of Figure 8.25. An important aspect of the test module `test_ram` is the requirement to declare the local signal to be connected to the bidirectional data port of the `ram` as a **wire** rather than a **reg**, as would normally be the case if it were purely an input.

The **wire** named `data`, declared and continuously assigned in lines 9 and 10, must be driven to the high-impedance state when the memory is being operated in read mode.

In order to achieve the above, the test module makes use of a single-bit **reg**, named `tri_cntr` (short for tri-state control), to control when the data to be written, `data_reg`, is driven onto the bus wire `data`. During write operations, `tri_cntr` is set high to enable the `data_reg` values to be written to the memory array, whereas during read operations `tri_cntr` is forced to logic 0 with the corresponding effect of making the `data` bus wire high impedance.

A 16-byte RAM is instantiated in the test module in lines 35–38, by overriding the address and data width parameters with the numbers 4 and 8 respectively. The **initial** sequential block, starting at line 11, performs a sequence of 10 writes to the address locations 0 to 9; the data being written is an alternating sequence containing the hexadecimal values `8'h55` and `8'hAA`. At the end of this sequence of writes the address is reset back to zero and the `data` bus wire is driven to the high impedance state by setting `tri_cntr` to logic 0 in line 26. The second **repeat** loop situated between lines 27 and 32 then performs 10 read operations from addresses 0 to 9, as above.

Figure 8.27 shows a block diagram to illustrate the structure of the test module described in Figure 8.26.

Simulation of the test-module results in the waveforms shown below the listing in Figure 8.26. As shown, the write operations occur as a result of the `webar` pulses being applied during the middle of each valid address and data value interval. The resulting stored values are then read out by disabling the `datareg` source by lowering `tri_cntr`, and then applying a sequence of `oebar` pulses while incrementing the address.

A ROM can be used wherever there is a need to store and retrieve fixed data during a simulation. For example, a set of test patterns could be stored in a ROM and subsequently used as test data (both stimulus and responses) for a module under test during the execution of a test module.

An embedded microcontroller may make use of an external ROM to store the fixed machine code program it will fetch and execute as part of a system-level simulation.

A simple Verilog model of a ROM, along with the corresponding symbolic representation, is given in Figure 8.28. In common with the RAM described above, the memory is designed to be scalable, having parameters to define the width of both the address bus and the data bus declared as part of the module header.

```
//test module for a 16-byte RAM

`timescale 1 ns/ 1 ns
module test_ram;

reg webar, oebar, csbar;
reg [7:0] datareg;
reg tri_cntr;    //data hi-z control
reg [3:0] address;
//three-state buffer for data input/output
wire [7:0] data = (tri_cntr == 1'b1)?
                            datareg : 8'bz;
initial
begin : test
     tri_cntr = 1'b1;    //make data available
     webar = 1'b1; oebar = 1'b1;
     csbar = 1'b1; datareg = 8'b01010101;
     address = 4'd0;
     #10 csbar = 1'b0;
     repeat (10) //perform 10 writes
     begin
          #10 webar = 1'b0;
          #10 webar = 1'b1;
          #10 address = address + 1;
          datareg = ~datareg;
     end
     address = 4'd0;
     tri_cntr = 1'b0;  //make data high impedance
     repeat (10)  //perform 10 reads
     begin
          #10 oebar = 1'b0;
          #10 oebar = 1'b1;
          #10 address = address + 1;
     end
     $stop;
end

ram #(.Awidth(4), .Dwidth(8))
          ram_ut(.web(webar),
               .oeb(oebar), .ceb(csbar),
               .data(data), .address(address));
endmodule
```

Figure 8.26 Test module and simulation results for the simple RAM.

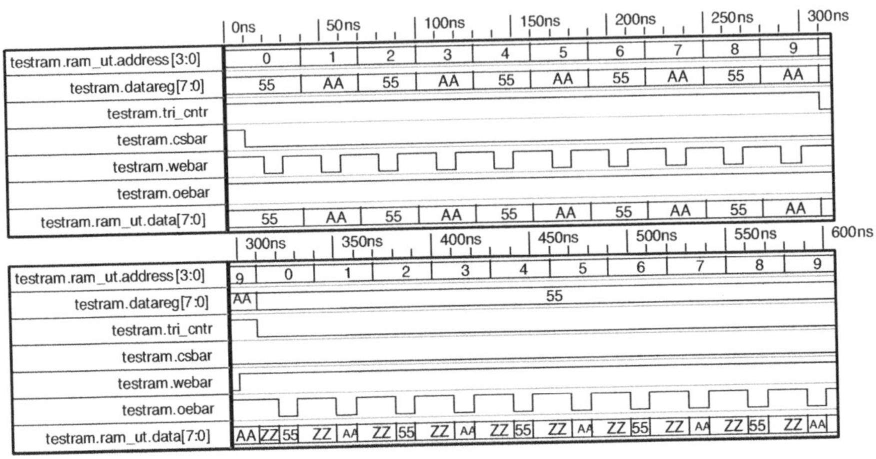

Figure 8.26 (*Continued*).

As in the case of the RAM module of Figure 8.25, the module `rom` in Figure 8.28 uses a **`localparam`** to calculate the length of the memory using the number of address bits at line 6, and then goes on to declare the actual memory array at line 7. The behaviour of the model is encapsulated in a single continuous assignment in line 8 of the listing in Figure 8.28; this statement assigns the contents of the memory array `mem`, indexed at location `address`, to the `data` output port, providing that the output enable control input `oeb` is asserted. Note that, in the case of the ROM, the data output port is of mode **`output`** rather than **`inout`**, since data are only ever read from the module. With the output enable control input at logic 1, the data output is set to the high-impedance state.

The actual contents of the ROM array `mem` are not specified anywhere in the Verilog description shown in Figure 8.28. For this type of ROM description, the stored data are defined externally, in an ASCII text file, and loaded into the memory array at the beginning of the

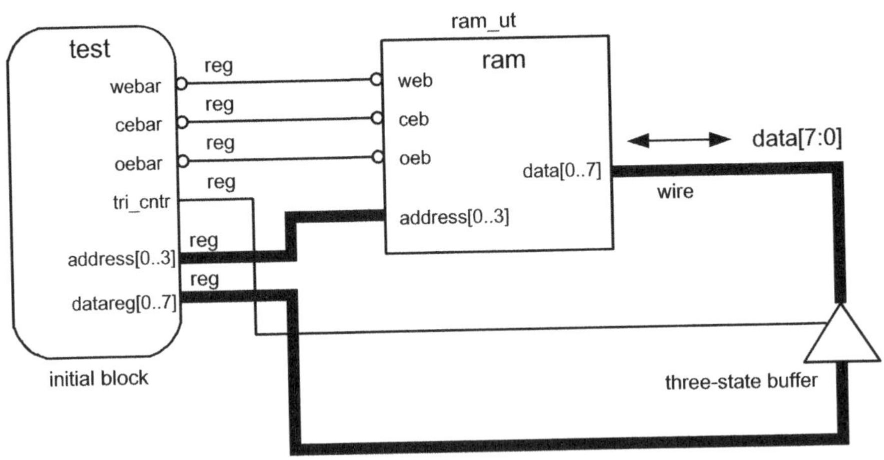

Figure 8.27 Block diagram of the module `test_ram`.

```
//a scalable read only memory module
module rom #(parameter Awidth = 8, Dwidth = 8)
                (input oeb,
                output [Dwidth-1:0] data,
                input [Awidth-1:0] address);

localparam Length = (1 << Awidth);

reg [Dwidth-1:0] mem[0:Length-1]; //memory array

assign data = (oeb == 1'b0) ? mem[address] : 'bz;
endmodule
```

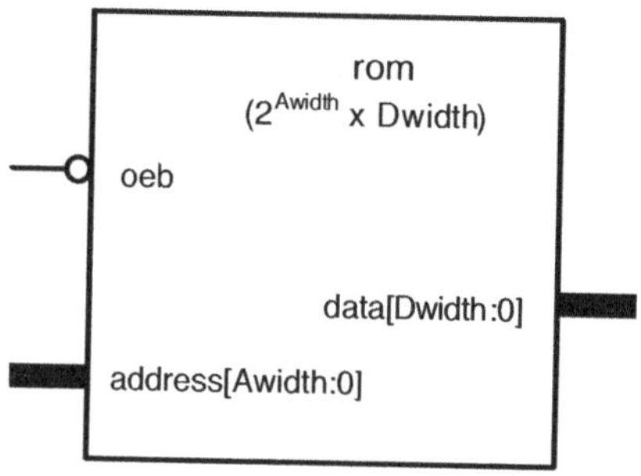

Figure 8.28 Verilog description of a ROM.

simulation. This method of initializing a ROM can also be used for a RAM, if required. It also provides a convenient way of loading a large amount of data into a memory from a file generated by a third-party tool, such as an assembler.

There are two 'system commands' that are available for loading a memory array from a text file:

- `$readmemb(''filename'', array_name);`
- `$readmemh(''filename'', array_name);`

The difference between the two functions lies in the format used to represent the stored data within the text file; the first function requires the data to be entered into the text file in *binary*, whereas the second makes use of a text file containing *hexadecimal* values.

Listing 8.2 shows the contents of an example text file containing binary data values for loading into a memory array. The first line specifies the numeric address, in hexadecimal format, of the starting location. This is usually equal to zero. Subsequent use of the `@hex_address` delimiter allows the memory to be initialized in discrete sections with different blocks of data.

```
@0
1010 0000 1111 1011 0010 1001 0110 1110
0111 1101 1011 1111 0000 0001 0010 0101
```

```
1010 0000 1111 1011 0010 1001 0110 1110
0111 1101 1011 1111 0000 0001 0010 0101
```

Listing 8.2 Contents of the file `rom_data.txt`.

The actual data values are listed in the order they will be stored in memory separated by white space, such as one or more space characters or the new-line character. If the number of values

```
`timescale 1 ns/ 1 ns
module Test_rom();

wire [3:0] Data;

reg [4:0] Address;

reg oebar;

initial //initialise rom with data from file
  $readmemb("rom_data.txt", dut.mem);


rom  #(.Awidth(5), .Dwidth(4))
          dut(.oeb(oebar),
          .data(Data),
          .address(Address));

initial
begin
  Address = 0;
  repeat (32) //read entire rom contents
  begin
    oebar = 1'b1;
    #25 oebar = 1'b0;
    #50 oebar = 1'b1;
    #25;
    Address = Address + 1;
  end
  $stop;
end
endmodule
```

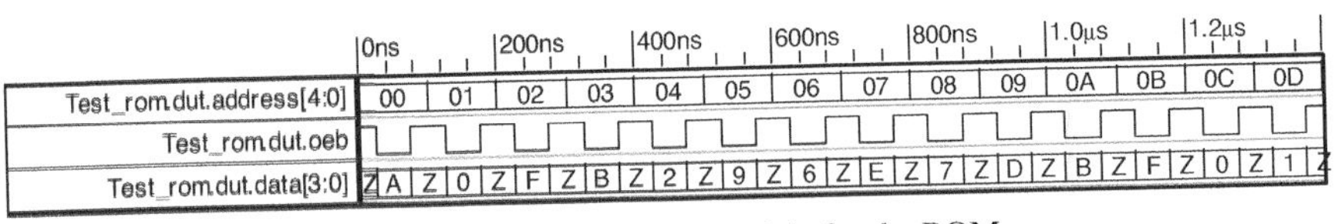

Figure 8.29 Verilog test-module for the ROM.

contained within the text file is less than the size of the memory array, then the remaining memory array locations are undefined.

The text file name field ``filename'' is a valid path name to the text file containing the data. The exact format used here depends on the operating system of the computer used to perform the Verilog simulation, but generally the name of the text file is all that is required if the file is in the same location (folder or directory) as the Verilog source files that make use of it.

The call to the system commands `$readmemb()` and `$readmemh()` may be made from within the actual memory module itself, in which case the `array_name` field refers to the name of the memory array defined within the enclosing module, e.g. `mem` in the listing shown in Figure 8.28.

In the present example, the initialization of the ROM memory array is performed within the test-module `Test_rom`, shown in Figure 8.29. Here, an **initial** block in lines 6 and 7, containing a single statement, loads the binary data shown in Listing 8.2 into the memory array:

```
$readmemb(''rom_data.txt'', dut.mem);
```

As shown above, the reference to `mem` must be preceded by the *instance name* of the `rom` being instantiated in lines 8–11 of the listing shown in Figure 8.29. The default values of the address and data widths of the ROM are overridden such that a '32 × 4' (32 words, 4-bits per word) memory is instantiated; this corresponds to the memory array values defined by the `rom_data.txt` file shown in Listing 8.2.

The remainder of the test module shown in Figure 8.29 corresponds to an **initial** block between lines 12 and 24 that reads each stored value out from the memory array, from location 0 to 31. The resulting simulation waveforms shown below the listing in Figure 8.29 illustrate this process; careful inspection of the data values output during the periods when `oebar` is asserted reveals that they are identical to those stored in the text file `rom_data.txt`.

The last example in this section on Verilog memories shows an alternative approach to describing a ROM. Listing 8.3 shows the source description of a module named `rom_case`. As the name suggests, this variation of a ROM makes use of the Verilog **case...endcase** sequential statement.

```
//read only memory using a case statement
module rom_case #(parameter Awidth = 8, Dwidth = 8)
                 (input oeb,
                 output [Dwidth-1:0] data,
                 input [Awidth-1:0] address);

reg [Dwidth-1:0] data_i;

always @(address)
begin
  case (address) //define rom contents
    0: data_i = 'h88;
    1: data_i = 'h55;
    2: data_i = 'haa;
    3: data_i = 'h55;
    4: data_i = 'hcc;
    5: data_i = 'hee;
```

```
16        6: data_i = 'hff;
17        7: data_i = 'hbb;
18        8: data_i = 'hdd;
19        9: data_i = 'h11;
20        10: data_i = 'h22;
21        11: data_i = 'h33;
22        12: data_i = 'h44;
23        13: data_i = 'h55;
24        14: data_i = 'h66;
25        15: data_i = 'h77;
26        default: data_i = 'h0; //use `'x' or '0'
27      endcase
28    end
29    //three-state buffer
30    assign data = (oeb == 1'b0) ? data_i: 'bz;

31    endmodule
```

Listing 8.3 Verilog description for the ROM using a **case** statement.

The module header is identical to that of the module shown in Figure 8.28; this is followed by the declaration of a **reg** named `data_i` having `Dwidth` bits. This object acts as a signal to hold the output of the **case** statement, prior to being fed through the 'three-state buffer' at line 30.

The **always** block in line 7 responds to events on the input address only; the enclosed **case** statement then effectively maps each address value to the appropriate data value. In this manner, the 'contents' of the memory are explicitly defined within the module itself, rather than being contained in an external file. This may restrict this approach to the description of relatively small memories, due to having to specify each value explicitly within the module text.

Where the number of data values is less than the capacity of the memory ($2^{\texttt{Awidth}}$), the **default** branch in line 26 must be included to cover the unused memory locations. A default value of `x` rather than zero will result in a smaller logic circuit if the ROM is to be implemented in the form of a combinational logic circuit, since an `x` is interpreted as a 'don't care' condition by a logic synthesis software tool.

8.7 DESCRIBING FINITE-STATE MACHINES

This section describes how the Verilog HDL can be used to create concise behavioural-style descriptions of FSMs. The underlying building block of many digital systems, the FSM is a vitally important part of the digital system designer's toolbox. The behavioural statements provided by Verilog facilitate the quick and straightforward creation of synchronous FSM simulation models, once the state diagram has been drawn. This, when combined with the wide availability of powerful logic synthesis software tools, makes the realization of state machines extremely efficient and rapid.

Figure 8.30 shows the block diagram structure of a general synchronous FSM. As shown in Figure 8.30, the FSM comprises two major blocks connected in a feedback configuration: the

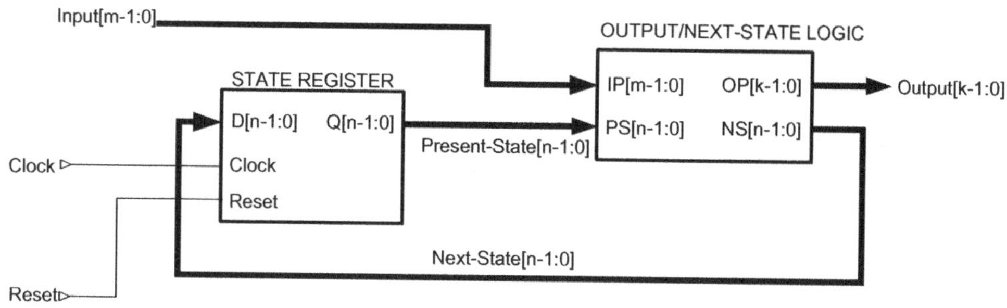

Figure 8.30 General FSM block diagram.

STATE REGISTER and the OUTPUT/NEXT-STATE LOGIC. There are several possible variations on the basic structure; however, the state register generally consists of a collection of n flip-flops (where 2^n must be greater than or equal to the number of FSM states), and the OUTPUT/NEXT-STATE LOGIC block contains the combinational logic that predicts the next state and the output values.

The general block diagram shown in Figure 8.30 represents the so-called *Mealy* FSM, where the `k` output bits depend both on the `n` state bits and the `m` input bits. Initialization of the FSM may be provided through the use of an asynchronous `Reset` input that forces all of the state flip-flops into a known state (usually zero). One possible disadvantage of the Mealy FSM architecture is the fact that the `Output` can change asynchronously, in response to asynchronous changes in the `Input`. This can be removed by making the outputs depend only on the `Present-State` signal, i.e. the output of the state register. This modified structure is better known as the *Moore* FSM. This section will present guidelines and examples on how to construct Verilog behavioural descriptions of both Mealy and Moore FSMs.

The starting point in the design of any FSM is the state diagram. This graphical representation provides a crucially important visual description of the machine's behaviour, allowing the designer to determine the number of states required and establish the logical transitions between them. Once the number of states has been determined, the next step is to assign a unique binary code to each state; this is known as the *state assignment*. In Verilog, the state assignment can be defined in a number of different ways, using:

- local parameters;
- parameters declared as part of the module header;
- the `` `define `` compiler directive.

The first of these is perhaps the most obvious choice, since the state values are likely to be a set of fixed codes referenced from within the module describing the FSM. The following line of Verilog illustrates how a set of state values is defined for an FSM having four states:

```
localparam s0 = 2'b00,
           s1 = 2'b01,
```

```
         s2 = 2'b10,
         s3 = 2'b11;
```

From the point of the above declaration, the symbolic names `s0...s3` can be used instead of the binary codes, making the description more readable.

Defining the state values as a set of in-line parameters within the module header provides the additional flexibility of being able to reassign them when the FSM module is instantiated, as shown below:

```
//module header with in-line parameters
module fsm #(parameter s0 = 0,
                       s1 = 1,
                       s2 = 2,
                       s3 = 3)
           (input clk, ..., output...);
//overriding default parameter values
fsm #(.s0(2),
      .s1(0),
      .s2(3),
      .s3(1))
   F1(.clk(CLK), ...);
```

The third approach makes use of the `` `define`` compiler directive in a similar manner to the way in which #define is used in the C/C++ languages to perform text substitution. The compiler directives come before the module header, as shown by the following example:

```
`define WAIT 4'b001
`define IDLE 4'b011
`define ACK1 4'b101
`define ACK2 4'b110

module fsm(...);
```

Within the body of the fsm module above, reference is made to the defined state values as follows:

```
//identifier must be prefixed by grave-accent character
Present-State <= `IDLE;
```

The STATE REGISTER block shown in Figure 8.30 is described by an **always** sequential block; therefore, the output signal it assigns to must be declared as a **reg**-type object, as shown below:

```
reg [n-1:0] Present-State; //number of states must be <= 2^n
```

The typical format of the state register sequential block is shown in Listing 8.4.

```
1   always @(posedge Clock or posedge Reset)
2   begin
3      if (Reset == 1'b1)
4         Present-State <= s0;
5      else
6         Present-State <= Next-State;
7   end
```

Listing 8.4 General format of state register **always** block.

As described in previous sections, the sequential block shown in Listing 8.4 describes synchronous sequential logic with active-high asynchronous initialization (active-low asynchronous initialization is equally possible).

On each 0-to-1 transition of the `Clock` signal, the `Present-State` is updated by the incoming `Next-State` value in line 6, the latter being produced by the OUTPUT/NEXT-STATE LOGIC block. Now the `Present-State` signal is an input to the OUTPUT/NEXT-STATE LOGIC block; therefore, it responds to this input change, combined with the current values of the inputs, by updating the `Next-State` output value. The feedback signal `Next-State` is now ready for the next positive edge of the clock to occur, thereby updating the `Present-State` in a cyclic manner.

It is good practice to split the OUTPUT/NEXT-STATE LOGIC block into two separate parts, one for the outputs and another for the next state. This results in a more readable and, therefore, maintainable description. Listing 8.5 shows the outline Verilog source description for the 'next-state' part of this block.

```
1   always @(Present-State, Input1, Input2, Input3...)
2   begin
3   //consider each possible state
4      case (Present-State)
5         s0: if (Input1 == 1'b0)
6               Next-State <= s1;
7            else
8               Next-State <= s0;
9         s1: ...;
10        s2: ...;
11        default: Next-State <= s0;
12     endcase
13  end
```

Listing 8.5 General format of next-state **always** block.

As shown in Listing 8.5, the next-state **always** block describes combinational logic; therefore, the guidelines discussed in Section 8.2 must be observed in order to ensure that `Next-State` is assigned a value under all possible conditions. (This is achieved in Listing 8.5 by means of the **default** branch in line 11.)

The **always** sequential block must be sensitive to changes in both the `Present-State` signal and all of the FSM inputs, as shown in line 1. The **case...endcase** statement, situated between lines 4 and 12 inclusive, considers each possible state and assigns the resulting `Next-State` depending on the input conditions. In this manner, the next-state part of the block describes the flow around the state machine's state diagram in terms of behavioural statements. The fact that the `Next-State` signal is assigned values by an **always** sequential block means that it must be declared in a similar manner to the `Present-State` signal, as follows:

```
reg [ n-1:0] Next-State; //output of combinational behaviour
```

The **default** branch (line 11) of the **case** statement is required to define the behaviour of the FSM for any *unused* states; these states result from the fact that the number of *used* states may be less than the number of *possible* states. If the FSM finds itself in an unused state, then the safest approach is to move it directly and unconditionally to the `reset` state, otherwise the designer may take the slightly more risky approach of treating all unused states as *don't care* states, in which case the **default** branch would be

```
default: Next-State <= 'bx;
```

The part of the OUTPUT/NEXT-STATE LOGIC block shown in Figure 8.30 that drives the FSM outputs may be described using either an additional **always** block or by means of continuous assignments. The choice between these approaches depends upon the complexity of the output logic. For Moore-type FSMs, the outputs depend only on the present state; therefore, the expressive capabilities of the continuous assignment are usually adequate. The potentially more complicated output logic of a Mealy FSM may require the use of a sequential block, in which case it is important to remember to qualify the outputs as being of type **reg**.

The following extract illustrates the use of the continuous assignment to describe the output logic of a simple Mealy FSM:

```
assign Output1 = ((Present-State == s0)
                 && (Input1 == 1'b0)) ||
                 ((Present-State == s2)
                 && (Input2 == 1'b1));
```

Here, the output `Output1` depends directly on both the present state and the inputs. A variation on the use of separate sequential blocks, for the state-register and next-state feedback logic, is to combine these in a single **always** block. This approach has the advantage of making the Verilog description more concise and involves combining the sequential logic behaviour shown in Listing 8.4 with the combinational logic behaviour shown in Listing 8.5, as shown in Listing 8.6.

```
...
reg[ n-1:0] state; //single state register
...
always @(posedge clock or posedge reset)
begin
if (reset == 1)
```

```
7       state <= State0;
8     else
9       case (state)
10        State1: if (Input1 == 0)
11              state <= State2;
12           else
13              state_reg <= read1one;
14        State2: if (Input2 == 1)
15              state <= State3;
16           else
17              state <= State2;
18        ...
19        default: state <= 3'bxxx
20      endcase
21    end
```

Listing 8.6 General format of combined state-register and next-state logic **always** block.

Another consequence of using a combined sequential block for the state register and next-state logic is the removal of the need for two separate **reg**-type signals for *present state* and *next state*. As shown in Listing 8.6, only a single declared **reg** named `state` is required in line 2; the behavioural description both assigns to (lines 7, 11, 15....) and reads from (line 9) this combined signal. The combined sequential block is triggered by positive edges on either the clock or reset input (assuming asynchronous active-high initialization is being employed). After testing for the reset condition in line 6, the behaviour is much the same as that of the next-state logic given in Listing 8.5, making use of the **case**...**endcase** statement to consider each state and input condition to implement the sequential behaviour described by the state diagram.

In effect, the statements between lines 9 and 20 of the source listing shown in Listing 8.6 describe a self-contained synchronous feedback logic system where the signal `state` is the output of a set of *D*-type flip-flops and the inputs of the flip-flops are described by the combination of the **case** and **if**...**else** statements.

The following example FSM designs serve to illustrate the points discussed above further. The first example is concerned with the description of an FSM to control the timers used by two people playing a game of timed chess, and the second looks at a simple combination lock with automatic locking mechanism.

8.7.1 Example 1: Chess Clock Controller Finite-State Machine

Figure 8.31 shows the block diagram of a system used by two chess players to record the amount of time taken to make their respective moves. The players, referred to as Player-A and Player-B, each have their own timer (TIMER-A and TIMER-B), the purpose of which is to record the total amount of time taken in hours, minutes and seconds for their moves since the commencement of the game.

The exact details of the timer internal operation are beyond the scope of this discussion, since we are primarily concerned with the description of the FSM that controls them. The timer control inputs, `en` and `rst`, shown in Figure 8.31, operate as follows:

- `rst` – when logic 1, resets the time to zero hours, zero minutes and zero seconds.

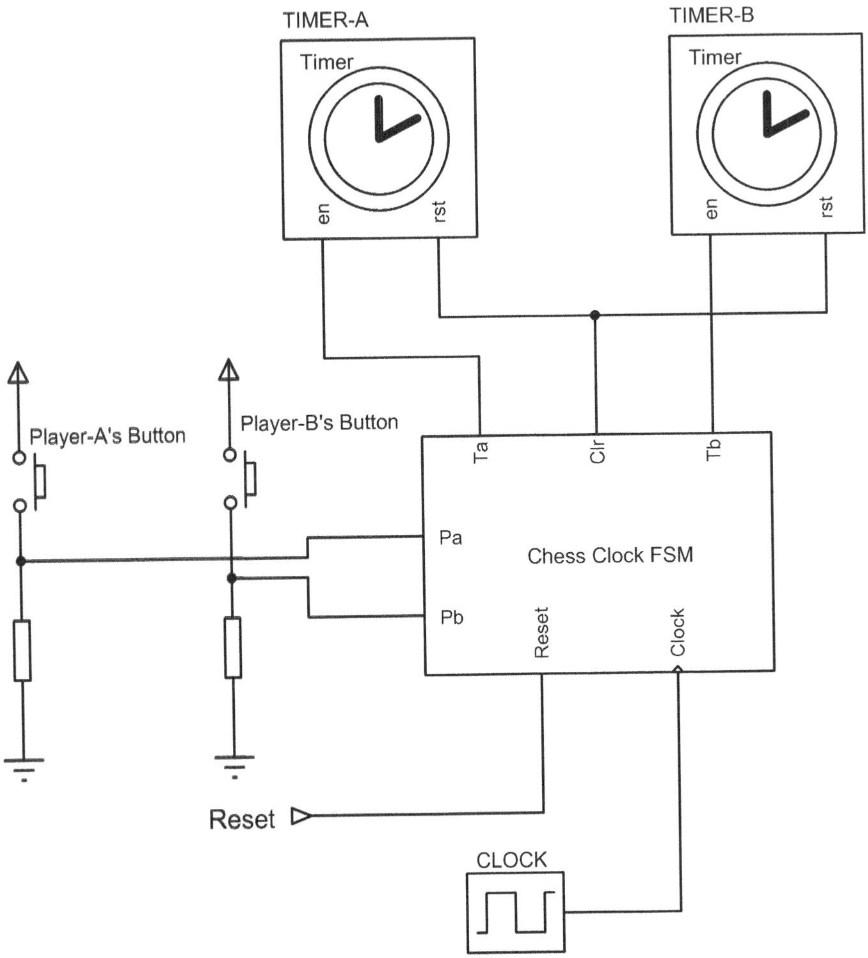

Figure 8.31 Block diagram of chess clock system.

- `en` – when logic 1, enables the time to increment from the current time value. When `en` is logic 0, the current elapsed time is held constant.

At the start of a new game, the `Reset` input is asserted to initialize the system and clear both timers to zero time. This is achieved by means of the `Clr` output of the Chess Clock FSM being driven high, thereby asserting the reset (`rst`) input of both timers. Each chess player has a push-button, which when pressed applies a logic 1 to their respective inputs, `Pa` and `Pb`, of the Chess Clock FSM. After resetting the timers, the player who is not making the first move presses their push-button in order to enable the other player's timer to commence timing.

For example, if Player-A is to make the first move, then Player-B starts the game by pressing their push-button. This has the effect of activating the `Ta` output of the Chess Clock FSM block shown in Figure 8.31, in order to enable TIMER-A to record the time taken by Player-A to make

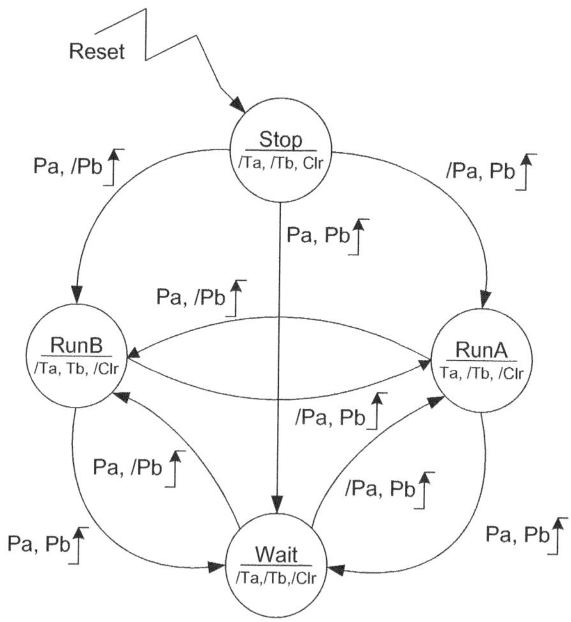

Figure 8.32 State diagram for chess clock controller FSM.

the first move. Once Player-A completes the first move, Player-A's button is pressed in order to stop their own timer and start Player-B's timer (`Ta` is negated and `Tb` is asserted).

For the purposes of this simulation, it is assumed that the `Pa` and `Pb` inputs are asserted momentarily for at least one clock cycle, and the potential problems resulting from switch bounce and metastability [3] may be neglected.

In the unlikely event that both players press their buttons simultaneously, the Chess Clock FSM is designed to disable both timers by negating `Ta` and `Tb`.

This will hold each player's elapsed time until play recommences in the manner described above, i.e. Player-A (Player-B) presses their push-button to re-enable TIMER-B (TIMER-A).

The state diagram for the Chess Clock FSM is shown in Figure 8.32. As shown, the FSM makes use of four states having the names shown in the upper half of the state circles. The states of the FSM outputs `Ta`, `Tb` and `Clr` are listed in the lower half of every state circle; those outputs preceded by '/' are forced to logic 0, whereas those without '/' are forced to logic 1. The presence of the output states within each of the state circles indicates that the Chess Clock FSM is of the Moore variety.

The values of the inputs, `Pa` and `Pb`, are shown alongside each corresponding state transition path (arrow) using a format similar to that used to show the state of the outputs. The movement from one state to another occurs on the rising edge of the `Clock` input. Where the number of transitions shown originating from a given state is less than the total number possible, the remaining input conditions result in a so-called sling, i.e. the next state is the same as the current state.

For example, the state named `RunA` in Figure 8.32 has two transitions shown on the diagram corresponding to the input conditions $\langle$`Pa, Pb`$\rangle = \langle 1, 0\rangle$ and $\langle 1, 1\rangle$. The remaining input conditions, $\langle$`Pa, Pb`$\rangle = \langle 0, 0\rangle$ and $\langle 0, 1\rangle$, cause the state machine to remain in the current state;

hence, there exists a sling in state RunA corresponding to the condition that the Pa input is at logic 0 and the Pb input can be either logic 0 or logic 1, the latter indicating the presence of a *don't care* condition for input Pb.

The asynchronous, active-high Reset input forces the FSM directly into the state named Stop, irrespective of any other condition.

The FSM depicted visually by the state diagram shown in Figure 8.32, is described in a behavioural style by the Verilog HDL listing given in Listing 8.7.

```
module chessclkfsm(input reset, Pa, Pb, clock,
                   output Ta, Tb, Clr);

//ascending state assignment
localparam RunA = 0, RunB = 1, Stop = 2, Wait = 3;

reg[ 1:0] state;

//combined state register and next state sequential block
always @(posedge clock or posedge reset)
  begin
    if (reset)
       state <= Stop;
    else
      case (state)
        RunA:
          casex ({ Pa, Pb})
           2'b0x: state <= RunA;
           2'b10: state <= RunB;
           2'b11: state <= Wait;
          endcase
        RunB:
          casex ({ Pa, Pb})
           2'bx0: state <= RunB;
           2'b01: state <= RunA;
           2'b11: state <= Wait;
          endcase
        Stop:
          case ({ Pa, Pb})
           2'b00: state <= Stop;
           2'b01: state <= RunA;
           2'b10: state <= RunB;
           2'b11: state <= Wait;
          endcase
        Wait:
          if (Pa == Pb)
           state <= Wait;
          else if (Pa == 1'b1)
           state <= RunB;
          else
           state <= RunA;
```

```
39         endcase
40     end

41 //Moore output assignments depend only on state
42 assign Ta = state == RunA;
43 assign Tb = state == RunB;
44 assign Clr = state == Stop;

45 endmodule
```

Listing 8.7 Verilog description of the Chess Clock FSM.

The module `chessclkfsm` makes use of a local parameter to define the state values. Each state name shown in the state diagram of Figure 8.32 is assigned a value in line 4. This is followed by the declaration of a 2-bit **reg** to hold the state of the FSM; this description makes use of the single **always** block approach outlined in Listing 8.6.

The sequential **always** block spanning lines 7–40 of the listing shown in Listing 8.7 describes the state register and next-state logic. The presence of a don't-care condition in one of the state transitions for states `RunA` and `RunB` suggests the use of a special variation of the **case** statement known as **casex**.

The use of **casex** instead of **case** in lines 14 and 20 allows the explicit use of the 'don't-care' value (`x`) within the literals specified in lines 15 and 21. In effect, this means that one or more of the inputs can be either logic 0 or logic 1, e.g. lines 14 and 15 are equivalent to the following:

```
14 case ({ Pa,Pb})
15    2'b00, 2'b01: state <= RunA;
16 ...
```

The **case** statement considers each possible value of `state;` in this example there is no requirement for a **default** branch, since the number of states is equal to a power of 2. State `Stop` has four unique next states, hence the need for a nested **case**...**endcase** statement with four branches, or limbs, situated in lines 27–30 inclusive. The **case** statement gives equal priority to each of the individual limbs or branches enclosed within the bounds of **case**...**endcase**; hence, the matching expressions must be *nonoverlapping* or *mutually exclusive*. As seen previously, multiple values may be specified on a single branch, so long as none of these values appears within any other of the branches within the statement.

The next-state behaviour of the `Wait` state is described using a nested **if**...**else** statement in order to illustrate the flexibility of the Verilog language. It is straightforward to appreciate that the semantics of the statement in lines 33–38 inclusive of the source description in Listing 8.7 are equivalent to the state transitions shown on the state diagram of Figure 8.32, bearing in mind that there is a sling condition corresponding to input values $\langle \texttt{Pa}, \texttt{Pb} \rangle = \langle 0, 0 \rangle$ and $\langle 1, 1 \rangle$.

It should be noted that, despite the priority implied by the nested **if**...**else**...**if** statement semantics, the circuitry resulting from synthesis of this description will not include any prioritized logic. This is due to the fact that the conditions specified in each part of the **if**...**else** statement are *mutually exclusive*.

The outputs Ta, Tb and Clr, of the Chess Clock FSM, are of the *Moore* variety, i.e. dependent on the state of the machine only. These are generated by means of the continuous assignments in lines 42–44 of the source description shown in Listing 8.7. Each output is generated by continuously comparing the value of the state-register state with the local parameter value corresponding to the state in which the output is asserted.

In this simple example, each output is asserted in only one state; therefore, the logic of the outputs amounts to little more than a single AND gate.

The output logic can be further simplified by encoding the states of the FSM with values that match the outputs. In the present example, the output values are unique for each state, so this would involve simply defining the state values to be the same as the output values, i.e. replace the local parameter declarations with those shown in lines 4–7 of Listing 8.8.

```
3   //state assignment matches outputs Ta, Tb, Clr
4   localparam RunA = 3'b100,
5          RunB = 3'b010,
6          Stop = 3'b001,
7          Wait = 3'b000;

8   reg [2:0] state; //no. of state bits = no. of outputs

    ...
39     default: state <= 3'bx;
40  endcase

    ...
41  //outputs are equal to state bits
42  assign Ta = state[2] ;
43  assign Tb = state[1] ;
44  assign Clr = state[0] ;
```

Listing 8.8 Alternative state assignment to match outputs.

The output continuous assignments, situated in lines 42–44 of the listing given in Listing 8.7, would be replaced by the corresponding lines shown in Listing 8.8. As shown, each output is now mapped directly to the corresponding bit of the state register.

Another consequence of modifying the state assignments, as described above, is the need to change the number of state bits to match the number of outputs. The replacement state-register declaration, in line 8 of Listing 8.8 now declares a register having 3-bits; therefore, the next-state behaviour must be modified by the addition of a **default** branch in line 39, so that the additional ($2^3 - 4 = 4$) unused states are covered by the **case** statement.

Simulation of the Chess Clock FSM module chessclkfsm is achieved by means of the simple test module shown in Figure 8.33. The resulting timing waveforms are also shown in Figure 8.33, where the relationship between the inputs, state and outputs can be seen to follow that defined by the state diagram. Most Verilog simulation tools provide a facility whereby the values of the state waveform can be displayed in terms of the state names used on the state diagram, as is the case here. This is a significant visual aid when attempting to analyse, understand and verify the behaviour of an FSM using simulation.

```
`timescale 1 ms / 1 ms
module Test_chessclkfsm();

reg RES, A, B, CLK;
wire Ta, Tb, Clrt;

//generate a 10 Hz clock
initial
begin
   CLK = 1'b0;
   forever
      #50 CLK = ~CLK;
end

//generate inputs
initial
begin
   RES = 1'b1; A = 1'b0; B 1'b0;
   #200 RES = 1'b0;
   #200;
   A = 1'b1;
   #550 A = 1'b0;
   #350 B = 1'b1;
   #750 B = 1'b0;
   #400;
   A = 1'b1; B = b1;
   #350;
   A = 1'b0; B = 1'b0;
   #450;
   A = 1'b1;
   #800;
   $stop;
end

//instantiate the FSM
chessclkfsm mut (.reset (RES),
                 .Pa (A), . Pb(B), . clock (CLK),
                 .Ta (Ta), .Tb (Tb), .Clr (Clrt));
endmodule
```

Figure 8.33 Test module and simulation waveforms for chess clock FSM.

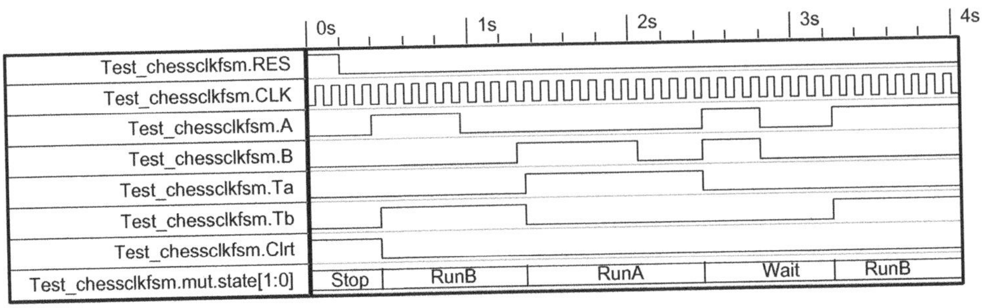

Figure 8.33 (*Continued*).

8.7.2 Example 2: Combination Lock Finite-State Machine with Automatic Lock Feature

The second example of an FSM-based design is a rather more complex system that makes use of several modules, both combinational and sequential. This example also serves to illustrate the interaction of an FSM with other synchronous sequential modules, all described in a behavioural style and clocked by a common clock signal.

Figure 8.34 shows the block diagram of a so-called 'digital combination lock' system. At the heart of the system there is an FSM, labelled CONTROLLER in the figure, the function of which is to detect when a user has entered the correct four-digit secret code via the Key Pad Switches, shown at the left-hand side of Figure 8.34.

The user sees a keypad with eight active-low push-button switches (SW[0]...SW[7]). The first four (SW[0...3]) are hardwired into the system via a four-to-one multiplexer; these represent the code switches. It is up to the user to connect the multiplexer inputs to the keypad switches corresponding to the secret code; in this manner, the secret access code is *hardwired* into the system.

The eight-input AND gate, connected to all of the switches in Figure 8.34, provides an output named `allsw` that goes to logic 0 if *any* switch is pressed. The output of the four-to-one multiplexer, named `mux_out`, will go to logic 0 if the switch being pressed corresponds to the multiplexer select address input `sel[0..1]`. In this manner, the multiplexer is able to select each switch in the code in sequence; the output `mux_out` will go low only if the correct switch has been pressed.

The input push-button switches are asynchronous inputs by nature, whereas the combination lock system operates entirely synchronously. It is impossible to predict for how long any push-button will be pressed; therefore, the duration of the logic 0 pulses coming into the system on signals `mux_out` and `allsw` is entirely unpredictable. If the aforementioned signals were fed directly into the FSM, then a single key depression lasting 0.5 s, for example, would be interpreted as a sequence of approximately n inputs, where

$$n = 0.5/\texttt{clock\_period}.$$

The above problem is overcome by means of the simple 'edge detector' circuit shown in Figure 8.35. The system makes use of two of these circuits, labelled DET1 and DET2 in Figure 8.34. As shown in Figure 8.35, the circuit is essentially a synchronous 2-bit shift

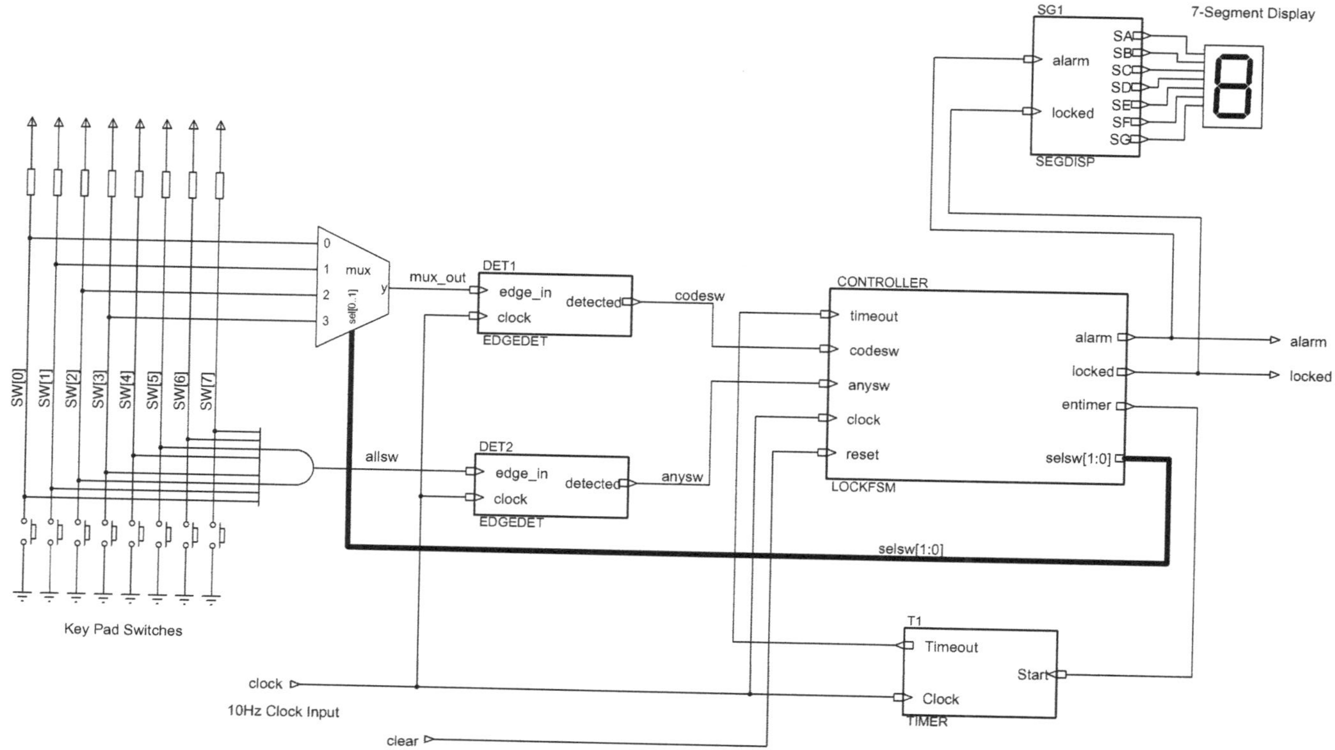

Figure 8.34 Block diagram of combination lock system.

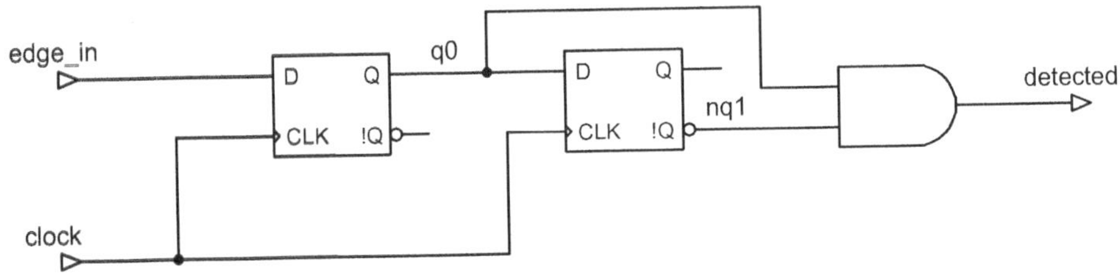

Figure 8.35 Logic diagram of edge detector `edgedet`.

register with the output of the first flip-flop ANDed with the inverse of the output of the second flip-flop.

This simple circuit performs both synchronization and edge detection, in that it produces a single clock-cycle-length logic 1 pulse at the output named `detected`, near to the point when the input, `edge_in`, undergoes a logic 1 to logic 0 transition, regardless of how long `edge_in` remains at logic 0.

Neglecting the usual problems of metastability [3], which result whenever there is a need to interface between asynchronous and synchronous domains, the logic circuit of Figure 8.35 provides an effective means of interfacing the push-button switches to the FSM.

The outputs, `codesw` and `anysw`, of the two edge detectors feed directly into the FSM LOCKFSM. The fact that the edge detectors and the FSM are clocked by the same signal ensures synchronization between the two separate modules such that if a key is pressed, and it is the correct key (i.e. the four-to-one multiplexer is selecting the key), the `lockfsm` receives a logic 1 pulse on both `codesw` and `anysw` during the same clock cycle. The arrival of the two pulses indicates the correct key was pressed and the FSM then advances to the next state.

The Verilog descriptions of the *D*-type flip-flop and the edge detector are shown in Listings 8.9 and 8.10 respectively.

```
module dff(output reg q, input d, clk);

always @(posedge clk) q <= d;

endmodule
```

Listing 8.9 Verilog source description of *D*-type flip-flop.

```
module edgedet(input edge_in,
               output detected,
               input clock);

wire q0, q1;

dff dff0(.q(q0), .d(edge_in), .clk(clock));
dff dff1(.q(q1), .d(q0), .clk(clock));

assign detected = q0 & ~q1;

endmodule
```

Listing 8.10 Verilog source description of edge detector.

The block diagram of Figure 8.34 includes a timer module (TIMER) labelled T1. This module interfaces with the FSM via signals `entimer` (enable timer) and `timeout` (timer timed out) and is clocked by the same master clock as the FSM and edge detectors, ensuring synchronization.

The function of the timer is to provide an automatic locking mechanism, returning the system to the locked state after a delay of 30 s subsequent to the system entering the unlocked state.

The master clock signal is intended to have a frequency of 10 Hz, so the timer implements the required delay by counting to 300_{10}, as shown in the Verilog source description shown in Listing 8.11.

```
module Timer(input Clock, Start, output Timeout);
//time delay value in clk pulses
localparam NUMCLKS = 300;

reg[8:0] q;
always @(posedge Clock)
begin
   if (!Start||(q == NUMCLKS))
      q <= 9'b0;
   else
      q <= q + 1;
end
//decode counter output
assign Timeout = (q == NUMCLKS);

endmodule
```

Listing 8.11 Verilog source description of automatic lock timer.

The `Timer` module behaviour is entirely synchronous: with the input named `Start` at logic 0, the timer is disabled and the count `q` held at zero.

The FSM starts the timer when it enters the unlocked state by asserting `entimer` (connected to timer input `Start`), this allows the count `q` to increment on each clock edge until it reaches the terminal value `NUMCLKS` (300_{10}), at which point the `Timeout` output of the timer goes high for one clock cycle and the count returns to zero.

The FSM responds to the logic 1 on its `timeout` input by returning to state s0, where the `locked` output returns high. By returning to state s0, the FSM also negates the `entimer` output, thereby disabling the timer until the next time it is required.

The remaining module, as yet not discussed, in the block diagram of Figure 8.34, is the seven-segment decoder named `SEGDISP`. This module is purely combinational and drives an active-low seven-segment display unit that displays the state of the system, based on the values of the `alarm` and `locked` outputs of the FSM: 'L' for locked, 'U' for unlocked and 'A' for alarm. The Verilog behavioural description of the module is given in Listing 8.12.

```
module segdisp(input locked,alarm,
               output SA,SB,SC,SD,SE,SF,SG);

reg[ 6:0] seg;

always @(locked or alarm)
begin
   if (alarm == 0)
     seg = 7'b0001000; //display 'A'
   else if (locked == 0)
     seg = 7'b1000001; //display 'U'
   else
     seg = 7'b1110001; //display 'L'
end

assign {SA, SB, SC, SD, SE, SF, SG} = seg;

endmodule
```

Listing 8.12 Verilog source description of seven-segment display decoder.

Figure 8.36 shows the state diagram for the `lockfsm` module at the heart of the combination lock system.

The FSM is initialized by asserting the asynchronous reset input, this forces it into state s0, where the `locked` and `alarm` outputs are both at logic 1, indicating the system is locked and not in a state of alarm (`alarm` is active-low). The 2-bit `selsw` output of the `lockfsm` is set to zero, thereby selecting the first input push-button in the sequence via the four-to-one multiplexer. The timer is disabled on account of `entimer` being at logic 0.

What happens next depends on which of the eight push-button switches is pressed. If the first switch in the code sequence is pressed (SW[0]), then the input signals `codesw` and `anysw` go high simultaneously, causing the FSM to move into state s1, where it remains until a subsequent key is pressed.

In state s1 the `selsw` output of the FSM is set to 1, thereby selecting the second input of the multiplexer, this being connected to the second switch in the code sequence, SW[1]. Pressing SW[1] in state s1 asserts both `codesw` and `anysw` again, advancing the FSM into state s2.

On entering state s2, the FSM changes `selsw` to 2, thereby selecting the third input of the multiplexer, this being connected to the third switch in the code sequence, SW[2].

In a similar manner to that described above, pressing switches SW[2] followed by SW[3] causes the `lockfsm` to enter the `unlock` state, having pressed all four keys (SW[0]...SW[3]) in the correct order. The `locked` output goes to logic 0 and the seven-segment display shows the letter 'U'.

As shown in Figure 8.36, the `entimer` output of the FSM is now asserted, thereby enabling the timer. The `lockfsm` will remain in the unlock state for as long as the `timeout` input remains at logic 0 (assuming the asynchronous reset input is not asserted).

As discussed above, this corresponds to a duration equal to 300_{10} clock periods or 30 s, whereupon the FSM will return to state s0 and reassert the `locked` output.

In any of the `lockfsm` states (s0, s1, s2 and s3), pressing the incorrect key pad switch will result in a pulse arriving from `anysw`, via the eight-input AND gate, but there will be no such

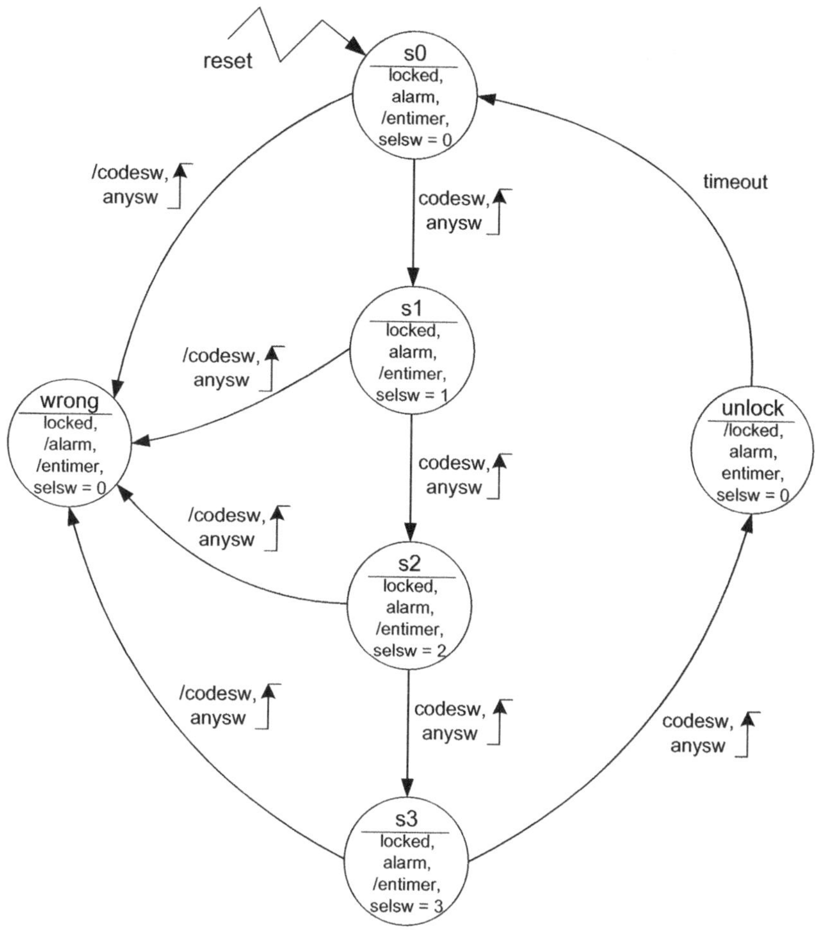

Figure 8.36 Combination lock FSM (`lockfsm`) state diagram.

pulse on `codesw`, due to the fact that the currently selected multiplexer input will not be asserted low.

The state diagram of Figure 8.36 shows that, under these circumstances, the FSM will move to state `wrong`, indicating that the incorrect key was pressed. In this particular state, the active-low `alarm` output is asserted and the display unit outputs the code for the letter 'A'.

The absence of any transitions leaving state `wrong` indicates the presence of an unconditional state transition leading from the `wrong` state back to itself (a 'sling'), i.e. the only way to exit the alarm state is to force an asynchronous reset. Needless to say, the `clear` input would, therefore, have to be located in a secure environment, enabling only a qualified operator to reset the alarm.

The Verilog behavioural description of the `lockfsm` module is shown in Listing 8.13.

```
module lockfsm(input clock, reset,
```

```
             codesw, anysw,
             output reg[1:0] selsw,
             output locked, alarm, entimer,
             input timeout);

localparam s0=3'b000, s1=3'b001, s2=3'b010,
             s3=3'b011,
             wrong=3'b100, unlock=3'b101;

reg[2:0] lockstate;

always @(posedge clock or posedge reset)
begin
if (reset == 1'b1)
   lockstate <= s0;
else
   case (lockstate)
     s0 : if (anysw & codesw)
          lockstate <= s1;
       else if (anysw)
          lockstate <= wrong;
       else
          lockstate <= s0;
     s1 : if (anysw & codesw)
          lockstate <= s2;
       else if (anysw)
          lockstate <= wrong;
       else
          lockstate <= s1;
     s2: if (anysw & codesw)
          lockstate <= s3;
       else if (anysw)
          lockstate <= wrong;
       else
          lockstate <= s2;
     s3: if (anysw & codesw)
          lockstate <= unlock;
       else if (anysw)
          lockstate <= wrong;
       else
          lockstate <= s3;
     wrong: lockstate <= wrong;
     unlock: if (timeout)
          lockstate <= s0;
       else
          lockstate <= unlock;
     default: lockstate <= 3'bx;
   endcase
end

always @(lockstate)
```

```
49 begin
50    case (lockstate)
51       s0: selsw = 0;
52       s1: selsw = 1;
53       s2: selsw = 2;
54       s3: selsw = 3;
55       wrong: selsw = 0;
56       unlock: selsw = 0;
57       default: selsw = 2'bx;
58    endcase
59 end

60 assign locked = (lockstate == unlock) ? 0: 1;
61 assign alarm = (lockstate == wrong) ? 0: 1;

62 assign entimer = (lockstate == unlock) ? 1: 0;

63 endmodule
```

Listing 8.13 Verilog source description of combination lock FSM.

In common with the previous example, this FSM is of the Moore type; therefore, the **always** sequential block starting at line 10 describes the state register and next-state behaviour only.

The output logic is captured by the combinational **always** block situated in lines 48–59 inclusive, and the continuous assignments on lines 60–62. The 3-bit state register `lockstate` is declared in line 9 and the six used states are assigned ascending numbers by means of a local parameter starting in line 6.

The two unused states are exploited as don't-care states by means of the **default** branches in lines 45 and 57 of the source shown in Listing 8.13.

All of the used states, with the exception of state `wrong`, make use of the **if...else** statement to describe the state transition logic defined by the state diagram of Figure 8.36. For example, the next-state behaviour for state s1 is repeated below:

```
s1 :  if (anysw & codesw)
          lockstate <= s2;
      else if  (anysw)
          lockstate <= wrong;
      else
          lockstate <= s1;
```

The first condition to be tested is the expression `anysw & codesw`; this will be true (logic 1) if both `anysw` and `codesw` are at logic 1. If this is the case, then the state of the FSM is moved to s2. If the first condition is false, then this leaves the possibility of either input being high or both inputs being low. The structure of the logic means that `codesw` cannot be high if `anysw` is low, so it is only necessary to test the state of `anysw` to see whether an incorrect key was pressed and, hence, move to the `alarm` state.

If no keys are pressed, then the FSM state remains the same, i.e. in this case s1. This is achieved by means of the final, and optional, **else** part of the above statement.

The complete combination lock system block diagram, shown in Figure 8.34, is described by the Verilog source given in Listing 8.14.

```
module comblock(input clock, clear,
            input[7:0] switches,
            output alarm, locked,
            output SA, SB, SC, SD, SE, SF, SG);

wire mux_out, anysw, codesw,
       allsw, entimer, timeout;

wire [ 1:0] selsw;

//4-to-1 multiplexor
assign mux_out = selsw == 0 ? switches[0] :
          (selsw == 1 ? switches[1] :
          (selsw == 2 ? switches[2] :
          (selsw == 3 ? switches[3] : 1'b0)));

//AND gate for all switches
assign allsw = &switches;

edgedet det1(.edge_in(mux_out),
            .detected(codesw),
            .clock(clock));

edgedet det2(.edge_in(allsw),
            .detected(anysw),
            .clock(clock));

Timer t1(.Clock(clock),
            .Start(entimer),
            .Timeout(timeout));

lockfsm controller(.clock(clock),
            .reset(clear),
            .codesw(codesw),
            .anysw(anysw),
            .selsw(selsw),
            .locked(locked),
            .alarm(alarm),
            .entimer(entimer),
            .timeout(timeout));

segdisp sg1(.locked(locked),
            .alarm(alarm),
            .SA(SA),
```

```
36              .SB(SB),
37              .SC(SC),
38              .SD(SD),
39              .SE(SE),
40              .SF(SF),
41              .SG(SG));
42 endmodule
```

Listing 8.14 Verilog source description of complete combination lock system.

The `comblock` module comprises instantiations of the modules discussed previously, along with two continuous assignments, situated in lines 9 and 14, to implement the four-to-one multiplexer and the eight-input AND gate respectively.

Simulation of the combination lock system is achieved with the use of a Verilog test module named `test_comblock`, shown in Listing 8.15.

```
`timescale 1 ms / 1 ms
module test_comblock();

// Inputs
reg clock;
reg clear;
reg[7:0] switches;

// Outputs
wire alarm;
wire locked;
wire SA, SB, SC, SD, SE, SF, SG;

// Instantiate the combination lock
comblock UUT(
            .clock(clock),
            .clear(clear),
            .switches(switches),
            .alarm(alarm),
            .locked(locked),
            .SA(SA), .SB(SB),.SC(SC),
            .SD(SD),.SE(SE),.SF(SF),. SG(SG)
            );

initial
begin
  clock = 1'b0;
  forever
    #50 clock = ~clock;
end

initial
begin
  clear = 1'b1;
```

```
switches = 8'b11111111;

repeat(3) @(negedge clock);
clear = 1'b0;

repeat(3) @(negedge clock);
switches[0] = 1'b0;

repeat(2) @(negedge clock);
switches[0] = 1'b1;

repeat(3) @(negedge clock);
switches[1] = 1'b0;

repeat(2) @(negedge clock);
switches[1] = 1'b1;

repeat(3) @(negedge clock);
switches[ 2] = 1'b0;

repeat(2) @(negedge clock);
switches[2] = 1'b1;

repeat(3) @(negedge clock);
switches[3] = 1'b0;

repeat(2) @(negedge clock);
switches[3] = 1'b1;

repeat(400) @(negedge clock); //wait for timeout

clear = 1'b1;

repeat(4) @(negedge clock);
clear = 1'b0;
repeat(3) @(negedge clock);
switches[0] = 1'b0;

repeat(2) @(negedge clock);
switches[0] = 1'b1;

repeat(3) @(negedge clock);
switches[5] = 1'b0;

repeat(2) @(negedge clock);
switches[5] = 1'b1;

repeat(3) @(negedge clock);
switches[2] = 1'b0;

repeat(2) @(negedge clock);
```

```
64      switches[2] = 1'b1;

65      repeat(3) @(negedge clock);
66      switches[3] = 1'b0;

67      repeat(2) @(negedge clock);
68      switches[3] = 1'b1;

69      repeat(4) @(negedge clock);
70      clear = 1'b1;

71      repeat(4) @(negedge clock);

72    $stop;
73  end

74  endmodule
```

Listing 8.15 Verilog source description of combination lock system test module.

The test-module generates a 10 Hz clock using an **initial** sequential block starting at line 21.

(a)

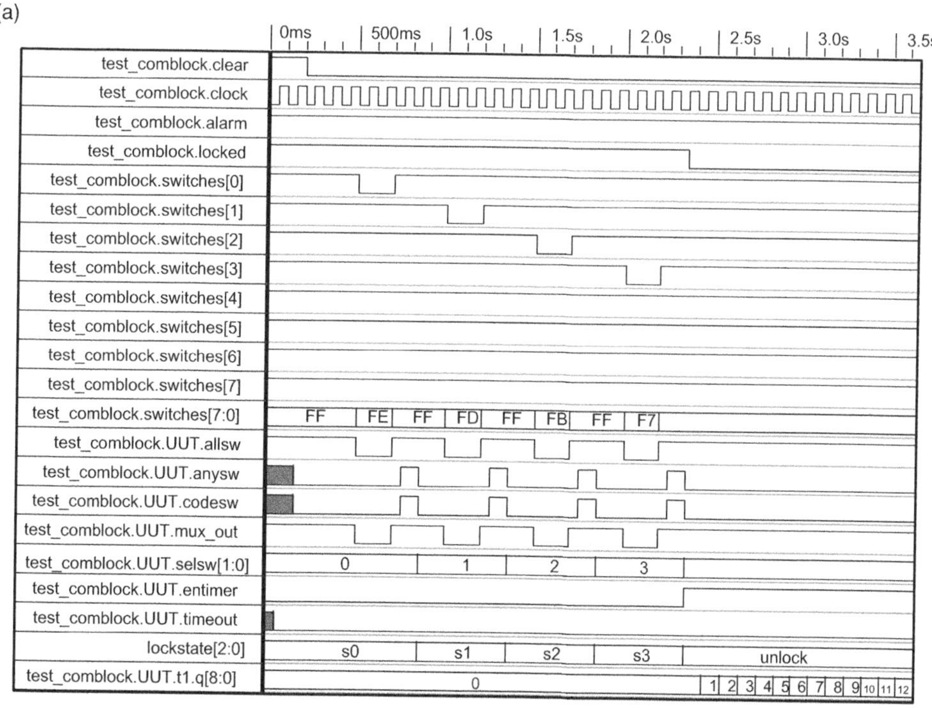

Figure 8.37 Combination lock simulation showing: (a) application of correct switch sequence; (b) automatic locking feature; (c) incorrect key input sequence.

(b)

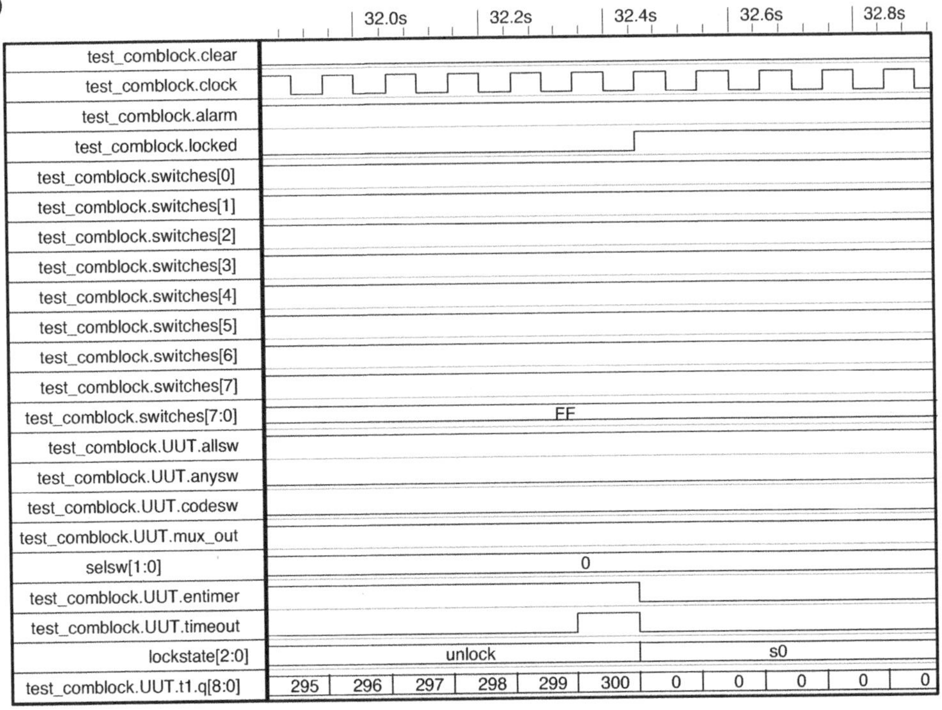

(c)

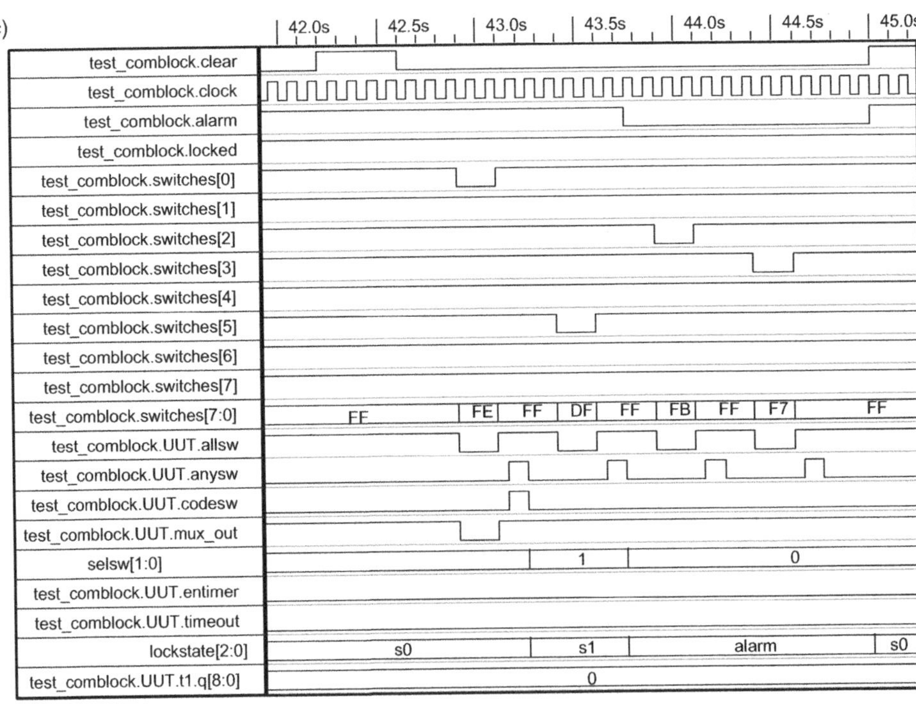

Figure 8.37 *(Continued).*

A second **initial** block, starting at line 27, exercises the combination lock by applying the correct sequence of switch inputs in order to reach the `unlock` state. This is followed by a 40 s delay, implemented using a **repeat** loop, to allow observation of the automatic lock feature. Finally, after resetting the system, an incorrect sequence of switches is applied in order to verify the operation of the `alarm` state.

Figure 8.37a–c shows a selection of simulation waveforms obtained as a result of running the test-module simulation.

REFERENCES

1. Ciletti MD. Modeling, Synthesis and Rapid Prototyping with the Verilog HDL. New Jersey: Prentice Hall, 1999.
2. Ciletti MD. Advanced Digital Design with the Verilog HDL. New Jersey: Pearson Education, 2003 (Appendix I – Verilog-2001).
3. Wakerly JF. Digital Design: Principles and Practices, 4th edition. New Jersey: Pearson Education, 2006 (Metastability and Synchronization, Section 8.9).

9

Asynchronous Finite-State Machines

9.1 INTRODUCTION

Most FSM systems are synchronous; that is, they make use of a clock to move from one state to the next. Using a clock to control the synchronous movement between one state and the next allows the FSM logic time to settle before the next transition and, hence, overcomes some logic delay problems that may arise. For this reason, synchronous systems are, by far, the most popular in digital electronics; and most HDLs used to define them are optimized for synchronous system design.

However, there is another kind of FSM, one that does not use a clock to instigate a transition between states. This is knows as the asynchronous FSM. In an asynchronous FSM, the transition between states is controlled by the event inputs, so that the FSM does not need to wait for a clock signal input. For this reason, asynchronous FSM are sometimes called 'event-driven' FSMs.

A typical event FSM is shown in Figure 9.1. In this FSM, the transition from state s0 to s1 will take place when input *s* is logic 1 AND input *c* is logic 0. On reaching state s1, the FSM will remain in this state until the input *c* goes to logic 1, at which point it will move to state s2. Here, it will remain until input *c* goes to logic 0 to move to state s3, before returning to state s0 when input *s* goes to logic 0.

In this example, the FSM will only change state when there is a change of input variable; hence, the event nature of the system.

Sometimes, it is desired to change state when there is no input signal change (as has been seen in clocked driven systems).

In Figure 9.2, the transition between s3 and s0 does not have an input term along the transitional line. This implies that when the FSM reaches state s3 (when input *x* became logic 1) the FSM will move to s0. The time taken for the FSM to move to s0, when it reaches state s3, will be determined by the propagation delay of the event logic used in the system. This will be as fast as the logic technology used to implement the design.

An important feature with event-driven FSM systems is that when the FSM is in a stable state (perhaps waiting for an input event to move to the next state) the power drain is very low in

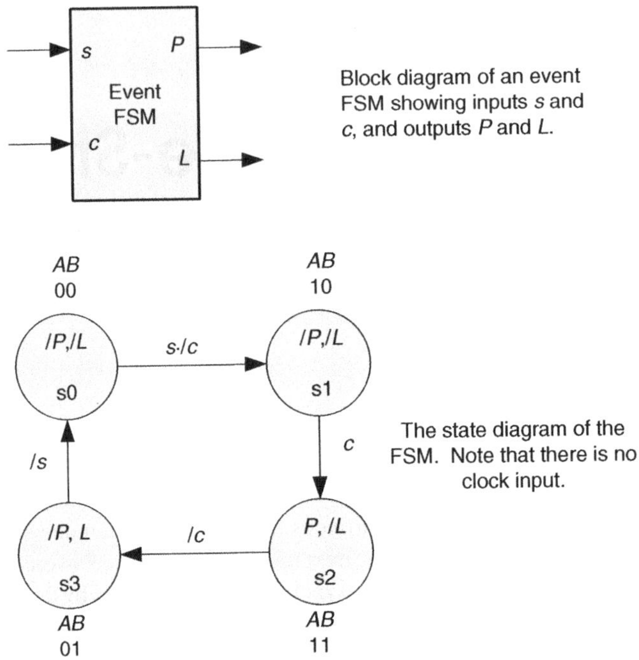

Figure 9.1 Example of an asynchronous FSM.

CMOS circuits, since there is no repetitive clock to consume power. This allows asynchronous (event) systems to be low power, while also being very fast. This latter point is due to the fact that the event FSM will move to the next state as soon as the relevant event input changes, and is only limited by the propagation delay for its event-driven logic.

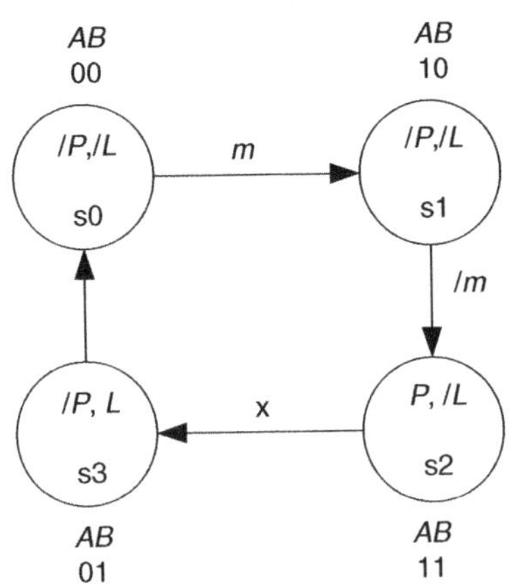

Figure 9.2 Transition without an input event.

9.2 DEVELOPMENT OF EVENT-DRIVEN LOGIC

From the previous section it is clear that an event state diagram can be developed in much the same way as a clocked driven state diagram. However, whilst with a clocked FSM, the implementation (synthesis) will make use of some type of flip-flop (D type, T type, or JK type), the event-driven system needs to make use of memory elements that do not require a clock input. This implies that perhaps SR latches are required. But in practice these latches may, in some cases, need multiple set (s) and multiple reset (r) inputs. What follows is the development of a set of equations that can be used to implement a general 'event-driven' cell for each particular application.

Consider Figure 9.3. This shows the block diagram for the proposed event cell. This cell has a 'turn-on set' input to set the cell output to logic 1, a 'turn-off set' to turn the cell output to logic 0, and a hold term input, derived from the cell output to hold the cell either in its set, or rest state.

In order to develop the logic equations for the event cell a table of required states for each input condition is required. This is shown in Table 9.1. In this table, the 'turn-on set' input is denoted as s, the 'turn-off set' is denoted as r, the current state of the cell output is Q_n, and the next state of the cell output is Q_{n+1}. The two inputs s and r, together with the current output state, are shown as a binary sequence. This defines all possible states for the cell. What is now required is to fill in the required state condition for each Q_{n+1} state.

- In row 1, $s=r=0$, and the cell is currently reset. Since our event cell is to remain in whatever state it happens to be in, when $s=r=0$, then $Q_{n+1}=Q_n=0$.
- In row 2, $s=r=0$, but the cell is currently set. Therefore, $Q_{n+1}=1$, since the cell must remain in the set state.
- In row 3, $s=0$ but $r=1$, implying a reset condition for the event cell. Since the cell in this row is currently reset, then $Q_{n+1}=0$ as well.
- In row 4, $s=0$, $r=1$ as before, but the cell is currently set, so the required action is that $Q_{n+1}=0$ to reset the cell.

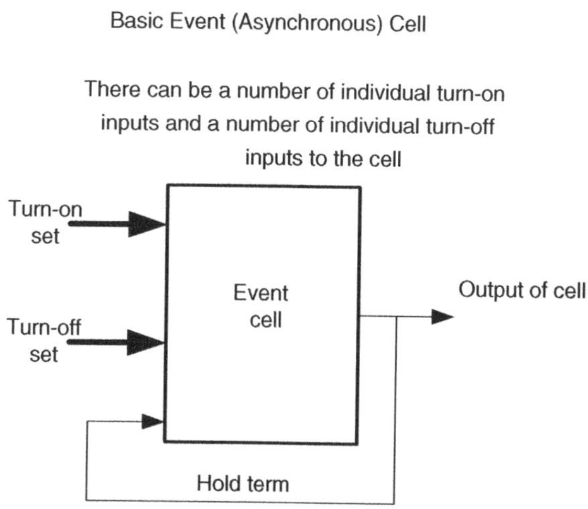

Figure 9.3 The event cell.

Table 9.1 State table for the event cell.[a]

Row	s	r	Q_n	Q_{n+1} event
1	0	0	0	No change
2	0	0	1	No change
3	0	1	0	Reinforced turn off
4	0	1	1	Turn off
5	1	0	0	Turn on
6	1	0	1	Reinforced turn on
7	1	1	0	Not allowed
8	1	1	1	Not allowed

[a] Q_n is the present state; Q_{n+1} is the next state of Q. Each row corresponds to a possible particular condition or state of the event cell. This can be used to determine the 'characteristic equation' for the event cell.

- Moving to row 5, $s = 1$ and $r = 0$. The cell is currently reset; thus, $Q_{n+1} = 1$ to set the cell.
- In row 6, $s = 1$ and $r = 0$ as before, and the cell is currently set; therefore, $Q_{n+1} = 1$ to hold the cell in its set state.
- In rows 7 and 8, both $s = 1$ and $r = 1$. This is not a very practical condition for the cell, since it implies that the cell inputs are ambiguous (i.e. set the cell and reset the cell at the same time!). Clearly, this is impossible. Here, our own common sense will prevail, and both rows 7 and 8 are defined as 'not allowed' states. What is meant here is that it is rather hoped that the input conditions defined by rows 7 and 8 'won't' happen. This is usually referred to as 'don't care' states. It is important that the 'don't care' does not happen, and this will be assumed in the design of asynchronous systems that use the corresponding equations being developed here. The input conditions $s = 1$ and $r = 1$ will not be allowed to occur. This is not too difficult to ensure, so one marks out the row 7 and 8 Q_{n+1} outputs with x.

Table 9.2 illustrates the completed table.

Now, an equation for Q_{n+1} can be developed from this table in terms of s, r, and Q_n:

$$
\begin{aligned}
Q_{n+1} &= /s/rQ_n + s/r/Q_n + s/rQ_n + srQ_n + sr/Q_n \\
&= /s/rQ_n + s/r + sr \\
&= /s/rQ_n + s.
\end{aligned}
$$

Table 9.2 Completed state table for the event cell.

s	r	Q_n	Q_{n+1}	
0	0	0	0	No change
0	0	1	1	No change
0	1	0	0	Reinforced reset
0	1	1	0	Turn off (reset)
1	0	0	1	Turn on (set)
1	0	1	1	Reinforced set
1	1	0	x	Don't care
1	1	1	x	Don't care

Applying the auxiliary rule and rearranging results in the following *sequential equation*:

$$Q_{n+1} = s + Q_n/r.$$

The sequential equation produced here represents the 'characteristic equation' for the event cell. Notice that in line 3 the 'don't care' terms $s/r + sr$ have been reduced to the term s.

The sequential equation can be stated as:

> The new output state for the event cell is equal to the condition of the set input s *or* the current state of the cell Q_n *and* the inverse of the reset input r.

This can be easily proved, as shown below, by defining initial states for s, r, and Q_n using the sequential equation to predict the new output Q_{n+1}. Note that, in these equations, the r term is $/r$, so $r = 0$ means $/r = 1$, and $r = 1$ means $/r = 0$.

Let $s = 1,\ r = 0,\ Q_n = 0$. Then $Q_{n+1} = 1 + 0 \cdot 1$
$= 1$; i.e. cell sets (output changes from 0 to 1).
Let $s = 0,\ r = 0,\ Q_n = 1$. Then $Q_{n+1} = 0 + 1 \cdot 1 = 1$;
cell remains set (output remains at logic 1).
Let $s = 0,\ r = 1,\ Q_n = 1$. Then $Q_{n+1} = 0 + 1 \cdot 0 = 0$;
cell is reset (output changes from 1 to 0).
Let $s = 0,\ r = 0,\ Q_n = 0$. Then $Q_{n+1} = 0 + 0 \cdot 1 = 0$;
cell remains reset (output remains at logic 0).

As it stands, the sequential equation is rather limited because it caters for only a single input s term and a single input r term. In real event-driven systems there may be a requirement for multiple set and multiple reset terms so that the cell can be set and reset under different conditions. But these will be OR terms, since the state diagram is sequential and can only deal with one set and one reset condition at a time. So the sequential equation can be modified by introducing the possibility of multiple set inputs as:

Sum of set inputs $\sum s = s_1 + s_2 + \cdots + s_n$, where s_n is the final set input term.

Sum of reset inputs $\sum r = r_1 + r_2 + \cdots + r_n$, where r_n is the final rest input term.

Thus, the sequential equation becomes:

$$Q_{n+1} = \sum s_Q + Q_n \cdot \sum /r_Q. \tag{9.1}$$

This is the final form of the sequential equation used to define the event cell. It is referred to as the NAND sequential equation [1].

Note that there is a corresponding equation called the NOR sequential equation that is defined as

$$Q_{n+1} = \left(\sum s_Q + Q_n\right) \cdot \sum /r_Q. \tag{9.2}$$

Equations 9.1 and 9.2 are reproduced from Page 19 in Chapter 1 'Basic Concepts in Logic Design', from 'Problems and Solutions in Logic Design' by Zissos, D. by permission of Oxford University Press.

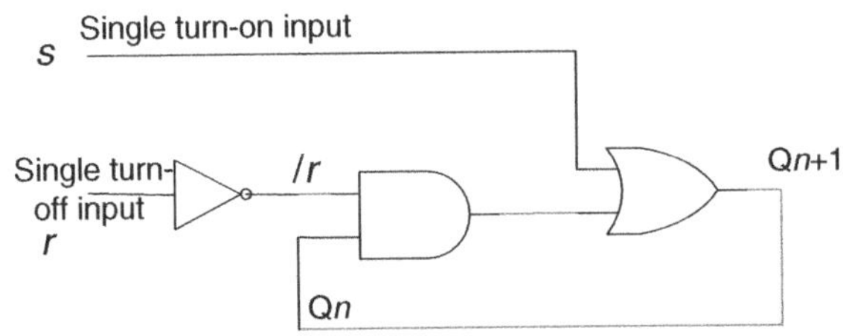

Figure 9.4 Basic event cell.

But it is not used much these days. You may like to try to see how this NOR sequential equation is obtained from the sequential equation. (*Hint*: AND $(\sum r_Q + \sum /r_Q)$ with the $\sum s_Q$ and expand.)

Equation (9.1) can be described for an event cell A as

$$A = \sum(\text{turn-on sets of } A) \cdot A + \sum /(\text{turn-off sets of } A)$$

and Equation (9.2) as

$$A = (\sum(\text{turn-on set of } A) + A) \cdot \sum /(\text{turn-off sets of } A).$$

Both these equations were used in the book *Problems and Solutions in Logic Design* by D. Zissos [1] (chapter 1: 'Basic concepts in logic design') and are repeated here by permission of Oxford University Press.

The next stage is to show how the sequential equation can be used to synthesize an event FSM. This will be followed by an example of how to design an event-driven FSM from a specification.

Returning to the sequential equation, Equation (9.1), a circuit can be produced. This is shown in Figure 9.4.

$$Q_{n+1} = s + Q_n \cdot /r.$$

This is the equation defining the circuit of Figure 9.4. This can be converted into NAND form by applying De Morgan's rule to obtain:

$$Q_{n+1} = /(/s \cdot /(Q_n \cdot /r)). \tag{9.3}$$

This is where the 'NAND' sequential equation name comes from. The event cell circuit is shown in Figure 9.5.

Either type can be used in practice, although with PLD and FPGA devices the AND/OR arrangement fits best.

9.3 USING THE SEQUENTIAL EQUATION TO SYNTHESIZE AN EVENT FINITE-STATE MACHINE

The event state diagram shown in Figure 9.6 will be used to synthesize an event system. The design process for event state diagrams will be dealt with later. The system is essentially able to determine a 0 to 1 transition on the c input.

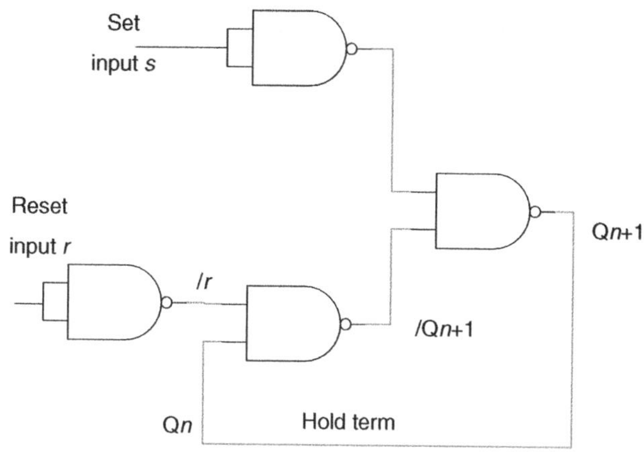

Figure 9.5 Event cell circuit.

In this system there are three inputs: st, *c*, and sp. There is a single output *P* that is logic 1 in state s3. Note that there are two event cells in this state diagram: event cell *A* and event cell *B*. These form the secondary state variables.

When the operator asserts input st, the system moves from state s0 to s1, where it waits for the input *c* (the incoming pulse) to become logic 0 (if *c* is logic 0 in state s0, then the FSM will simply move through s1 to s2). When the FSM reaches s2 the system waits for *c* going high. In this way,

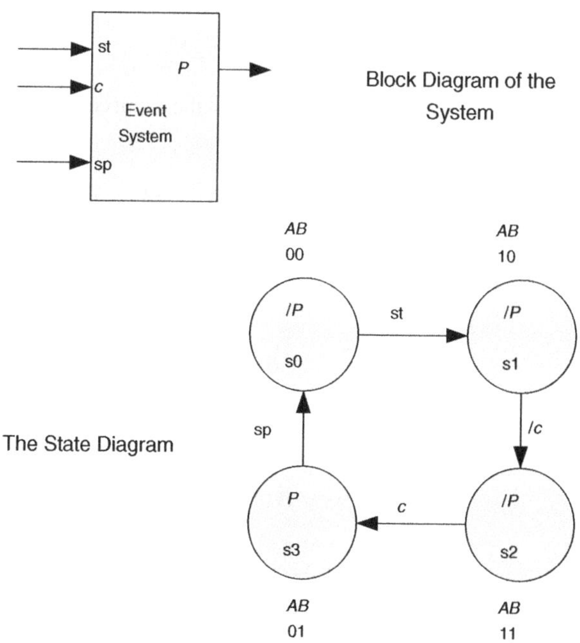

Figure 9.6 The basic event-driven system.

the event-driven system is able to catch the positive-going transition on input c. The P output will remain high until the sp input is asserted. In this way, the P output acts as a memory of the transition event on c.

When the operator asserts input sp, the FSM will move back to state s0.

This system can be left unattended, since it will indicate the c 0 to 1 transition, and asserting sp will allow the operator to return the system to its initial state again. Note that the system can be reset to s0 via its reset input as well (not shown).

First, the turn-on set of conditions to set the A event cell must be determined.

The $\sum s_A$ is found by looking for the state where A_n goes from 0 to 1. This is state s0, for $A_n = 0$ in state s0, and $A_n = 1$ in state s1. There is an input along the transitional line between s0 and s1, so this input st is included in the turn on set for A_n. Therefore:

$$\sum s_A = \mathrm{s0} \cdot \mathrm{st} + \mathrm{s1} \cdot \mathrm{st}. \tag{9.4}$$

The reason why s1 · st is needed is because the input st must still be logic 1 (active) when the FSM reaches state s1 to ensure that it will remain in this state.

Now:

$$\sum s_A = /A_n \cdot /B_n \cdot \mathrm{st} + A_n \cdot /B_n \cdot \mathrm{st} = (/A_n \cdot /B_n + A_n \cdot /B_n) \cdot \mathrm{st} = /B_n \cdot \mathrm{st}$$

due to the application of the logical adjacency rule. *Note*: this has effectively led to the removal of the A_n term in the equation for the turn-on set for event cell A_n.

Now, looking for the turn-off condition, this occurs in state s2 when A_n is changing from 1 to 0. Therefore:

$$\sum r_A = \mathrm{s2} \cdot c + \mathrm{s3} \cdot c, \tag{9.5}$$

since the c input must be held true in state s3 to ensure that the event cell hold reset. In terms of the state variables:

$$\sum r_A = A_n B_n \cdot c + /A_n \cdot B_n \cdot c = (A_n \cdot B_n + /A_n \cdot B_n) \cdot c = B_n \cdot c.$$

The A_n term is removed by the logical adjacency rule to leave the B_n and c terms. This results in the turn-off term

$$\sum r_A = B_n \cdot c.$$

The complete sequential equation can now be written thus:

$$A_{n+1} = \sum s_A + A_n \cdot \sum /r_A$$
$$A_{n+1} = /B_n \cdot \mathrm{st} + A_n \cdot /(B_n \cdot c).$$

This represents the required behaviour for the event cell A. It is the sequential equation for the event cell A originally called the NAND sequential equation by Professor D. Zissos in his book *Problems and Solutions in Logic Design* [1].

The sequential equation for the event cell B can be obtained in the same way:

$$B_{n+1} = \sum s_B + B_n \cdot \sum /r_B.$$

The turn-on set for B is

$$\sum s_B = \text{s}1 \cdot /c + \text{s}2 \cdot /c = A_n/B_n \cdot /c + A_n \cdot B_n \cdot /c = A_n \cdot /c. \tag{9.6}$$

Note here that the application of the logical adjacency rule has removed the B_n term in the same way that the A_n term in the turn on set equation for A was dropped.

The turn-off set for B is

$$\sum /r_B = /(\text{s}3 \cdot \text{sp} + \text{s}0 \cdot \text{sp}) = /((/A_n \cdot B_n \cdot \text{sp} + /A_n \cdot /B_n \cdot \text{sp}) = /((/A_n \cdot \text{sp}). \tag{9.7}$$

Likewise, the B_n term is dropped using the logical adjacency rule. So now the logic to specify the behaviour of the event cells is complete.

The complete sequential equation for cell B is thus

$$B_{n+1} = \sum s_B + B_n \cdot \sum /r_{B.} = A_n \cdot /c + B_n \cdot /((/A_n \cdot \text{sp}).$$

The two sequential equations

$$A_{n+1} = \sum s_A + A_n \cdot \sum /r_A = /B_n \cdot \text{st} + A_n \cdot /(B_n \cdot c)$$
$$B_{n+1} = \sum s_B + B_n \cdot \sum /r_{B.} = A_n \cdot /c + B_n \cdot /((/A_n \cdot \text{sp})$$

represent the behavioural logic for the two event cells.

The final equation is the output equation for the signal P. This, like clock-driven systems, is based on the state, in this case state s3:

$$P = \text{s}3 = /A_n \cdot B_n.$$

Note that in the $\sum s_A$ set the $/A_n$ state variable has disappeared and in the $\sum /r_A$ set the A_n terms have disappeared.

Likewise, the $/B_n$ and B_n terms have disappeared from the respective $\sum s_B$ and $\sum /r_B$ sets.

9.3.1 Short-cut Rule

This is always going to be the case since the logical adjacency rule will always be applied to the state variable for the cell.

Thus, it is possible to apply a short-cut where in the event cell X the turn-on set $\sum s_x$ will have the $/x$ term removed, and in the turn-off set $\sum /r_x$ the x term will be removed as a result of applying the logical adjacency rule in Equations (9.4)–(9.7).

This allows the equations to be written thus:

$$\sum s_A = \mathrm{s0} \cdot \mathrm{st} = /A_n \cdot /B_n \cdot \mathrm{st} = /B_n \cdot \mathrm{st};$$

i.e. drop the /A state variable in the 0 to 1 term. This means you do not need to write down the second term s1 · st in Equation (9.4).

$$\sum r_A = \mathrm{s2} \cdot c = A_n B_n \cdot c = B_n \cdot c;$$

i.e. drop the A state variable in the 1 to 0 term. This means you do not need to write down the second term s3 · c in Equation (9.5).

$$\sum s_B = \mathrm{s1} \cdot /c = A_n/B_n \cdot /c = A_n \cdot /c;$$

i.e. drop the /B state variable in the 0 to 1 term. The s2 · c term is not required in Equation (9.6).

$$\sum /r_B = /\mathrm{s3} \cdot \mathrm{sp} = /((/A_n.B_n \cdot \mathrm{sp}) = /((/A_n \cdot \mathrm{sp});$$

i.e. drop the B state variable in the 1 to 0 term and you do not need to write down the term s0 · sp in Equation (9.7).

This provides a rapid way to obtain the sequential equations direct from the state diagram. The easiest way to remember this rule is to simply 'drop' the state variable term in the equation for that state variable. Therefore, in the equation for A, drop the /A state variable in the $\sum s_A$ 0 to 1 transition term. In the equation for B, drop the A state variable in the $\sum r_A$ 1 to 0 transition term. From now on, the short-cut rule will be applied.

Having established the equations, they can now be implemented using a PLD or FPGA.

9.4 IMPLEMENTING THE DESIGN USING SUM OF PRODUCT AS USED IN A PROGRAMMABLE LOGIC DEVICE

To do this the NAND part of the equations might want to be replaced to turn them into sum of product terms:

$$A_{n+1} = \sum s_A + A_n \cdot \sum /r_A = /B_n \cdot \mathrm{st} + A_n/(B_n \cdot c)$$
$$B_{n+1} = \sum s_B + B_n \cdot \sum /r_B = A_n \cdot /c + B_n \cdot /((/A_n \cdot \mathrm{sp}).$$

In the equation for A_{n+1}, for example, the term $/(B_n \cdot c)$ can be converted using De Morgan's rule. The De Morgan rule used here is

$$/(X \cdot Y) == /X + /Y$$

to produce

$$/(B_n \cdot c) == /B_n + /c.$$

This results in

$$A_{n+1} = \sum s_A + A_n \cdot \sum /r_A = /B_n \cdot \mathrm{st} + A_n \cdot (/B_n + /c).$$
$$= /B_n \cdot \mathrm{st} + A_n \cdot /B_n + A_n \cdot /c.$$

And for the term $/(/A_n \cdot \mathrm{sp})$, using De Morgan's rule results in $A_n + /\mathrm{sp}$. The final equation is

$$B_{n+1} = \sum s_B + B_n \cdot \sum /r_B = A_n \cdot /c + B_n \cdot /(/A_n \cdot \mathrm{sp}) = A_n \cdot /c + B_n \cdot (A_n + /\mathrm{sp})$$

and

$$B_{n+1} = A_n \cdot /c + A_n B_n + /\mathrm{sp} \cdot B_n.$$

Using these two sequential equations, the final event cell circuits can be synthesized.

9.4.1 Dropping the Present State n and Next State n + 1 Notation

Up to now the sequential equations used have been of the form:

$$A_{n+1} = \sum s_A + A_n \cdot \sum /r_A,$$

where A_{n+1} is the next state of the event cell. However, it could be written as

$$A = \sum s_A + A \cdot \sum /r_A,$$

where A on the left is taken to be the next state and A on the right the present state of the event cell. This is, in effect, a recursive equation.

This notation will be used from now on. This can be clearly seen in Figure 9.7, where the outputs A and B are fed back to inputs. Figure 9.7 illustrates the final circuit for the system. This could be synthesized using a PLD device such as the 22V10.

9.5 DEVELOPMENT OF AN EVENT VERSION OF THE SINGLE-PULSE GENERATOR WITH MEMORY FINITE-STATE MACHINE

The clock-driven single-pulse generator circuit that was developed in Chapter 1 when dealing with synchronous (clock-driven) systems will now be revisited. This time it will be developed as an event-driven system.

In the clocked version, use was made of a system clock to control the timing of the single pulse produced when the input p was asserted. However, in an event version, there is no system clock, so an input (named the c input) will be used for that purpose (it can also be used to set the pulse duration). The event-driven system will make use of this input as an event input that happens to be changing state at a regular interval, but it will be seen by the event system as 'an event' input.

Figure 9.8 illustrates the final system. Looking at the state diagram, it can be seen that the system starts when input s is asserted, but the FSM will not move from state s0 until both s is logic

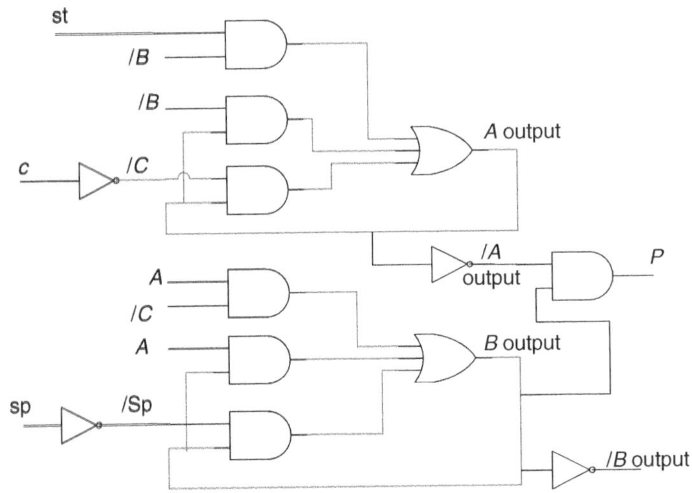

Event cells and output for the system

Figure 9.7 Final circuit for the system.

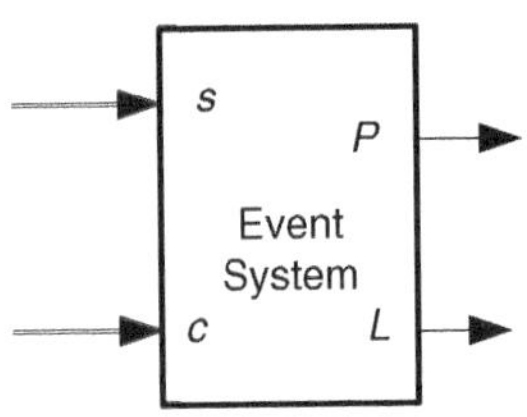

$A = \Sigma s_A + A \cdot /\Sigma r_A$

$B = \Sigma s_B + B \cdot /\Sigma r_B$

Equations:

$A = /B \cdot s \cdot /c + A \cdot /(B \cdot /c)$
$\quad = /B \cdot s \cdot /c + A \cdot /B + A \cdot c$

$B = A \cdot c + B \cdot /(/A \cdot /s)$
$\quad = A \cdot c + A \cdot B + B \cdot s$

$P = s2 = A \cdot B$

$L = s2 + s3 = B$

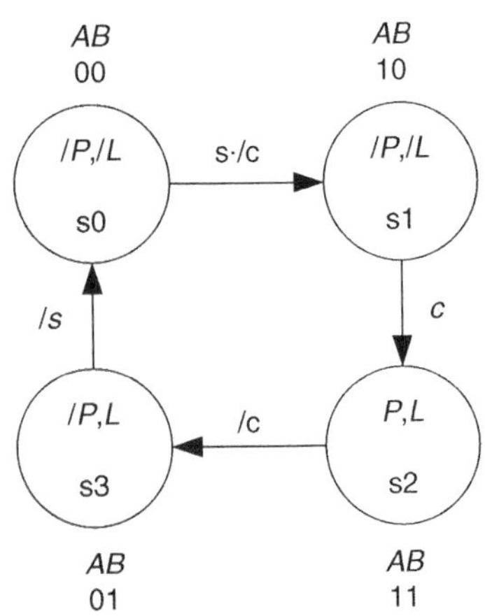

Figure 9.8 Event-driven single-pulse system with memory showing block diagram, state diagram and equations.

1 *and* the c input is at logic 0. The reason for this is that the transition of c from 0 to 1 is to be used to assert the outputs P and L to logic 1 (beginning of the output pulse).

Therefore, when the FSM moves from state s0 to s1 it waits in s1 for the c input to go to logic 1, then moves to state s2 where P and L are made logic 1. This will happen on the 0 to 1 transition of the c input. The FSM will remain in state s2 until the c input again drops to logic 0, and the FSM will move to state s3 where the output P will resume its logic 0 state while the output L remains at logic 1.

At this point, the FSM will remain in state s3 until the input s reverts back to logic 0, ready for the next single-pulse generation. This will also cancel the output L. In this design, the output L is being used as a pulse indicator, since the pulse duration is dependant upon the width of the c pulse and may not be seen by the user.

In this system, the actual width of the P output pulse can be controlled by the logic 1 period of the c input.

Turning to the equations, the two-event cell equations can be obtained in the same way as in the previous example, by first obtaining the turn-on set and then the turn-off set for each equation, then inserting them into the sequential equations. However, a little thought shows that each sequential equation can be written down directly using the short-cut method, more or less as has been done in Figure 9.8:

$$A = \sum s_A + A \cdot \sum /r_A = \text{s0} \cdot s/c + A \cdot /(\text{s2} \cdot /c) = /B \cdot s \cdot /c + A \cdot /(B \cdot /c)$$
$$B = \sum s_B + B \cdot \sum /r_{B.} = \text{s1} \cdot c + B \cdot /(\text{s3} \cdot /s) = A \cdot c + B \cdot /(/A \cdot /s)$$

with outputs

$$P = \text{s2} = A \cdot B$$

and

$$L = \text{s2} + \text{s3} = A \cdot B + /A \cdot B = B.$$

The two-event cell equations can now be converted so that they can be implemented with sum of product logic (typically found in PLD devices):

$$A = /Bs/c + A \cdot /B + A \cdot c$$
$$B = Ac + B \cdot A + B \cdot s$$
$$P = A \cdot B$$
$$L = B.$$

The circuit is illustrated in Figure 9.9. Notice the Reset line (thick line) to initialize the event cells to zero. This is essential in order to ensure that the system is reset to state s0. In operation, this Reset line will be at logic 1. During reset it will be at logic 0, thus clearing both event cells to zero. Clearly, the reset line is ANDed with the turn-on/turn-off logic of the event cells:

$$A = (/Bs/c + A/B + Ac) \cdot \text{Reset}$$
$$B = (Ac + AB + B \cdot s) \cdot \text{Reset.}$$

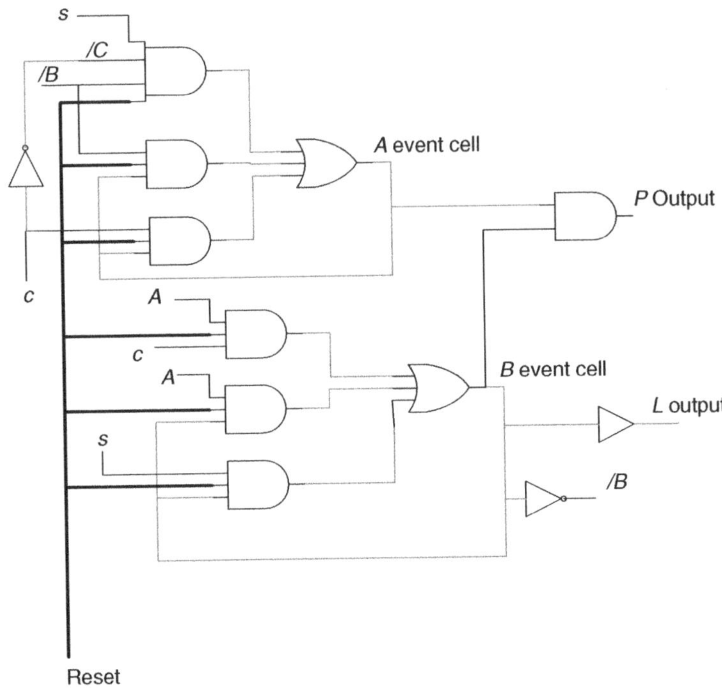

Figure 9.9 Circuit for the event-driven FSM system.

For clarity, the Reset input line will be left out of the event cell equations, but remember to add it in when implementing each design, otherwise the circuit will not simulate, since it will not be able to initialize. At this stage the reader might like to revisit Figure 9.7 and add a reset connection to allow this circuit to reset to state s0.

9.6 ANOTHER EVENT FINITE-STATE MACHINE DESIGN FROM SPECIFICATION THROUGH TO SIMULATION

In this next example, an event FSM will be developed from its written specification through to a Verilog HDL description of the FSM (as described in Chapter 6). This is then simulated using the Syncad™ simulator system.

The idea here is to illustrate how a complete design can be implemented. Later, the Verilog file could be used to program a PLD device and, hence, realize the design in physical hardware.

9.6.1 Important Note!

Since the Verilog behavioural level is not optimized for an event-driven system, as yet, the Verilog description is at the Boolean equation level. This is fine for our purposes, since it will provide a one-to-one correspondence with the system equations. It is also possible to implement the event-driven system using the gate level direct. The Boolean equation level, however, is useful for quick simulation and verification. On the other hand, simulating in terms of the logic gates allows the designer to experiment with different gate delay values to ensure that the circuits will not maloperate due to violation of the 33.3% gate tolerance rule (see Section 9.12.3 and Reference [1] for details).

9.6.2 A Motor Controller with Fault Current Monitoring

This is an event-based FSM used to control a motor. An external device (possibly based on a Hall-effect transducer) is used to monitor the motor current. This will be set so that normal start current is allowed, but if the motor current exceeds some defined limit a fault signal will be sent to the FSM to switch off the motor and light up a fault LED indicator. The details of the Hall-effect fault circuitry and the power circuit to switch the motor on and off are excluded from the diagram of Figure 9.10a.

Figure 9.10b shows the state diagram for the FSM controller. The motor can be switched on by asserting input st, and off by disasserting input st. If a fault is encountered by the Fault Detection Unit its output signal ms will go high thus causing the FSM to move into state s2 where the motor will be switched off and the Fault indicator L turned on (an active-low signal). The system will remain in s2 until the start input st is disasserted to move the FSM into state s3 turning off the Fault indicator LED L. The FSM can return to its initial state s0 on reaching state s3 if input *t* is logic 0.

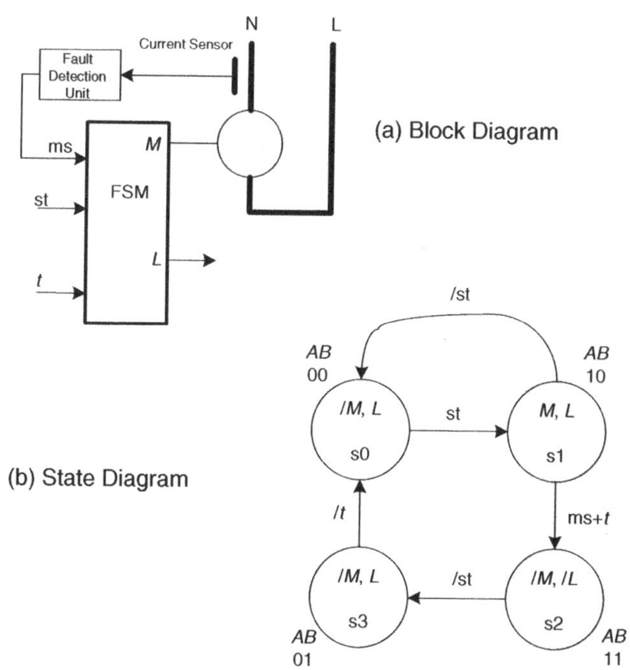

Figure 9.10 The block diagram and state diagram for the motor controller.

The system can be tested in the absence of a fault by pressing the test input t. Note that $t = 1$ will hold the FSM in state s3. The equations for the event cells can now be developed:

$$\begin{aligned} A &= \sum s_A + A \cdot \sum /r_A \\ &= \mathrm{s0} \cdot \mathrm{st} + A \cdot /(\mathrm{s1} \cdot /\mathrm{st} + \mathrm{s2} \cdot /\mathrm{st}) \\ &= /B \cdot \mathrm{st} + A \cdot /(/B \cdot /\mathrm{st} + B \cdot /\mathrm{st}) \\ &= /B \cdot \mathrm{st} + A \cdot //\mathrm{st} \\ &= /B \cdot \mathrm{st} + A \cdot \mathrm{st} \end{aligned}$$

$$\begin{aligned} B &= \sum s_B + B \cdot \sum /r_B \\ &= \mathrm{s1} \cdot (\mathrm{ms} + t) + B \cdot /(\mathrm{s3} \cdot /t) \\ &= A \cdot \mathrm{ms} + A \cdot t + B \cdot /(/A \cdot /t) \\ &= A \cdot \mathrm{ms} + A \cdot t + A \cdot B + B \cdot t \end{aligned}$$

$$M = \mathrm{s1} = A \cdot /B \quad \text{and} \quad L = /\mathrm{s2} = /(A \cdot B) = /A + /B.$$

The schematic diagram of the design is illustrated in Figure 9.11. This has the test input included so that the system can be tested in the absence of a fault input.

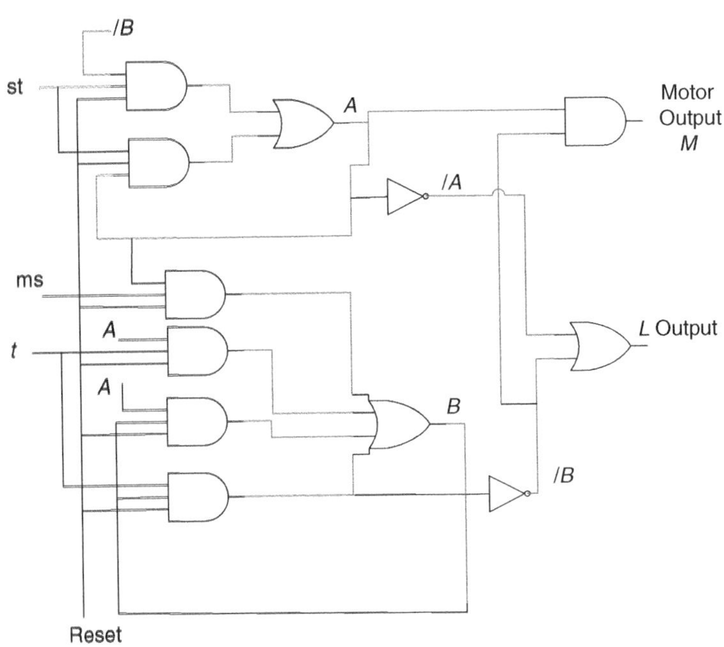

Figure 9.11 Schematic circuit diagram for the FSM controller.

Although it is not necessary to draw a circuit diagram, it is useful to see the circuit of the FSM. Note that the essential interface buffering between the low-voltage FSM circuit and the high-voltage motor circuit is not shown.

This design can be developed as a Verilog module and this is illustrated in Listing 9.1.

```
/////////////////////////////////////////
module fsm(rst,st,ms,t,M,L,A,B);
  output M,L,A,B;
  input rst,st,sp,ms,t;

assign
  A= (~B&st | A&st)&rst,
  B= (A&ms | A&t | A&B | B.t)&rst,
  M=A&~B,
  L=~A | ~B;
endmodule
/////////////////////////////////////////
```

Listing 9.1 FSM module.

Note that the module inputs and outputs are defined outside of the parentheses, as was usual in older style Verilog modules. This is still supported in later versions of the Verolog compiler tools. Chapter 6 shows the more recent way to define the inputs and outputs.

In the Verilog file, the event equations have been implemented using an assign with blocking statements. The equations also cater for the test *t* input to test the system in the absence of a fault.

The Verilog code in Listing 9.2 is a test bench that is used to test the design. A test bench provides an instance of the FSM, along with a set of test signals to be used in the simulation in order to verify the design.

```
module test;
reg st,ms,t,rst;

fsm uut(rst,st,ms,t,M,L,A,B);

initial
   begin
   $dumpfile("motflt.vcd"); // to get a printout of waveforms.
   $dumpvars;
   rst=0;
   st=0;

   ms=0;
   t=0;
// Note it is important to ensure signals change in
// proper sequence. Also to ensure ms and sp are
```

```
// mutually exclusive.
//----------------- remove reset
   #20 rst=1;
//----------------- move to s1
   #20 st=1;
//----------------- stay in s1
//----------------- move to s0
   #20 st=0;
//----------------- move to s1 again
   #20 st=1;
//----------------- move to s2
   #20 ms=1;
//-----------------
   #30 ms=0;
//----------------- move to s3
   #20 st=0;
//----------------- move back to s0
   #20 st=0;
//----------------- move to s1
   #20 st=1;
//----------------- stay in s1
//----------------- move to s2
   #20 t=1;
//----------------- move to s3
   #20 st=0;
//----------------- move to s0
//----------------- end of tests.

$stop(60); // stop the simulation.
   end
endmodule
```

Listing 9.2 Test-bench module.

The FSM module is very simple and, apart from the input and output defines, consists of only an assign block. The event cells are defined individually within this block, together with the output equations.

The test bench module is also seen to be quite simple. One point to note is that the signals must change one at a time, and with a time delay. This is mandatory, since the event cells can respond to potential static 1 or 0 hazards (glitches). This will be a necessary requirement with all event-driven designs.

Finally, Figure 9.12 illustrates the timing waveforms from the simulation.

Comparing this with the test bench module sequence, it can be seen that the state diagram has been traversed twice: once with a fault signal and next with a test signal.

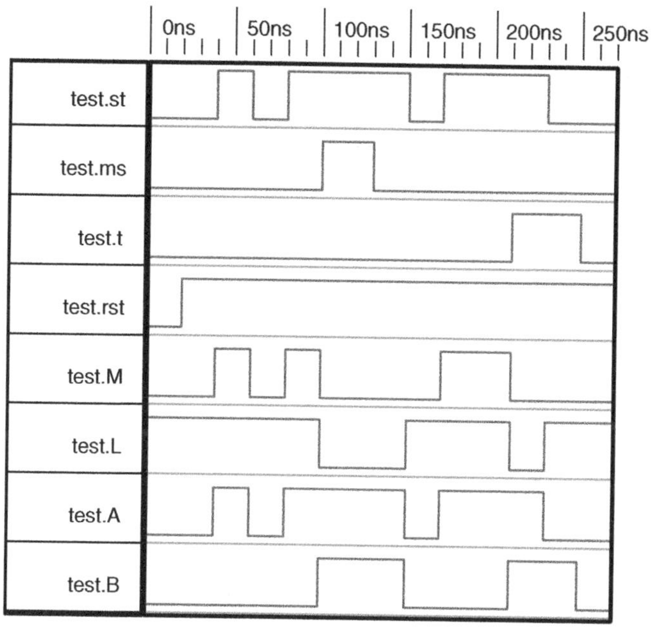

Figure 9.12 Verilog simulation of the design.

Note that, in the case of a fault, the transition from s2 to s3 to s0 is very fast and the s3 state is not apparent in the simulation. In the case of the test, the FSM stops in state s3 until the *t* input is returned to its low state. In this way, the operator can test the complete state sequence (particularly if the state variables are available as LED indicator outputs).

9.7 THE HOVER MOWER FINITE-STATE MACHINE

A hover-type lawnmower usually uses a mechanical interlock to prevent the motor from starting unless the user presses a button before operating the on/off lever. By replacing the mechanical mechanism with an electronic equivalent, the safety mechanism can be made easier to manufacture.

9.7.1 The Specification and a Possible Solution

A hover lawnmower has a safety button sf that must be pressed before operating the start lever st. When the safety button is pressed, an LED indicator *P* is lit; when the start lever st is operated after this, the motor will turn on. The motor can be stopped by releasing the start lever. The safety button sf must be pressed before the motor can be restarted with the start lever.

A block diagram with a suitable state diagram for the system are illustrated in Figure 9.13a and b. The specification is a typical one that might be given as a specification for a product. Looking at the original state diagram 1 in Figure 9.13b (with four states), it can be seen that a number of safety features have been added. This was done during the development of the state diagram as the true nature of the control sequence was revealed.

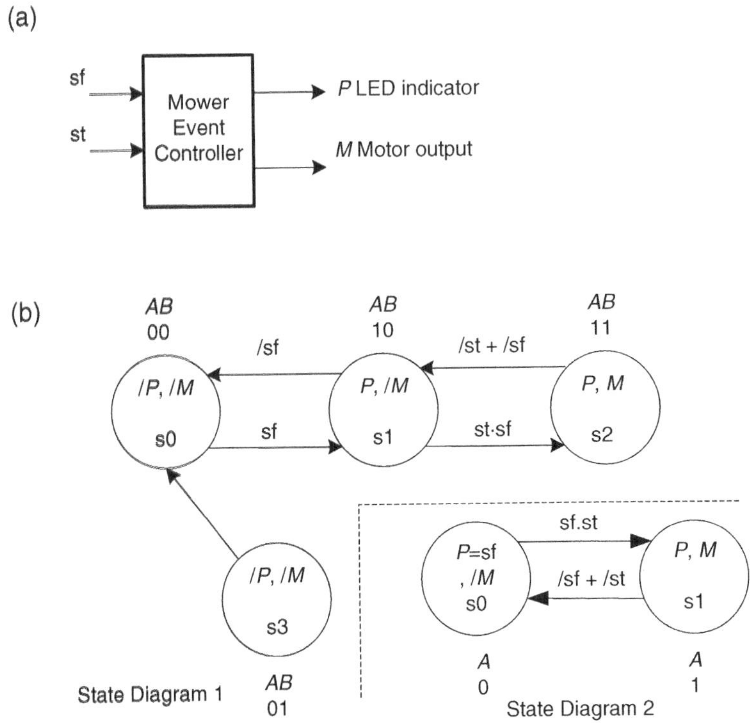

Figure 9.13 (a) Block diagram of the mower FSM controller. (b) Two possible state-diagram solutions.

The state diagram controls the sequence of the controller by ensuring that only if the safety button is pressed before the start lever is operated will the motor operate. The *P* LED will remain on in s1 if the start button is released. If the safety button is released in either states s1 or s2 the FSM will move back to state s0.

Note that the FSM will return to s1 when the start lever st is released or the safety button is released. This ensures that the operator's hands are on both the safety button and the start lever in order to start the motor. The operator must see the LED *P* turn on before the start lever can be used to turn on the motor. Finally, note that the unused state s3 has been returned to s0. This ensures that the system will fall into s0 should a glitch cause it to get into this unused state.

Returning to Figure 9.13, state diagram 2 (with only two states) is an alternative solution, where combinational logic is used on the inputs (along the transitional lines between s0 and s1). The logic equations can be deduced in the usual way:

$$\begin{aligned} A &= \sum s_A + A \cdot / \sum r_A \\ &= \mathrm{s0} \cdot \mathrm{st} \cdot \mathrm{sf} + A \cdot /(\mathrm{s1} \cdot (/\mathrm{st} + /\mathrm{sf})) \\ &= \mathrm{st} \cdot \mathrm{sf} + A \cdot /(/\mathrm{st} + /\mathrm{sf}) \\ &= \mathrm{st} \cdot \mathrm{sf} + A \cdot //(\mathrm{st} \cdot \mathrm{sf}) \\ &= \mathrm{st} \cdot \mathrm{sf} + A \cdot \mathrm{st} \cdot \mathrm{sf} \\ P &= \mathrm{s0} \cdot \mathrm{sf} = \mathrm{sf}. \end{aligned}$$

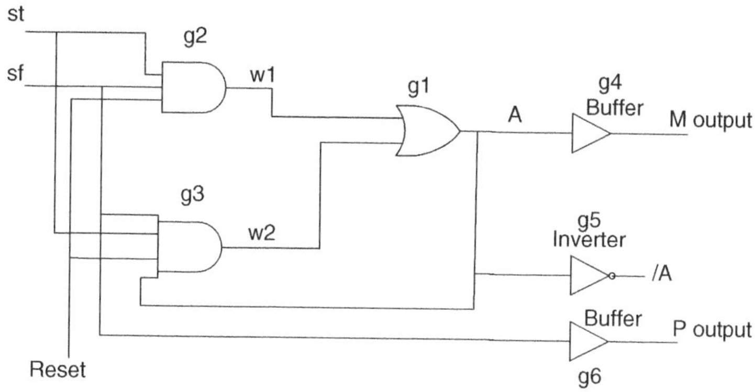

Figure 9.14 Schematic circuit diagram of the mower FSM.

This latter equation can be seen by noting that the *P* indicator can be on in s0 (Mealy output) and also in s1 as a result of getting into s1 via inputs sf · st.

$$M = \mathrm{s1} = A.$$

This leads to more simplified logic requiring only three logic gates: two AND gates and one OR gate. Buffers would, of course, be required for outputs *P* and *M*.

Figure 9.14 illustrates the circuit for the mower FSM of state diagram 2 in Figure 9.13. Additional buffers have been added to provide appropriate power levels. In particular, the motor output *M* would need to be connected to a relay (static or electromechanical) to isolate the FSM from the mains electrical supply.

The FSM of state diagram 2 is implemented in Verilog using a gate-level module. This allows individual gates to be given propagation delay values. This is shown in Listing 9.3.

```
module mowerfsm(st,sf,P,M,A,rst);

   input st,sf,rst;
   output P,M,A;
   wire na,nb,w1,w2;


   or #5 g1(A,w1,w2);
   and #5 g2(w1,sf,st,rst);
   and #5 g3(w2,A,st,sf,rst);
   //---------------
   buf #5 g4(M,A);
   //---------------
   not #5 g5(na,A);
   buf #5 g6(P,sf);
   //---------------

endmodule
```

Listing 9.3 Mower FSM module.

The test bench module is illustrated in Listing 9.4.

```
module test;
  reg rst, st, sf;
  mowerfsm uut(st,sf,P,M,A,rst);

  initial
    begin
      $dumpfile("mower.vcd");
      $dumpvars;
      rst=0;
      st=0;
      sf=0;
      #20 rst=1;
      #20 sf=1;
      #20
      #20 st=1;
      #20
      #20 st=0;
      #20
      #20 st=1;
      #20
      #20 sf=0;
      #20
      #20 st=1;
      #20
      #20 st=0;
      #10 $finish;
 end
 endmodule
```

Listing 9.4 Test-bench module.

Figure 9.15 illustrates the simulation of state diagram 2 in Figure 9.13. This follows the test-bench sequence of Listing 9.4.

The simulation starts by activating the sf input. The *P* indicator turns on. This is followed by the st input going high, which starts the mower motor. The start input is released and the motor stops. It can be started again with the start input because the sf input is still activated. The sf input is then deactivated (with the start input st still asserted) and the motor turns off.

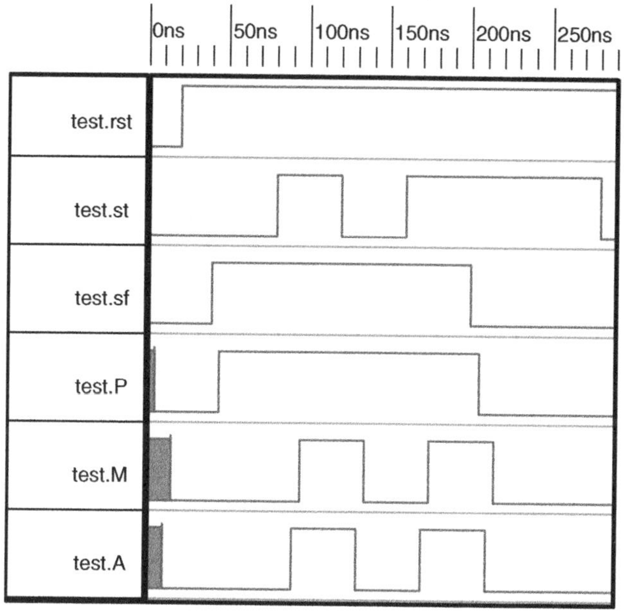

Figure 9.15 Mower FSM simulation.

Returning to the problem again, and after a little thought, the control of the mower can be reduced to a combinational one requiring

$$M = \mathrm{sf} \cdot \mathrm{st}$$
$$P = \mathrm{sf}.$$

This final solution is now obvious when seen, and perhaps you saw this at the beginning of this example. This is effectively back to a mechanical switch design!

The original solution based on state diagram 1 in Figure 9.13 is correct but requires three states and two event cells. The second attempt provides an equally working solution with fewer states using state diagram 2 in Figure 9.13. Finally, the combinational solution provides the simplest solution. It pays to look at the problems carefully to see whether they can be simplified. The sequential nature of the specification can easily lead to this kind of overdesign from the designer.

9.8 AN EXAMPLE WITH A TRANSITION WITHOUT ANY INPUT

Now consider the next example in Figure 9.16; in this example, the transition between s3 and s0 does not have any input.

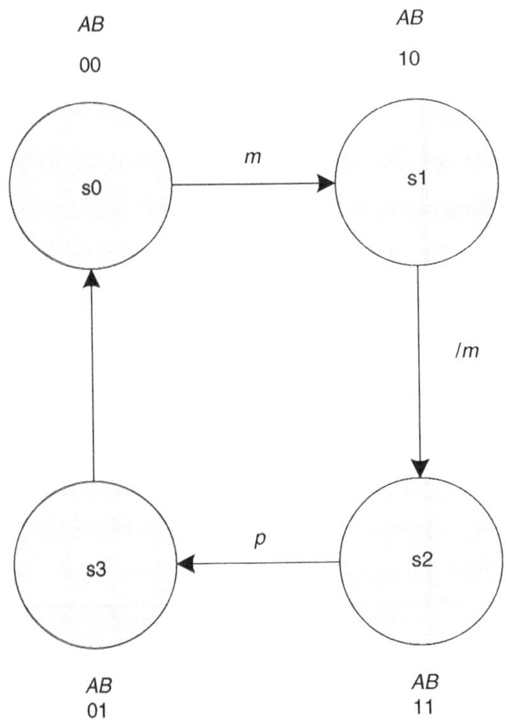

Figure 9.16 State diagram with no input along a transition.

Here are the equations for this example:

$$\begin{aligned} A &= \sum s_A + A \cdot \sum /r_A \\ &= \mathrm{s0} \cdot m + A \cdot /(\mathrm{s2} \cdot p) \\ &= /B \cdot m + A \cdot /(B \cdot p) \\ &= /B \cdot m + A/B + A/p. \end{aligned}$$

The equation for B will be obtained by not using the short-cut rule:

$$\begin{aligned} B &= \sum s_B + B \cdot \sum /r_B \\ &= \mathrm{s1} \cdot /m + \mathrm{s2}/m + B \cdot /(\mathrm{s3} + \mathrm{s0}) \\ &= A \cdot /B \cdot /m + A \cdot B \cdot /m + B \cdot /(/A \cdot B + /A \cdot /B) \\ &= A \cdot /m + B \cdot /(/A) \\ &= A \cdot /m + B \cdot A. \end{aligned}$$

In the equation for B, the $\sum /r_A$ term is (by the short-cut method) //s3 which is $//A$ because there is no input term along the transitional line.

This example does not have any output (something that most FSMs would have), but it is only an academic example.

Remember: add the reset input before trying to simulate the design.

9.9 UNUSUAL EXAMPLE: RESPONDING TO A MICROPROCESSOR-ADDRESSED LOCATION

Now here is an unusual example. Suppose one has an FSM-based event controller chip (PLD/FPGA) that is required to synchronize with a microprocessor. A possible solution follows.

In the system shown in Figure 9.17, an address 380h is produced by the address decoding logic. This might be implemented on the PLD/FPGA chip. The output of this is the signal 380h, which is to be used to operate the FSM.

The FSM will respond to this signal when *c* is low by moving to its state s1, where it will wait for *c* to go high.

At this point the FSM will move to s2 to assert the ACK signal to signal to the microcontroller that it has seen the 380h signal. The FSM will return to its initial state when the signal *c* goes low again via state s2 and s3. The signal *c* is derived from the system clock.

Note that this signal is used by the event FSM to control the return to initial state and thus provide a clearly defined ACK pulse width. If this were not done, the width of ACK signal would be dictated by the propagation time of the event logic only.

The state diagram is shown in Figure 9.18. Here, one can see the turn-on and turn-off terms, derived from the address decoder and *c* clock signals. This example shows how a simple event-driven FSM can be used to provide a control action without having to add a lot of logic to the system.

In a microprocessor system, the address decoding logic might well be already available; in a microcontroller system the FPGA could provide the decoding logic as well as the FSM, although it is using up a lot of I/O pins. The ACK signal, as implied in Figure 9.17, could be used to cause an interrupt in the microprocessor system, thus avoiding the need to provide an input port bit.

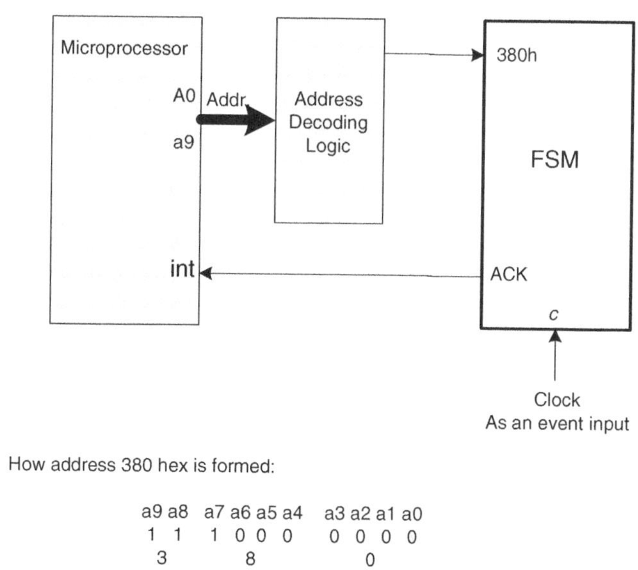

Figure 9.17 Block diagram of the basic address-activated system.

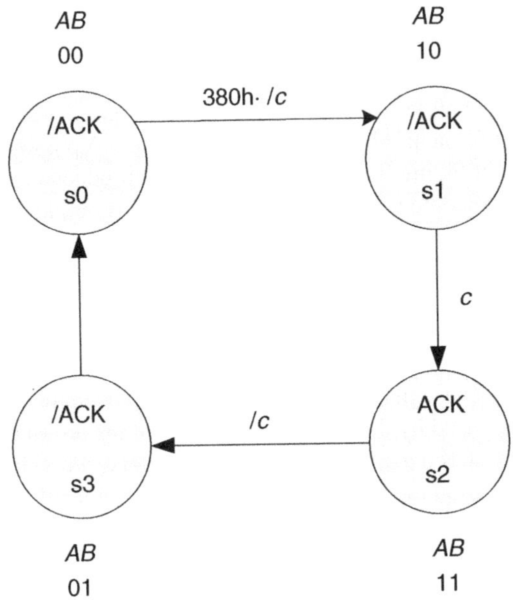

Figure 9.18 State diagram and equations for address-activated FSM.

This system will work correctly if the clock signal c connection between the microprocessor and the FSM is short to avoid lead delays. It is also assumed that the clock period is much greater than the largest propagation delay in the FSM, so as to allow the FSM time to settle.

The equations are

$$\begin{aligned} A &= \sum s_A + A \cdot \sum /r_A \\ &= \text{s0} \cdot \text{380h} \cdot /c + A \cdot /(\text{s2} \cdot /c) \\ &= /B \cdot \text{380h} \cdot /c + A \cdot /(B \cdot /c) \end{aligned}$$

$$\begin{aligned} B &= \sum s_B + B \cdot \sum /r_B \\ &= \text{s1} \cdot c + B \cdot /\text{s3} \\ &= A \cdot c + B \cdot //A \\ &= A \cdot c + B \cdot A \end{aligned}$$

$$\begin{aligned} \text{ACK} &= \text{s2} \\ &= A \cdot B. \end{aligned}$$

9.10 AN EXAMPLE THAT USES A MEALY OUTPUT

Sometimes it is useful to have an output that is a function of one or more inputs, but only in particular states. You might remember that during the programmed learning sections a Mealy FSM was defined as one in which some of the outside world inputs were fed into the outside world decoder. This next example illustrates this.

9.10.1 Tank Water Level Control System with Solutions

In the example shown in Figure 9.19, a pump is used to fill the tank (by making P1 = 1 and P2 = 0). The idea is to fill the tank so that the liquid level is between the level sensors Sh and Sl. When this is the case, thc outlet flow from the tank is balanced by the inlet flow to the tank via the pump.

If the liquid level falls below level sensor Sl (l1 asserted), the pump is to be switched to high-speed mode where P1 = 0 and P2 = 1. This is important to avoid air locks in the outlet part of the system.

Should the liquid level rise to level Sh (l2 asserted), the pump is to switch off.

This system will work continuously to maintain the liquid level. It can, of course, be switched on, or off via the relevant switches st and sp, which could be replaced with a single on/off switch if desired.

Table 9.3 shows the relationship between the level sensor inputs l1 and l2, and the outputs to the pump P1 and P2 can be constructed as illustrated below. Note that the last row of Table 9.3 is

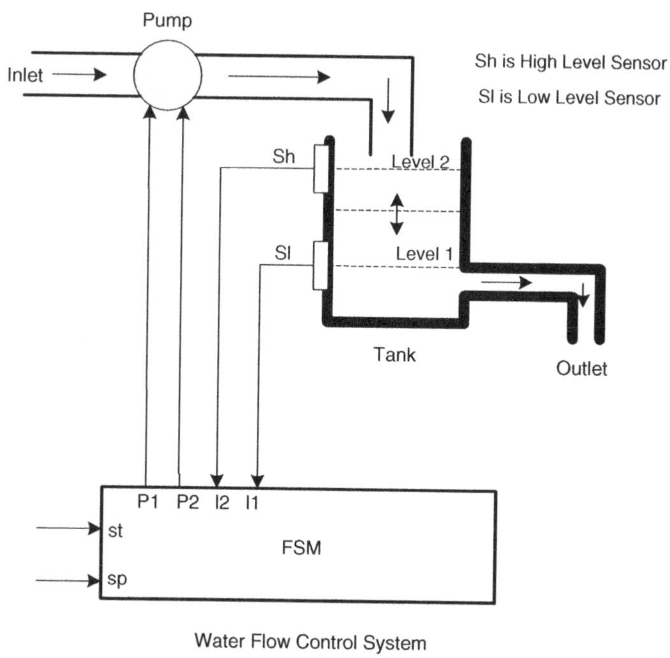

Figure 9.19 Block diagram of the FSM-based Mealy pump control system.

Table 9.3 Relationship between level sensor inputs and outputs to the pump.

l1	l2	P1	P2	Comment
1	1	0	0	Pump off, as water is in danger of overflowing tank
1	0	1	0	Pump at normal speed; water between sensors
0	0	0	1	Pump at high speed; water below sensors.
0	1	0	0	Impossible situation; pump off

dictated by the practical arrangement of the system. Clearly, the water level cannot be at the high setting in the tank and there be no water at the lower setting.

From this information, a state diagram can be developed that will meet the required specification. Figure 9.20 illustrates the state diagram. In this design, the system resets in to its idle state and waits for a start signal. Once obtained, the system moves into s1, the dormant state. It will stay in this state while the water level is above the level 1 sensor. The level sensors will now dictate when the system will move to s3. This will only occur if sensor l1 is zero, so the P2 input can start the pump in high mode to pump water above the lower level sensor. Once in state s3 the FSM will move between s3 and s2 to maintain the water level between the two level sensors.

Note that the system can be stopped at any time and the FSM will fall back to state s0. Note also that the P2 output will be disabled in state s3 if stop is activated, thus preventing the pump speed changing on a transition from s2 to s3 to s1 to s0. The water level would then fall to empty once the system was turned off. If the tank is empty when the system is turned on, then the FSM will move from s0 to s1, then straight to s3 to fill the tank to a level between l1 and l2.

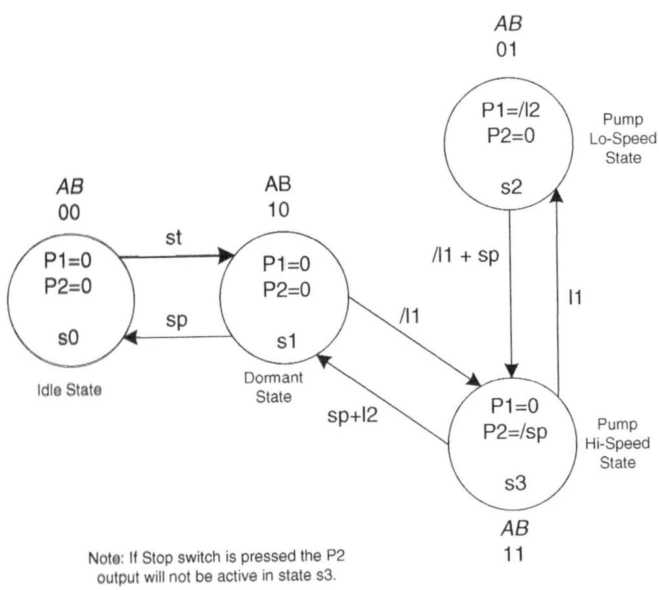

Figure 9.20 First attempt at a solution: four-state FSM with Mealy output.

This solution can now be developed into a practical system by assigning a set of secondary state variables. In this example, possible assignments could be

$$s0 = /A/B \quad s1 = A/B \quad s2 = /AB \quad s3 = AB$$

or perhaps

$$s0 = /A/B \quad s1 = /AB \quad s2 = A/B \quad s3 = AB.$$

Looking at this solution, one may wonder if it could be made simpler. In fact, looking at the table of sensor inputs and pump outputs, there is a combinational equation that can be formed using the level sensor inputs l1 and l2, and the two pump outputs P1 and P2. This is because the physical liquid movement forms a natural sequence for the problem. Look back to Table 9.3 with the impossible situation of l1 not active but l2 active, in which the pump should be held off. The equations for P1 and P2 are

$$P1 = l1 \cdot /l2$$
$$P2 = /l1 \cdot /l2.$$

However, these on their own are not enough, since there is the start and stop switch inputs to consider. Assuming that these two switches are push buttons, an event memory cell is needed to allow the system to occupy the two states.

The final system is illustrated in Figure 9.21. Here, the system is only able to operate when it is in state s1. In state s0 it is disabled.

The two equations for P1 and P2 are only true when the FSM is in state s1. Therefore, the two equations are written in the form

$$P1 = s1 \cdot /l2 \cdot l1$$
$$P2 = s1 \cdot /l2 \cdot /l1.$$

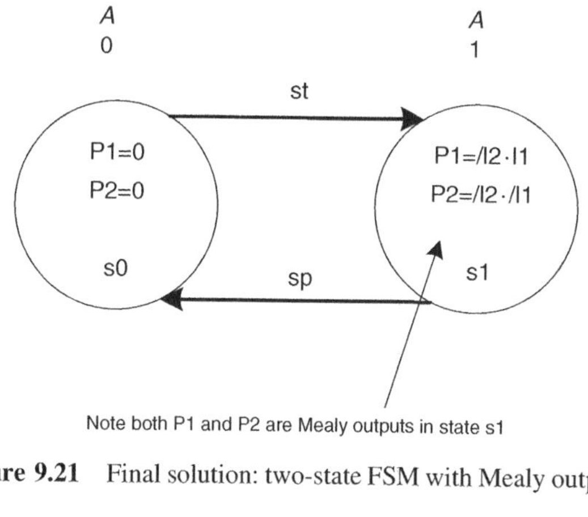

Figure 9.21 Final solution: two-state FSM with Mealy outputs.

To obtain the event cell (there is only one in this state diagram)

$$A = \sum s_A + A \cdot \sum /r_A.$$

Therefore:

$$A = \mathrm{s0} \cdot \mathrm{st} + A \cdot /(\mathrm{s1} \cdot \mathrm{sp}).$$

Replacing s0 and s1 with the secondary state variables gives

$$A = /A \cdot \mathrm{st} + A \cdot /(A \cdot \mathrm{sp})$$

The /*A* in /*A* · st and the *A* in *A* · sp need to be dropped (short-cut method), leaving

$$A = \mathrm{st} + A \cdot /\mathrm{sp}.$$

This is because when the /*A* term in $/A \cdot \mathrm{st}$ is dropped the result is effectively 1 · st, since $/A \cdot 1 = /A$.

In a similar way, *A* · sp is 1 · *A* · sp which is 1 · sp = sp. Therefore, the final set of equations for this example is

$$\begin{aligned} A &= \mathrm{st} + A \cdot /\mathrm{sp} \\ \mathrm{P1} &= A \cdot \mathrm{l1} \cdot /\mathrm{l2} \\ \mathrm{P2} &= A \cdot /\mathrm{l1} \cdot /\mathrm{l2}. \end{aligned}$$

Finally, before leaving this example, it is possible to reduce this particular problem to a combinational logic circuit that does not require an event cell. This is possible owing to the physical nature of the problem. The water in the tank creates a sequential operation for the water level sensors.

$$\begin{aligned} \mathrm{P1} &= \mathrm{l1} \cdot /\mathrm{l2} \cdot \mathrm{st} \cdot /\mathrm{sp} \\ \mathrm{P2} &= /\mathrm{l1} \cdot /\mathrm{l2} \cdot \mathrm{st} \cdot /\mathrm{sp}. \end{aligned}$$

This is only possible if the design uses switches that remain open or closed when released. If the system uses push switches that release when one leaves go of them, then the event cell is needed to remember the switch action.

9.11 AN EXAMPLE USING A RELAY CIRCUIT

The event sequential equations can be used to implement a design using relay logic. This might seem to be an outdated way to implement an FSM, but, in some cases, old-style electromechanical relays might be a more preferred solution. Alternatively, semiconductor static relays could be used in place of electromechanical relays. Both could be designed to operate at high voltage or high current levels.

In this next example, the design will be implement using logic gates and then relay logic.

Consider the following specification, which is very similar to the motor controller problem of Section 9.6.

A motor can be started by pressing the start button st, provided the stop button sp is not pressed. It can be stopped by pressing the stop button provided the start button is not pressed. If the stop button is pressed while the start button is still pressed, then the motor is to stop and an indicating LED turned on. The system can only leave this state and return to its initial state via a manual reset-key-activated switch. The reset key switch can also be used to deactivate the system regardless of the state of the start and stop buttons.

The state diagram in Figure 9.22 is developed to implement the specification. In this state diagram, the motor can be started by pressing the start button st thus moving the FSM to state s1, but only if sp = 0. The motor can be stopped by pressing the stop button to move the FSM back to state s0, provided st = 0. Pressing the stop button while the start button is still pressed will cause the FSM to move to s2, which is an invariant state (from which the FSM cannot leave without a system reset). The idea is to allow the reset input to move the FSM back to state s0.

From this, a set of equations can be derived, resulting in

$$A = /B\mathrm{st} \cdot /\mathrm{sp} + AB + A/\mathrm{sp} + A \cdot \mathrm{st}$$
$$B = A \cdot \mathrm{st} \cdot \mathrm{sp} + B \cdot /0.$$

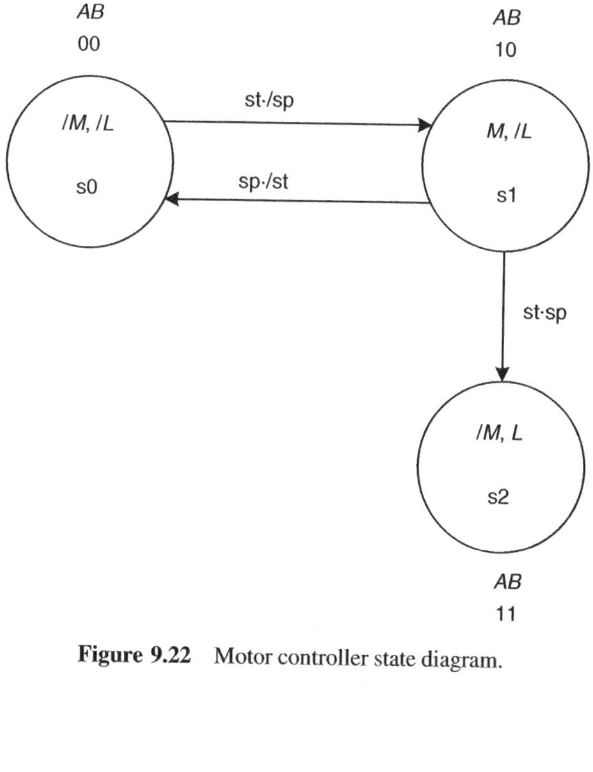

Figure 9.22 Motor controller state diagram.

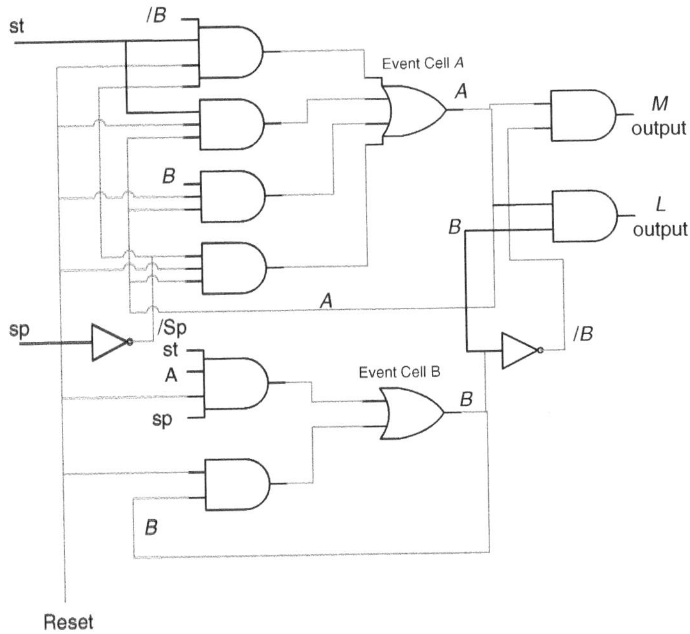

Figure 9.23 Logic circuit for the motor controller FSM.

Note, there is no turn-off term in the B equation, so the negated term is $/0$, which of course is 1. The outputs are:

$$M = \mathrm{s1} = A/B$$
$$L = \mathrm{s2} = AB.$$

These equations are in a suitable form for implementing the design using either a PLD or relays.

A circuit schematic is drawn in Figure 9.23. This circuit uses AND/OR/NOT logic, so is suitable to be implemented using a PLD device. Note that an AND gate is needed in the feedback loop for event cell B so that the reset can be used to reset the cell back to its zero state.

However, a little thought will reveal that such a circuit needs a 5 V power supply, and this would need to be obtained from the mains supply via a transformer and rectifier circuit. The transformer could be replaced with a mains resistor dropper and single diode and capacitor, but this still requires these overheads.

An alternative design could be based upon electromechanical or static relays. These have the advantage that they can be used with a very rough power supply direct from the mains (using relays that can be operated at mains voltage of course). The relay circuit is obtained directly from the sequential equations.

The circuit in Figure 9.24 is the final result. In this diagram, the relay contacts are shown with the relays not operated. This circuit will use a simple half-wave rectifier in series with a suitable capacitor to obtain a rough DC voltage for the relays A and B. The resistor across the capacitor provides a discharge path when the supply is disconnected (by a reset for example).

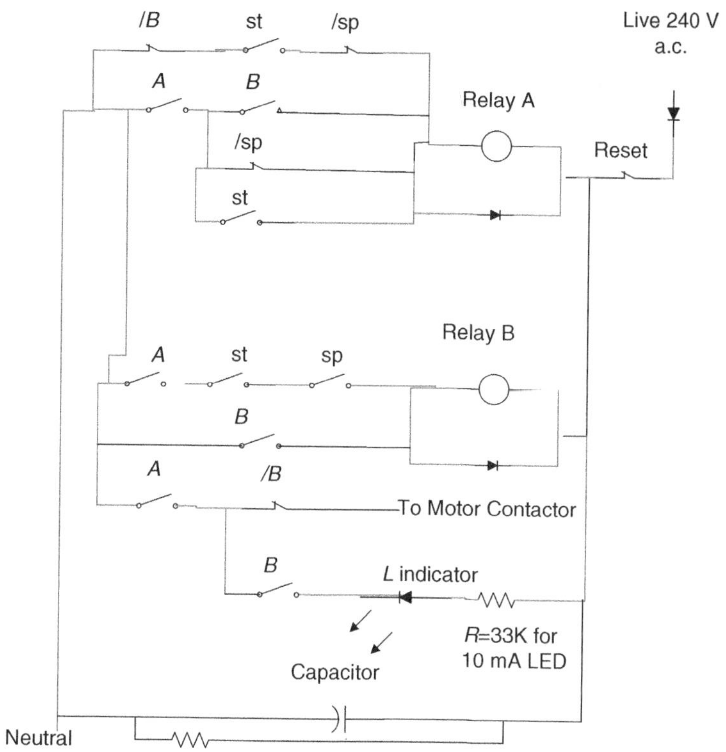

Figure 9.24 Relay logic for the motor controller FSM.

The circles A and B represent the relay operating coils (or control input to static relays). The diodes across each coil are needed to provide a path for relay current when the contacts open; otherwise, the large EMF across the coils could damage the switch and relay contacts. These are usually referred to as 'catching diodes'.

Note: the reset switch is in series with the supply. This can reset both relays and turn off both the motor and indicator LED.

Before moving on to look at more asynchronous (event-driven) examples, one needs to consider the effects of race hazards in event-driven types of FSM.

9.12 RACE CONDITIONS IN AN EVENT FINITE-STATE MACHINE

In this section some of the problems that can occur in asynchronous (event) FSM systems will be discussed, with suggestions on how they can be eliminated.

In an event FSM there are three types of potential race condition:

- race between primary inputs;
- race between secondary state variables (the event cells themselves);
- race between primary and secondary variables.

This is particularly important, since one needs to be aware of potential problems that can occur in event-driven systems in order to avoid making design errors. These are used with permission from Oxford University Press from their publication *Problems and Solutions in Logic Design* [1].

9.12.1 Race between Primary Inputs

This is when two signals both happening at the same time on the same transition of a three-way branch state are expected to cause the FSM to move to one particular state. Clearly, one cannot guarantee that two (or more) input signals will change at the same time, since there are always delays in the paths from two or more signals.

Note: to avoid this type of condition, do not try to look for more than one input changing at the same time.

In the example of Figure 9.18 there are two signals 380h · /c along a transitional line, but in this case the FSM was looking for the condition 380h AND /c, and in the next state *c* was to be seen to go high before a state change (it must have been low to get to this state). So, this is a very different situation, where the inputs have a dictated sequence and cannot cause confusion if they happen at the same time.

9.12.2 Race between Secondary State Variables

This is when the designer has not followed a unit distance coding for the secondary state sequence (A, B, event cells for example). The use of a none unit distance coding can result in the FSM falling into a state different to the one intended as a result of unequal propagation delays between event cells.

Consider the earlier state diagram of Figure 9.18 with the following secondary state assignment:

$$\text{s0} = /A/B, \quad \text{s1} = AB, \quad \text{s2} = A/B, \quad \text{s3} = /AB.$$

If, in state s0, the 380h input is 1 and *c* is 0, with *A* changing to 1 before *B*, then the resulting transition might be s0 to s2, and in s2, since *c* is still logic 0, a further transition to s3. Since there is no input along the transitional line between s3 and s0, the FSM would move back to s0! This sort of behaviour is unpredictable, since if it was *B* that changed first in state s0 then the transitional path could be s0 to s3, back to s0.

Remember, in an asynchronous (event-driven) FSM there is no synchronizing clock to introduce a delay to allow signals to settle.

Solution: always use a unit distance code for asynchronous (event) FSM systems.

9.12.3 Race between Primary and Secondary Variables

The final race condition to look at is also the most complex. There are more details to be found in Reference [1].

Essentially, as the heading suggests, this is a race between the primary (outside world) inputs to the FSM and its event cell operation (secondary state variables). This is caused if the propagation delay through to the primary input path to set the cell is greater than the secondary delay path (cell output to cell input to cause the cell to set or clear). It can result in the cell maloperating.

To prevent this kind of race from occurring, ensure that the primary delay Tp is less than the secondary delay Ts at all times, i.e.

$$Tp < Ts.$$

More specifically:

$$Tp_{max} < Ts_{min}.$$

This leads to the identity defined in Reference [1], repeated here by permission of Oxford University Press:

$$Tp_{max}/Ts_{min} < 1,$$

where Tp_{max} is the maximum possible propagation delay for a primary input path and Ts_{min} is the minimum possible delay for a secondary delay path (total gate delays between *A* and *B* outputs, for example).

The event cell structure used in the asynchronous designs in this book (and, indeed, in the Zissos book [1]) meet these requirements if the gate tolerances are within 33.3% of each other. This is not difficult to achieve in modern integrated circuits, particularly PLD and FPGA devices.

There is also a somewhat dated paper on gate tolerances in Reference [2] that is worth studying.

9.13 WAIT-STATE GENERATOR FOR A MICROPROCESSOR SYSTEM

Some microprocessor systems have a feature that allows the processor to introduce 'wait states' into a particular memory cycle.

Figure 9.25 shows the basic input or output (I/O) cycle timing (simplified for this example, but accurate in its sequence to produce a working design). In this example, it is assumed that each memory or I/O cycle consists of four *T* states created by the system clock *c*. T1 is address setup time, T2 read or write setup time, T3 a wait state to allow the data bus time to settle, and T4 used to read or write data. In this figure, the event FSM controller monitors the chip enable signal ce, which will go low when a slow I/O device is selected by the microprocessor software. This will occur in the T1 timing slot for the particular I/O cycle. There are four T slots per I/O access. During the T2 period of the clock, either the input/output write *w* or the input/output read *r* signal lines will be taken low by the microprocessor.

During the T2 period, an output signal from the FSM (IORDY), which is a special input signal to the microprocessor, can be taken low, and if the microprocessor detects this during the T2 period it will insert an additional *T* period Tw between T3 and T4.

This extra period is known as a wait state (Tw) and it effectively increases the T3 period used to allow slow devices time to settle before the T4 period that is used to perform the data

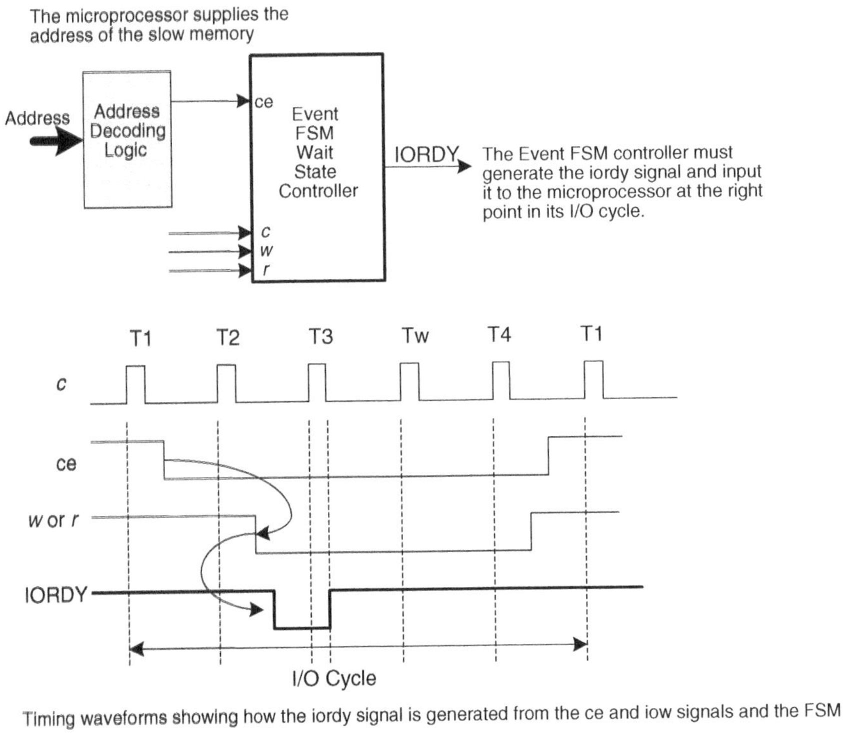

Figure 9.25 Showing the block diagram and memory/IO cycle timing.

transfer. In this way, a slow I/O device can have its chip enable ce signal monitored by the FSM controller and used to generate a wait state. To be sure, the particular microprocessor will need to be consulted to find out how to activate a wait state, but this is usually available in the data sheets for the microprocessor.

The purpose of the event FSM controller is to identify when to send the IORDY signal line low, and when to return it high again. In effect the event FSM is being used to detect the point in the timing diagram of Figure 9.25 at which to generate the IORDY signal to be sent to the microprocessor.

Using the timing diagram as a guide, the required state diagram can be developed as seen in Figure 9.26. As can be seen from Figure 9.26, the state diagram follows the sequence by detecting ce and either *w* or *r* going low in state s0 to turn on the IORDY (active-low) signal in T2. Then, it detects when the clock *c* goes low in state s1 in order to identify when it goes high in state s2 (to identify entry into T3 state). The FSM must then determine when the clock signal *c* goes low again, indicating the point at which IORDY must go high again.

Note that fast memory cycles will not activate the wait-state generator because those chip select signals will not be connected into the wait-state event FSM controller.

Finally, Figure 9.27 illustrates the sequential equations and output equation for the system. This example has illustrated how an event-driven FSM can be used to track points in a sequential sequence of signals. This example could easily be adapted for a particular microprocessor. However, one must determine the correct sequence from the microprocessor data sheet, since different microprocessors use their own signals and sequences to control access to slower memory.

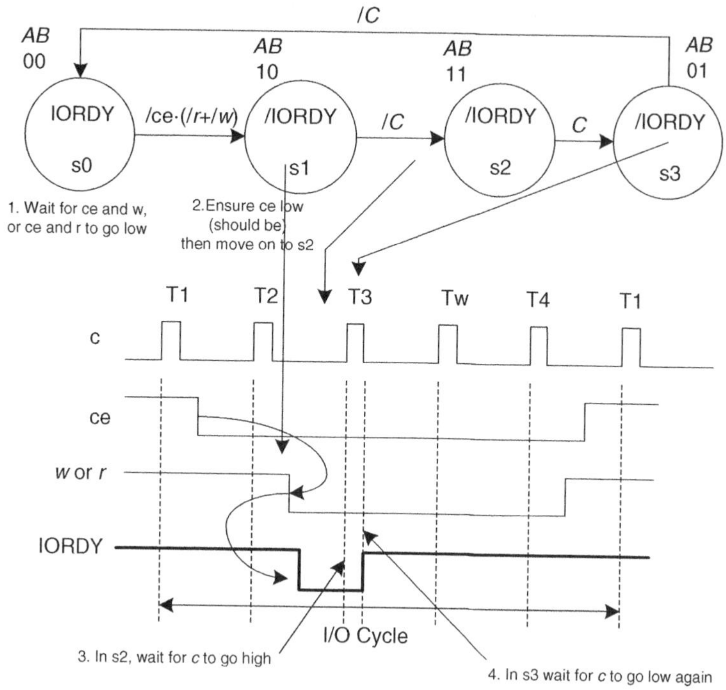

Figure 9.26 The state diagram and how it was derived from the timing waveform.

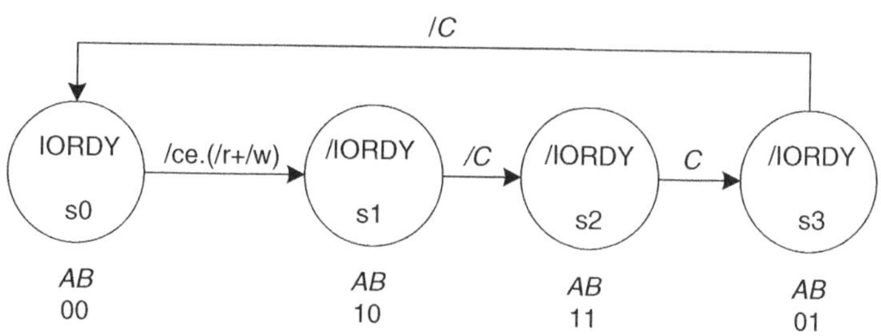

Event Cell Equations:

$A = \sum s_A + A \cdot / \sum r_A$

$\boldsymbol{A = /B \cdot /ce \cdot (/r + /w) + A \cdot /B + A \cdot /C}$

$B = \sum s_B + B \cdot / \sum r_B$

$B = A \cdot /C + B \cdot /(/A \cdot /C)$

$\boldsymbol{B = A \cdot /C + A \cdot B + B \cdot C}$

<u>Output:</u>

IORDY = s0 = $/A \cdot /B$

Figure 9.27 The sequential equations for the memory/IO FSM controller.

9.14 DEVELOPMENT OF AN ASYNCHRONOUS FINITE-STATE MACHINE FOR A CLOTHES SPINNER SYSTEM

Figure 9.28 illustrates the system. There is a spin motor to spin the clothes drum at high speed so as to remove excess water from the clothes by centrifugal force. The water released from the clothes into the drum is removed by the pump. There is a water level sensor to detect whether the water level is too high before turning on the spin motor to avoid excess load on the latter.

The user loads wet clothes into the clothes drum and presses the start button st. This starts the pump on release of the start button. When the water level is below the water level sensor, the spin motor is started and a timer (not shown here) is started.

In due course, the timer times out and the system stops both the spin motor and the pump. A done indicator is illuminated to indicate to the user that the spin cycle is complete. The user must press the stop button sp before another spin cycle can commence. This system does not have a sensor to test that the door is closed. You might like to add this to the system and modify the state diagram to include this feature.

A suitable state diagram is illustrated in Figure 9.29. In this state diagram, on pressing the start button a test is made to determine whether the water level is above or below the sensor on the drum. If above the sensor, the FSM moves to s2 via s1 and starts the pump.

Note, the pump can only start if the start button has been released. Once the water level has dropped below the sensor, the FSM moves to s3 to turn on the spin motor, as well as start the timer. At time out, the FSM moves to s4 to turn off both spin motor and pump as well as turn on the done indicator *D*. Note that the FSM cannot leave s4 via any transition. In fact, the stop input acts as a reset input and can stop the system in any state.

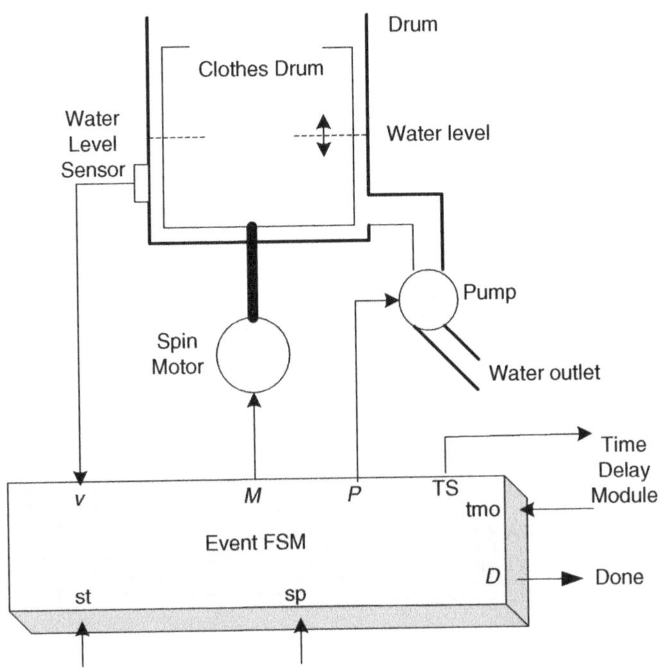

Figure 9.28 Basic system showing a clothes spin system with FSM.

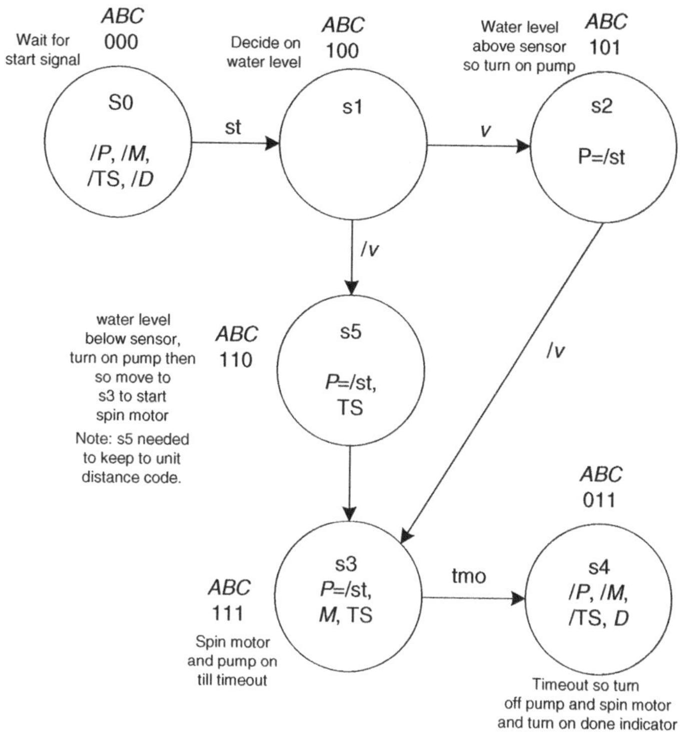

Figure 9.29 State diagram of a possible solution for clothes spinning system.

If, on starting the system, the water level in the drum is below the water level sensor, the FSM will move from s0 to s1, to s5, then directly to s3. State s5 is needed to allow a unit distance code to be used for the state machine; s5 is in fact a dummy state.

Note that there is no input along the transitional line connecting s5 to s3. This implies that when the FSM moves into s5, it will immediately move on to state s3, the delay being that of the propagation delay of the logic used to implement the event cells B then C.

The equations for the design are

$$\begin{aligned} A &= \sum s_A + A \cdot / \sum r_A \\ &= /B \cdot /C \cdot \mathrm{st} + A \cdot /(B \cdot C \cdot \mathrm{tmo}) \\ &= /B \cdot /C \cdot \mathrm{st} + A \cdot /B + A \cdot /C + A \cdot /\mathrm{tmo} \end{aligned}$$

$$\begin{aligned} B &= \sum s_B + B \cdot / \sum r_B \\ &= A \cdot /C \cdot /v + A \cdot C \cdot /v + B \\ &= A \cdot /v + B \end{aligned}$$

$$\begin{aligned} C &= \sum s_c + C \cdot / \sum r_c \\ &= A \cdot /B \cdot v + A \cdot B + C \\ &= A \cdot v + A \cdot B + C. \end{aligned}$$

The outputs are

$$P = s2 \cdot /st + s3 \cdot /st + s5 \cdot /st$$
$$= A \cdot C \cdot /st + A \cdot B \cdot /st$$
$$M = s3 = A \cdot B \cdot C \quad TS = s5 + s3 = A \cdot B \quad D = s4 = /A \cdot B \cdot C.$$

The stop input sp will be logically ANDed to each equation A, B, and C to allow the FSM to return to $ABC = 000$ when sp is made logic 0.

The Verilog module follows in Listing 9.5. In this module, the equation level is seen commented out and replaced with a gate-level description.

```
/////////////////////////////////////////////////////
              // Spin motor and pump Asyhchronous FSM //
/////////////////////////////////////////////////////
module smpfsm(st,sp,v,tmo,P,M,TS,D,A,B,C);

   input st,sp,v,tmo;
   output P,M,TS,D,A,B,C;
wire w1,w2,w3,w4,w5,w6,w7,w8,w9;
// equation level description. Used in Figure 9.31.
//assign
  //A = (~B&~C&st | A&~B | A&~C | A&~tmo)&sp,
  //B = (A&~v | B)&sp,
  //C = (A&v | A&B | C)&sp,

    // alternative gate level description Used in Figure 9.32.
    // each gate has been given a delay of 5 time units.

       or #5 g1(A,w1,w2,w3,w4);
       and #5 g2(w1,~B,~C,st,sp);
       and #5 g3(w2,~B,A,sp);
       and #5 g4(w3,~c,A,sp);
       and #5 g5(w4,~tmo,A,sp);
       //-------------------
       or #5 g6(B,w5,w8);
       and #5 g7(w5,A,~v,sp);
       and #5 g11(w8,B,sp);
       //-------------------
       or #5 g8(C,w6,w7,w9);
       and #5 g9(w6,A,v,sp);
       and #5 g10(w7,A,B,sp);
       and #5 g12(w9,C,sp);
       //---------------------
```

```
P = A&C&~st | A&B&~st,
M = A&B&C,
TS = A&B,
D = ~A&B&C;

endmodule
//////////////////////////////////////////////////
```

Listing 9.5 Verilog module for clothes spin FSM.

The test bench module is illustrated in Listing 9.6.

```
  `timescale 1ns / 10ps
module test;
   reg st,sp,tmo,v;
   smpfsm uut(st,sp,v,tmo,P,M,TS,D,A,B,C);
   initial
     begin
       sp=0;
       st=0;
       v=0;
       tmo=0;
       ///////
       #10 sp=1; // remove reset.
       #10
       #10 v=1; // water in drum.
       #10
       #10 st=1; //start system
       #10 //should move to s1 then s2.
       #10 st=0;
       #10 // starts pump to empty drum.
       #10 // wait for drum empty.
       #10 v=0; // signal that drum empty.
       #10 // should move to s3 and turn on spin motor
       #10
       #10 //waiting for timer to stop spn motor.
       #10 tmo=1; // signal to stop spin motor.
       #10 // should have moved to s4.
       #10 tmo=0;
       #20 st=0; //return start to off state.
       #10 sp=0; //stop system and return to s0.
       #20
       #20 sp=1; // release reset buton.
```

```
        #10 st=1; //start system with empty drum.
        #20
        #20 st=0;
        #20 // should move to s1 then s5 then s3.
        #10 tmo=1; // time out. should move to s3.
        #20 //waiting for user to press stop.
    #10 $stop;
    end
endmodule
```

Listing 9.6 Verilog test-bench module.

Finally, the simulation is shown in Figure 9.30 using the *equation-level description.* In the simulation, the event cells A, B, and C appear to be changing state at the same time in some parts of the simulation, but in fact the transitions are so fast that the actual transitions cannot be seen. However, care must be taken to ensure that propagation timing satisfies the 33.3% rule discussed in Section 9.12.3.

In Figure 9.31, the simulation using the *gate-level description* is seen. Here, each gate has been given a delay value of 5 time-units so that the state transitions can be clearly seen. In

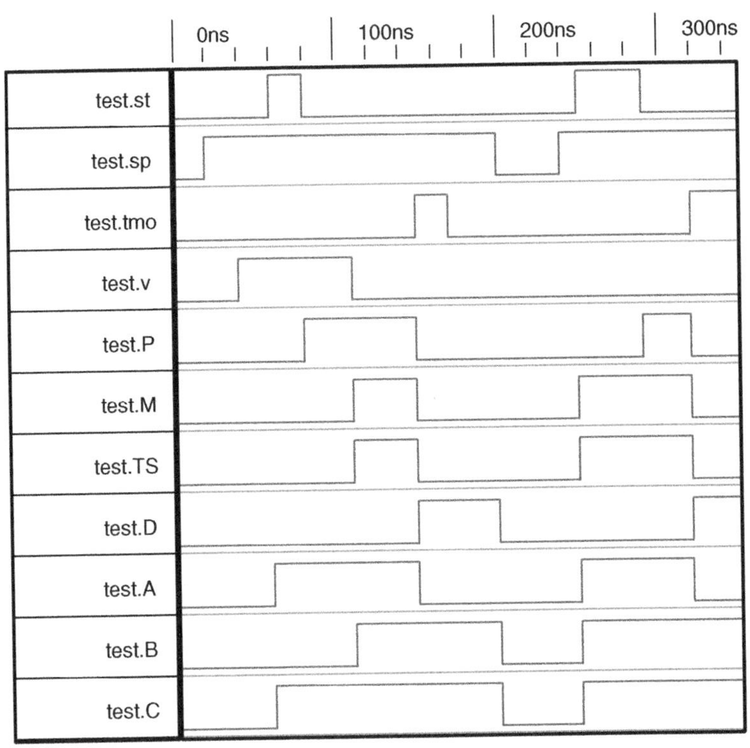

Figure 9.30 Simulation of a clothes spinner system using equation-level description.

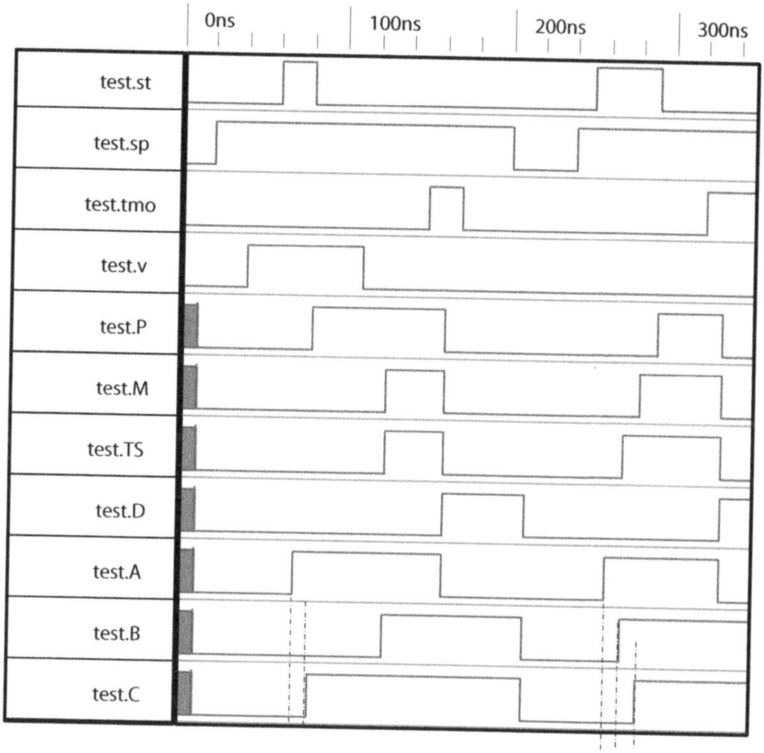

Figure 9.31 Gate-level simulation of a clothes spin system.

Figure 9.31, the delays between the gates allow the state transitions to be seen clearly. For example, the transitions between s1 ($ABC = 100$) and s2 ($ABC = 101$), and the transitions between s1 ($ABC = 100$) to s5 ($ABC = 110$), then on to s3 ($ABC = 111$). The dashed lines help to identify these transition points.

9.15 CAUTION WHEN USING TWO-WAY BRANCHES

In the state diagram of Figure 9.10 there is a two-way branch in state s1with /st along one transitional line and ms + *t* along the other. These inputs must be mutually exclusive, otherwise the FSM could maloperate. If this cannot be guaranteed, then the design will need to be changed so that the state diagram can only change from one state to the next on a single input change.

Figure 9.32 illustrates a possible alternative design (without the test input *t*). In this arrangement, the FSM can move from s1 to s2 if either the start input st is returned to logic 0 and/or if the fault input ms becomes logic 1. On reaching s2 from a fault, the motor is turned off and the fault indicator *L* turned on (active-low). If the st input is now returned to logic 0, then the fault indicator can be turned off but the FSM can only return to s0 if the fault input ms returns to its logic 0 level.

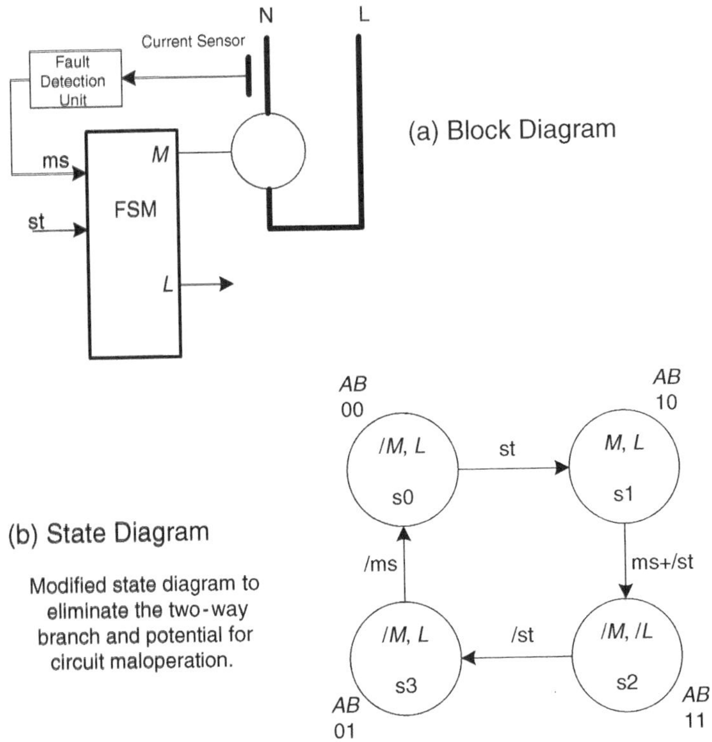

Figure 9.32 Modified state diagram for the motor controller of Section 9.6.2.

The equations for A and B are

$$
\begin{aligned}
A &= \sum s_A + A \cdot \sum /r_A \\
&= \mathrm{s0} \cdot \mathrm{st} + A \cdot /(\mathrm{s2} \cdot /\mathrm{st}) \\
&= /B \cdot \mathrm{st} + A \cdot /(B \cdot /\mathrm{st}) \\
&= /B \cdot \mathrm{st} + A/B + A \cdot \mathrm{st} \\
B &= \sum s_B + B \cdot \sum /r_B \\
&= \mathrm{s1} \cdot (\mathrm{ms} + /\mathrm{st}) + B \cdot /(\mathrm{s3} \cdot /\mathrm{ms}) \\
&= A \cdot \mathrm{ms} + A \cdot /\mathrm{st} + AB + B \cdot \mathrm{ms}.
\end{aligned}
$$

The output equations are the same as those for Figure 9.10.

Other examples using two-way branches in this chapter are as follows.

In Section 9.10.1, Figure 9.20, there are two possible two-way branches: one in state s1 and the other in state s3. In each case there are different inputs along each transition path that could result in maloperation; therefore, this design could fail. However, the alternative design in Figure 9.21 overcomes this problem.

In Section 9.11, Figure 9.22, there is a two-way branch in state s1. If input sp is logic 1 in state s1, then the FSM can move to either s0 if st = 0, or to s2 if st = 1. If, however, inputs st and sp were to change at the same time from logic 0 to logic 1 in state s0, then it is possible that the sequence shown below could occur:

st	sp	
0	0	in state s0
0	1	sp reaches logic 1 before st; stay in s0 while signals still changing
1	1	stay in state s0

or

st	sp	
0	0	in s0
1	0	st reaches logic 1 before sp; move s0 to s1
1	1	move to s2

The latter example appears to work correctly.

In general, however, changes in two or more input signals can result in circuit maloperation due to propagation delays between input signal changes producing static or dynamic hazards. The best way to handle this situation is to allow only one input to affect the FSM. Figure 9.33 shows how this could be done.

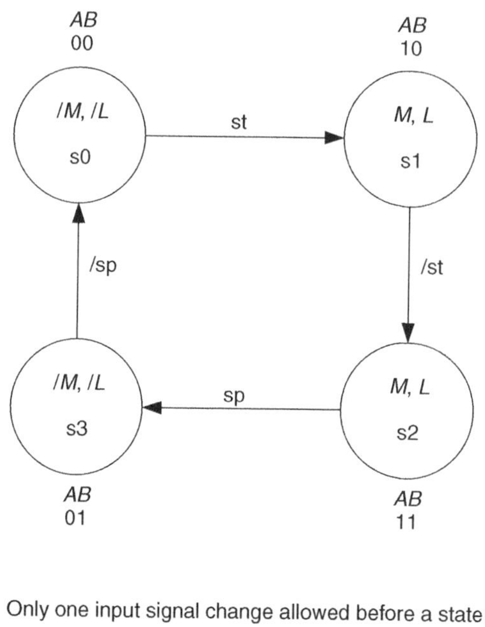

Figure 9.33 Modification to the state diagram of Figure 9.22 to avoid maloperation.

This, of course is not what the original specification for this FSM was designed to do. In fact the idea of trying to produce an event FSM to meet the specification in Figure 9.22 is not very practical.

Designing an asynchronous FSM to work correctly under multiple changing inputs is not easy and is beyond the scope of this book. Reference [3] is a good source that covers in detail and in a formal manner how to develop complex asynchronous FSMs using both Huffman and Muller circuits. In particular, the C gate is used to decouple the set terms and reset terms. This can reduce the potential for static and dynamic hazards when two or more inputs are changing.

9.16 SUMMARY

This chapter has introduced the idea of asynchronous (event-driven) FSMs and how to design them for implementation in devices such as PLD and FPGSs, as well as relay circuits. Also, the simplest method to simulate the designs has been considered, using the Verilog HDL at the equation and basic gate levels. This allows designs to be implemented directly at either the equation or logic gate level, and avoids the problems that most HDL systems can introduce at the behavioural level when implementing event-driven controllers. A number of simple FSM designs have been considered, showing how the event FSM can be used. In addition, the potential race problems associated with event-driven FSMs have been discussed, with ways to avoid these conditions from happening.

REFERENCES

1. Zissos D, Duncan FG. Problems and solutions in logic design, 2nd edn. Oxford University Press, 1979.
2. Duncan FG, Zissos D. Gate tolerance in sequential circuits. Proc. IEE 1971;118(2):317–320.
3. Myers C. Asynchronous circuit design. John Wiley & Sons, Ltd, 2001.

10

Introduction to Petri Nets

10.1 INTRODUCTION TO SIMPLE PETRI NETS

The Petri net is a state diagram that can be used to describe the behaviour of both sequential and parallel systems. It was initially conceived by Karl Petri in the 1960s and has had a good following of academics ever since. There is a website devoted to all things Petri at http://www.informatik.unihamburg.de/TGI/PetriNets/.

Petri nets are often used as a tool to study the behaviour of parallel and concurrent systems (not necessarily electrical). They have also been used to study parallel and concurrent programming methods. In recent years, researchers have shown [1] how the Petri net can be used to develop and synthesize electronic FSM systems, in a similar way to how synchronous and asynchronous systems can be developed and synthesized. The main reason for employing Petri nets is the ability to create parallel systems. The following method makes use of material with permission from [1].

Figure 10.1 illustrates a two-state diagram and its Petri net equivalent. In a Petri net, the 'state' is represented by a 'placeholder' and the 'transitional lines' between states are represented by 'arcs' that connect the placeholder (P1 and P2) to transition points (T1 and T2). The inputs along the transitional lines of a state diagram are placed against the transition points along the connecting arcs that link one placeholder to another in a Petri net.

The Petri net uses a memory element to represent each placeholder (rather like in a One Hot state diagram – as illustrated in Figure 10.1). However, in Petri nets used to represent parallel systems, there can be more than one active placeholder (whereas in a state diagram only one state can be active at any one time). For this reason, a Petri net needs some way to show which of its placeholders are active. This is done by using a 'token' to represent an active placeholder and by placing a 'dot' in the placeholder that is active.

In Figure 10.1, placeholder P1 is active, since it has a token, and placeholder P2 is not active and, hence, does not have a token.

A brief explanation of the behaviour of the Petri net in Figure 10.1 follows.

Initially, a token is in placeholder P1 (via initialization logic to be explained later). When the input x becomes active ($x = 1$) the transition T1 will fire, and the token will move (following the

---

All Petri Net Equation generations are reproduced from 'VHDL generation from hierarchical Petri net specifications of parallel controllers' by JM Fernandes, M Adamski and A J Proenca, (IEE Proceedings- Computers and Digital Techniques, Vol.144, No.2 March 2007) with permission from IET.

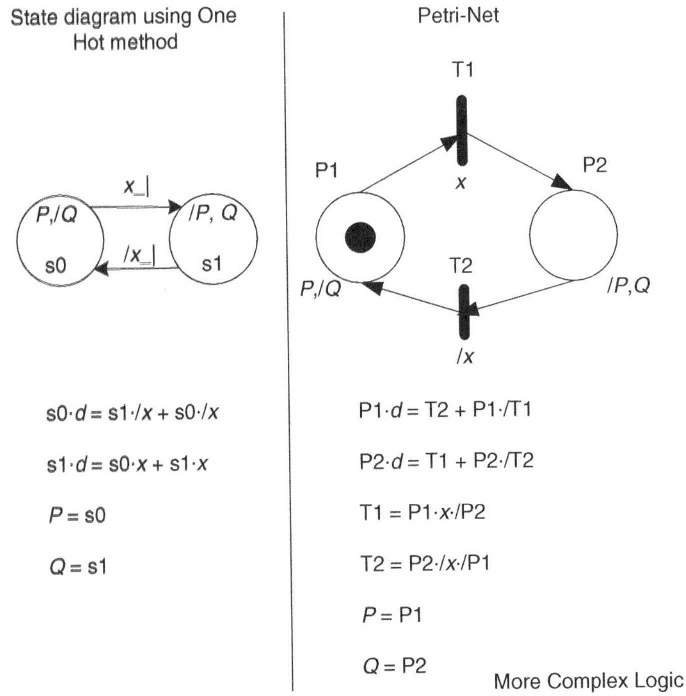

Figure 10.1 Comparison between a state diagram and Petri net with respective equations.

arc path) to placeholder P2, where it will remain (because T2 is not able to fire since x is still 1), as illustrated in Figure 10.2.

It should be noted that transition T1 will only fire when $x = 1$ and a clock pulse occurs. Note also that outputs $P = 0$ and $Q = 1$ in P2, so outputs are following a Moore-type model. When $x = 0$ and the next clock pulse occurs, the token will pass back to P1, as shown in Figure 10.1.

The syntheses for this Petri net are based upon the equations shown in Figure 10.1. There are three basic equation types:

- placeholder equations;
- transient equations;
- output equations.

The placeholder equations follow the same format as the sequential equations for an event-driven state machine. This is best described in terms of the Petri net in Figure 10.1, shown in Equation (10.1). The Petri-net equations define the input to a D-type flip-flop, hence the '$P \cdot d$' on the left-hand side.

$$\mathrm{P1} \cdot d = \mathrm{T2} + \mathrm{P1} \cdot /\mathrm{T1}. \tag{10.1}$$

This is interpreted as: for P1 to get a token, T2 must have fired; *or*, to hold on to the token, a token must be in P1 *and* T1 must *not* have fired.

For P2:

$$\mathrm{P2} \cdot \mathrm{d} = \mathrm{T1} + \mathrm{P2} \cdot /\mathrm{T2}. \tag{10.2}$$

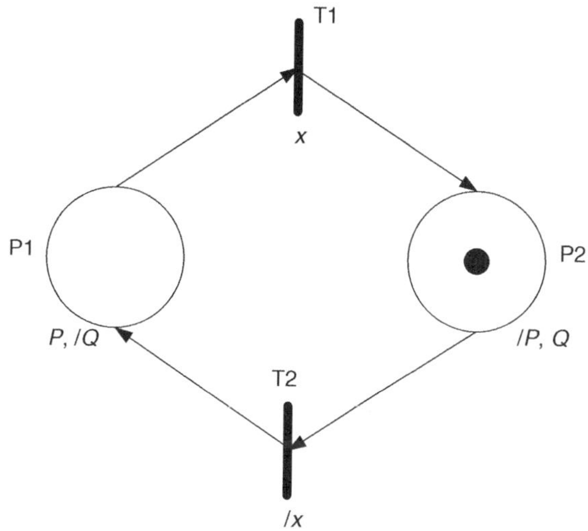

Figure 10.2 Token moved to P2 after T1 fired ($x = 1$).

The first term T1 on the right-hand side of Equation (10.1) for P1 is, in effect, a turn-on condition for the placeholder P1. The product term $\text{P1} \cdot \text{T1}$ is a hold term for the placeholder.

The transition equations are made up of the conditions necessary for the transition to fire. In the Petri net of Figure 10.1 it can be seen that T1 will only fire if P1 has the token *and* P2 does *not* have the token *and* the input $x = 1$,. Hence:

$$\text{T1} = \text{P1} \cdot x \cdot /\text{P2}. \tag{10.3}$$

In the same way:

$$\text{T2} = \text{P2} \cdot /x \cdot /\text{P1}. \tag{10.4}$$

There is more to these rules when describing more complex Petri nets, which will be explained later.

Since the placeholder equations of Equations (10.1) and (10.2) are equal to $\text{P1} \cdot d$ and $\text{P2} \cdot d$ respectively, they define the D inputs to D-type flip-flops. This is illustrated in Figure 10.3.

In future examples, the distinction between the left-hand side of a placeholder equation $\text{P}n \cdot d$ will not be made and will take on the appearance of a recursive equation, as in

$$\text{P1} = \text{T2} + \text{P1} \cdot /\text{T1}$$
$$\text{P2} = \text{T1} + \text{P2} \cdot /\text{T2}.$$

This implies that the left-hand side is the input to the flip-flop. Reference [1] uses this approach.

Figure 10.3 illustrates the cycle of design from Petri net to equations, and finally synthesized circuit. It implies that once a Petri net has been developed, the synthesization is a systematic application of the rules.

Of course, a PLD device or FPGA could be used and the equations used directly, or the Petri net could be written at the behavioural level in VeriLog HDL.

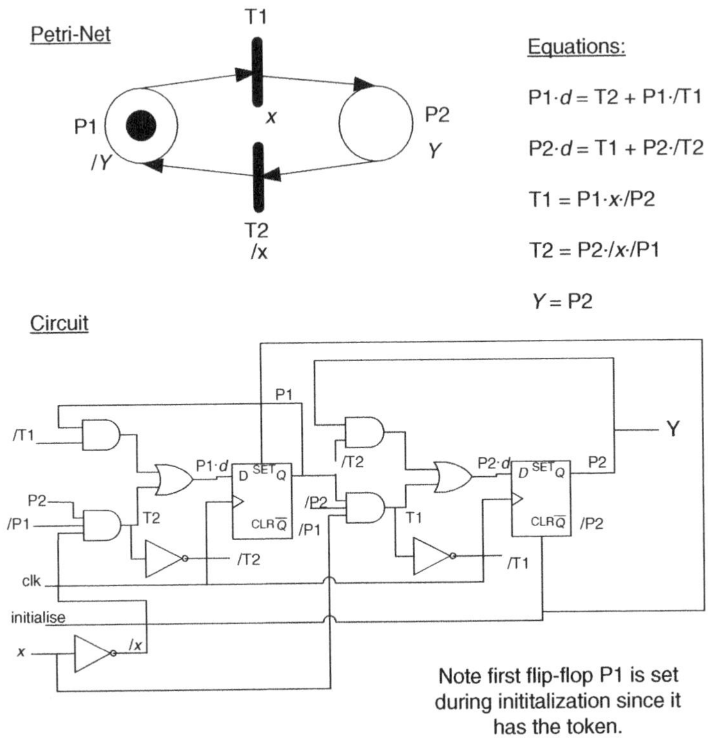

Figure 10.3 Full cycle of design from Petri net to circuit.

In the circuit schematic of Figure 10.3, note the initialization arrangement. This is the same as that used in the One Hot design of state machines. Also note the topological arrangement for the gate logic. The flip-flop output P1 is connected back into the turn off *and* gate logic; and likewise for the P2 flip-flop. This provides the hold term required to keep the placeholder active.

From this diagram, and the foregoing description of the equations, it can be seen that the flip-flops provide memory for the placeholder element and that a set flip-flop is equivalent to a placeholder with a token and a reset flip-flop is equivalent to a placeholder without a token.

The topological structure of the Petri net can be seen in Figure 10.4:

$$\text{Pn} = \text{Tin} + \text{Pn} \cdot /\text{Tout}. \tag{10.5}$$

Tin is the turn-on input, and the feedback from output Pn to the input of the AND gate forms the hold term. The term Tin in Equation (10.5) is of the form:

Tin = input placeholder AND input enable AND NOT output placeholder.

Tout is the turn-off term, which is negated in Equation (10.5). When Tout becomes asserted high, the /Tout input will go low so as to open the feedback hold term to allow the *D* flip-flop to reset (Tin will not be active at this point).

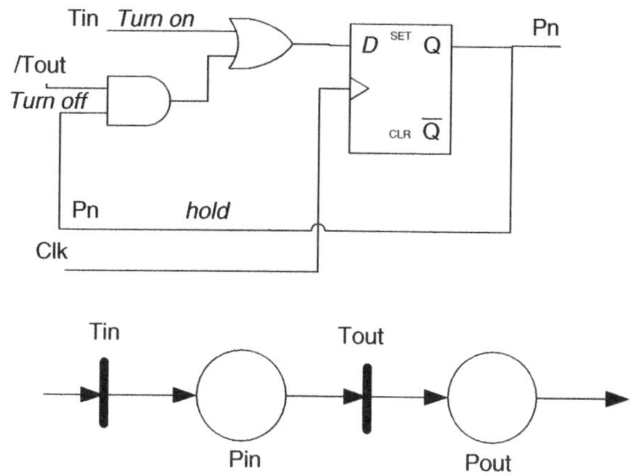

Figure 10.4 Basic topological structure of the Petri net.

A close look at Figure 10.4 shows that the gate logic of the AND and OR gates themselves with the feedback loop would form an asynchronous event cell if the D flip-flop were removed. This is illustrated in Figure 10.5. It can be seen that the Petri net can be synthesized as either a clocked or unclocked (event-driven) system.

Note that if an unclocked (event-driven) system is to be designed, then the gate propagation delays would need to be considered. This is similar to the effects on asynchronous (event-driven) FSMs discussed in Chapter 9.

Pn = Tin + Pn·/Tout

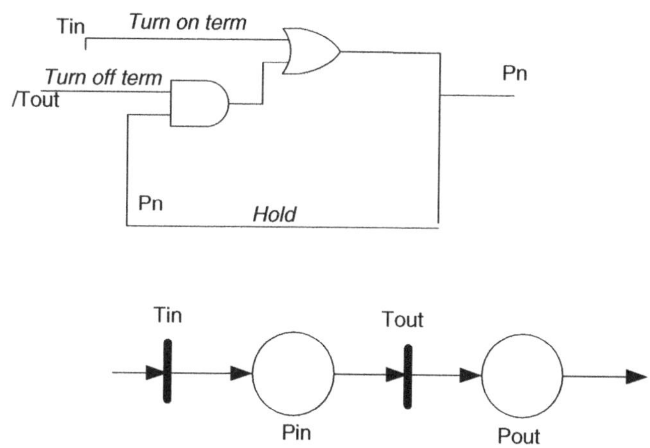

The Tin term equation is of the form:

Tin = input placeholder AND input enable AND NOT output placeholder

Figure 10.5 Asynchronous (event-driven) Petri net structure.

In Petri nets:

- synchronous designs are clock driven with the *D* flip-flop elements;
- asynchronous designs are event driven with the *D* flip-flop elements removed.

There is much research work being carried out on asynchronous Petri nets at a number of universities. You might wish to do a web search using the key words 'Petri nets' and 'C gate' to obtain further information.

The remainder of this chapter will deal with synchronous clock-driven systems.

To consolidate the ideas discussed so far, a sequential Petri-net controller example will be considered.

10.2 SIMPLE SEQUENTIAL EXAMPLE USING A PETRI NET

A sequential Petri-net controller example is illustrated in Figure 10.6. In this example, a pump *P* can be turned on by asserting st high to fire T1. After sensor *v* becomes high, T2 will fire to turn on the motor. Pressing the stop button sp will cause T3 to fire and return the system to placeholder P1, where both motor and pump are turned off.

The equations for this design are shown below, but you might want to cover them up and try to produce them. The equations are illustrated in Figure 10.7, which shows the circuit diagram of the system; initialization circuitry is also shown, with flip flop P1 being set while flip flops P2 and P3 are cleared.

To make this system event driven, the *D* flip-flops can be removed and the feedback loops completed from the OR gate outputs to the two input AND gates so as to form the event cells for P1, P2, and P3.

Sequential Petri-net pump – spin motor problem

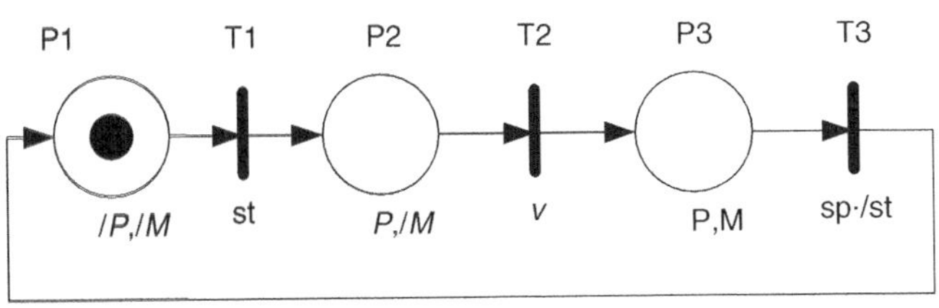

Produce the Petri-net equations for this controller.

T1 =
T2=
T3 =

P1 =
P2 =
P3 =

Figure 10.6 Another sequential Petri net design.

Equations for the design:

T1 = P1·st·/P2
T2 = P2·v·/P3
T3 = P3·sp·/st·/P1·

P1 = T3 + P1·/T1
P2 = T1 + P2·/T2
P3 = T2 + P3·/T3

P = P2 + P3

M = P3

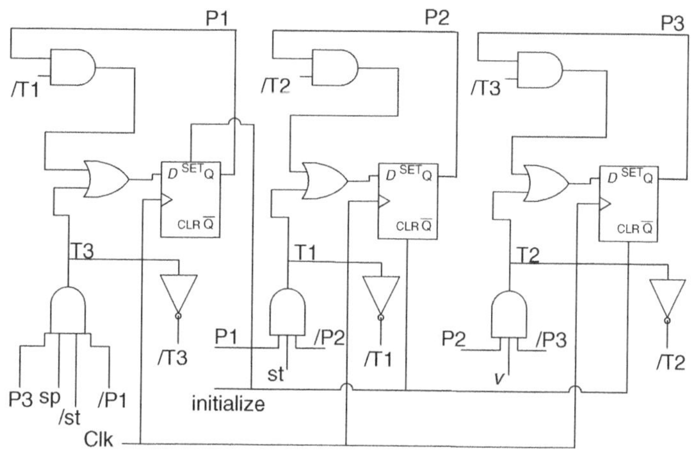

Figure 10.7 Circuit diagram of the Petri net design.

10.3 PARALLEL PETRI NETS

Up to this point, only sequential Petri nets have been considered. However, the main point of using the Petri net is to allow parallel systems to be developed. Therefore, parallel Petri nets will now be discussed.

A parallel Petri net will have parallel paths containing sequences. Figure 10.8 illustrates such a Petri net. In this Petri net there are three parallel paths between the T2 and T5 transitions. P1 and P2 form a sequential path. At T2, they 'fork' into three parallel paths. At T5 these parallel paths 'join' to form a sequential path again.

When the token reaches P2 and the syn1 input becomes active (high), the token will transfer to P3, P4, and P5, as illustrated in Figure 10.9. The system will now have three event cells (and *D* flip-flops) set at the same time.

Suppose input *p* becomes active (high) but input *q* is not yet active (high). The result will be as shown in Figure 10.10. If, at this point, syn2 were to go active (high), then transition T5 would not fire because the token has not yet reached P7.

A requirement for a Petri net is that all the placeholders merging into a transition (P6, P4, and P7 into T5) must have a token before the transition can fire.

Eventually, when input $q = 1$, T4 will fire and the token in P5 will move to P7.

In Figure 10.11, all placeholders merging into T5 have tokens; so, whenever syn2 $= 1$, T5 will fire and the tokens will 'join' and P1 will obtain the token again.

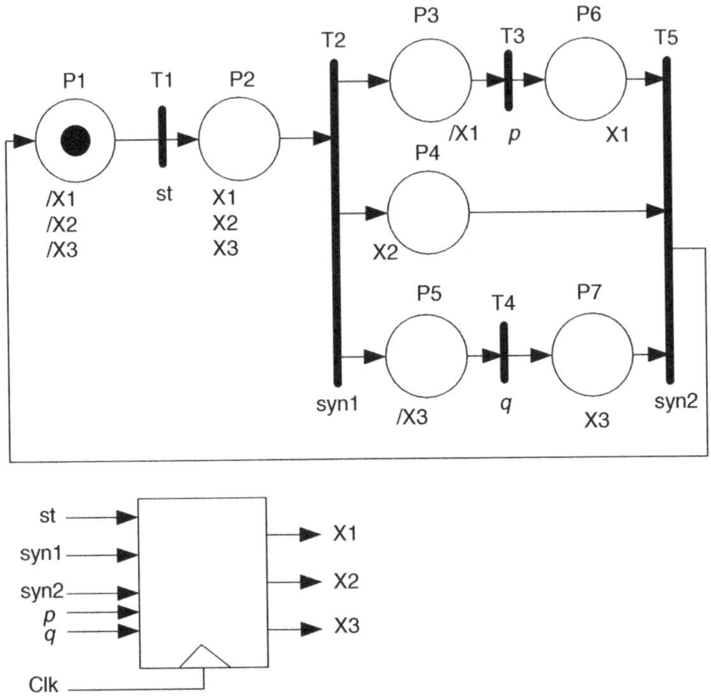

Figure 10.8 Petri net with parallel paths.

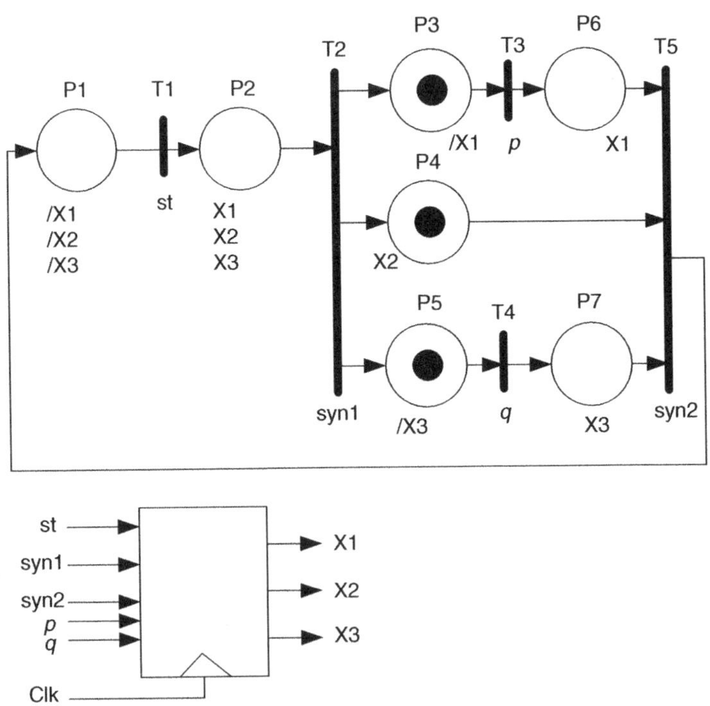

Figure 10.9 Tokens moved into three parallel paths (fork).

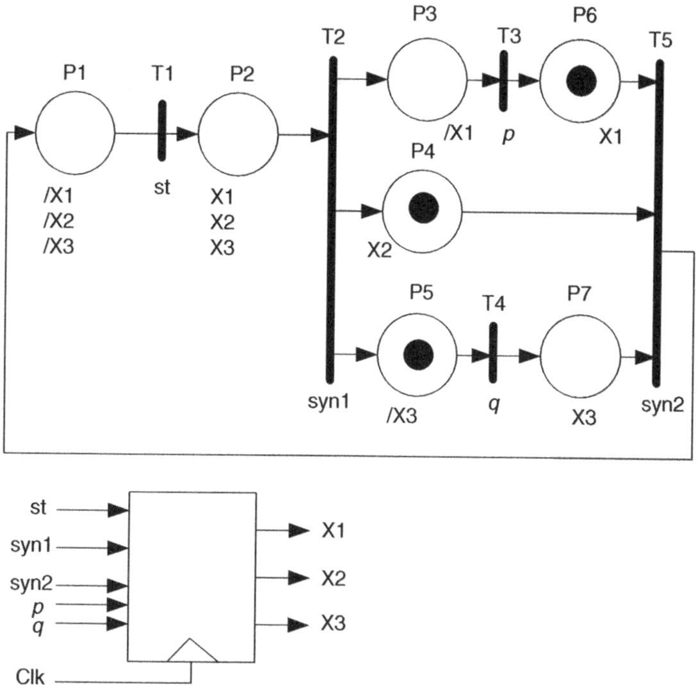

Figure 10.10 Input $p = 1, q = 0$ with P5, P6, and P4 active, but not P7.

The above discussion has described a mechanism in which sequential flow can become parallel flow and merge back into sequential flow again. Most parallel systems behave in this manner, and the Petri net can be used to model such behaviour. This has been one of the principle uses for Petri nets in the past.

In the example illustrated in Figures 10.8–10.11, the transitions T2 and T5 act as synchronizing points; syn1 (controlling the firing of T2) is used to synchronize the point of 'fork', and syn2 (controlling the firing of T5) is used to synchronize the point of 'join'. So, in a hardware system, the two signals syn1 and syn2 act as synchronizing points.

However, the Petri net is self-regulating, since all placeholders converging onto a transition must have tokens before the transition can fire.

The equations will now be developed for this example.

First the placeholder terms:

$$\mathrm{P1} = \mathrm{T5} + \mathrm{P1} \cdot /\mathrm{T1}$$

$$\mathrm{P2} = \mathrm{T1} + \mathrm{P2} \cdot /\mathrm{T2}$$

$$\mathrm{P3} = \mathrm{T2} + \mathrm{P3} \cdot /\mathrm{T3}$$

$$\mathrm{P4} = \mathrm{T2} + \mathrm{P4}/\mathrm{T5}$$

$$\mathrm{P5} = \mathrm{T2} + \mathrm{P5} \cdot /\mathrm{T4}$$

$$\mathrm{P6} = \mathrm{T3} + \mathrm{P6} \cdot /\mathrm{T5}$$

$$\mathrm{P7} = \mathrm{T4} + \mathrm{P7} \cdot /\mathrm{T5}.$$

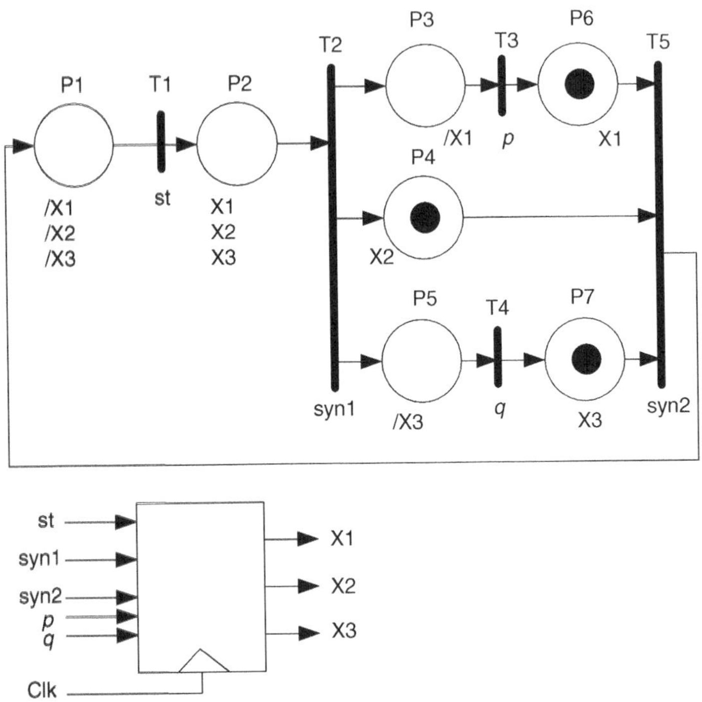

Figure 10.11 T5 can fire whenever input syn2 becomes active (high).

Now the transition terms:

$$T1 = P1 \cdot st \cdot /P2$$
$$T2 = P2 \cdot syn1 \cdot /P3 \cdot /P4 \cdot /P5.$$

Note here that for T2 to fire there must be a token in placeholder P2, signal syn1 must be active, but *none* of the P3, P4, or P5 placeholders must be active.

$$T3 = P3 \cdot p \cdot /P6$$
$$T4 = P5 \cdot q \cdot /P7$$
$$T5 = P6 \cdot P4 \cdot P7 \cdot syn2 \cdot /P1.$$

Here, all parallel path placeholders merging onto T5 must have a token. The equations for T2 and T5 need to be noted.

Finally, the outputs can be written as

$$X1 = P2 + P6$$
$$X2 = P2 + P4$$
$$X3 = P2 + P7.$$

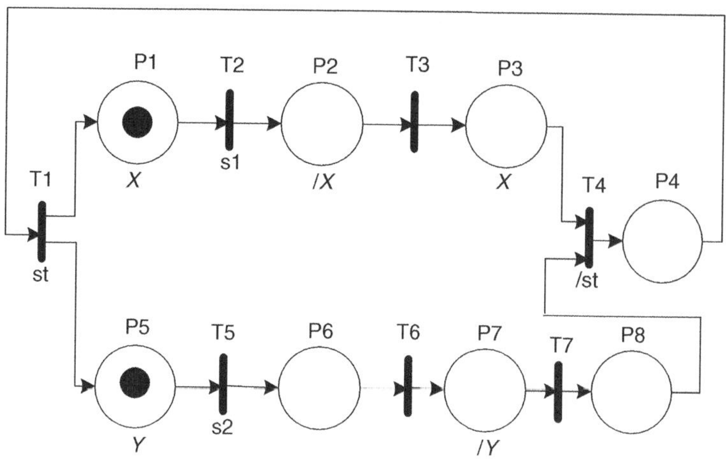

Figure 10.12 Another parallel Petri net example.

10.3.1 Another Example of a Parallel Petri Net

Figure 10.12 illustrates another Petri net example. You might like to try to write down the equations for this one and check the solution with the equations below. The results should be as follows.

The placeholder terms are

$$
\begin{aligned}
P1 &= T1 + P1 \cdot /T2 \\
P2 &= T2 + P2 \cdot /T3 \\
P3 &= T3 + P3 \cdot /T4 \\
P4 &= T4 + P4 \cdot /T1 \\
P5 &= T1 + P5 \cdot /T5 \\
P6 &= T5 + P6 \cdot /T6 \\
P7 &= T6 + P7 \cdot /T7 \\
P8 &= T7 + P8 \cdot /T4.
\end{aligned}
$$

The transitional terms are

$$
\begin{aligned}
T1 &= P4 \cdot st \cdot /P1 \cdot /P5 \\
T2 &= P1 \cdot s1 \cdot /P2 \\
T3 &= P2 \cdot /P3 \quad \text{there is no input against the transitionT3} \\
T4 &= P3 \cdot P8 \cdot /st \cdot /P4 \\
T5 &= P5 \cdot s2 \cdot /P6
\end{aligned}
$$

$$T6 = P6 \cdot /P7$$
$$T7 = P7 \cdot /P8.$$

The outputs are

$$X = P1 + P3 + P4$$
$$Y = P5 + P6.$$

10.4 SYNCHRONIZING FLOW IN A PARALLEL PETRI NET

In the example in Section 10.3, use was made of synchronizing inputs syn1 and syn2 to synchronize the flow from sequential to parallel, and from parallel to sequential. Sometimes, however, there is a need to synchronize between two separate Petri nets. Consider the example in Figure 10.13.

This clearly cannot be done without having some shared communication. It is a classical problem in parallel programming systems. However, in a parallel programming system, a share variable might be considered appropriate. This is dangerous, since this variable could be written to by either of the two parallel entities.

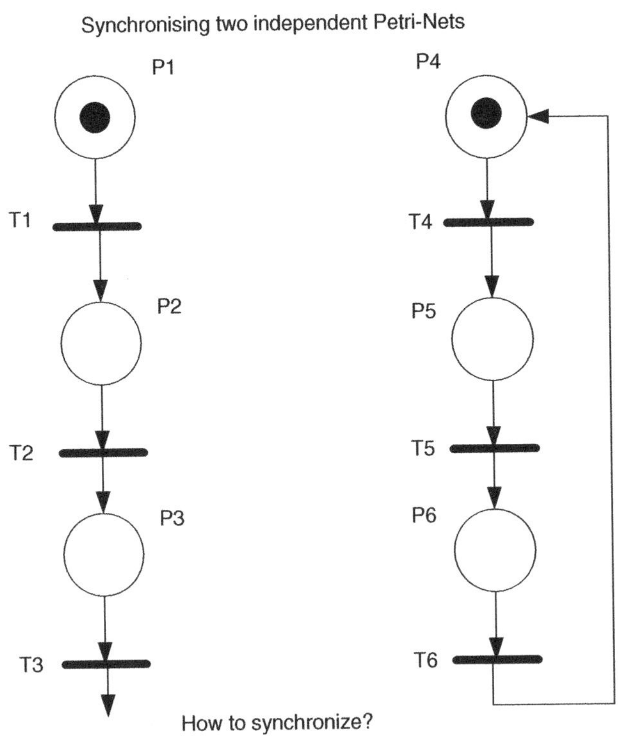

Figure 10.13 Synchronizing two independent Petri nets?

10.4.1 Enabling and Disabling Arcs

In the Petri net there is a way to overcome this problem, using either

- an enabling arc, or
- a disabling arc.

Consider, first, the action of an enabling arc. As can be seen from Figure 10.14, the process made up from P1 to P3 and the process made up from P4 to P6 are totally independent. However, the dashed line from P2 to T5 indicates that there must be a token in P2 in order to enable T5. However, T5 must also have a token in P5 *and* its go signal must be active (high). So, the condition for T5 to fire will be

$$T5 = P2 \cdot P5 \cdot go \cdot /P6 \quad \text{transition equation with enabling arc.}$$

This arrangement ensures that both Petri nets are at a particular state in their sequence (P2 and P5) before T5 can fire.

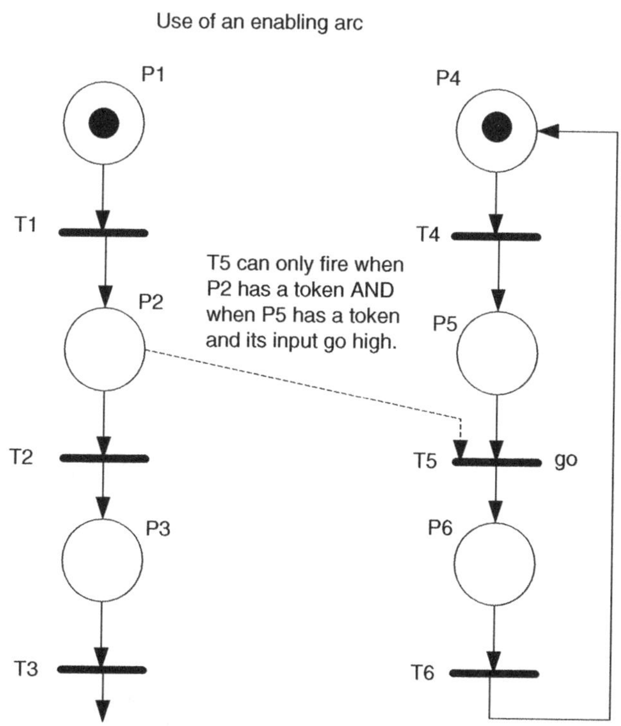

Figure 10.14 The enabling arc.

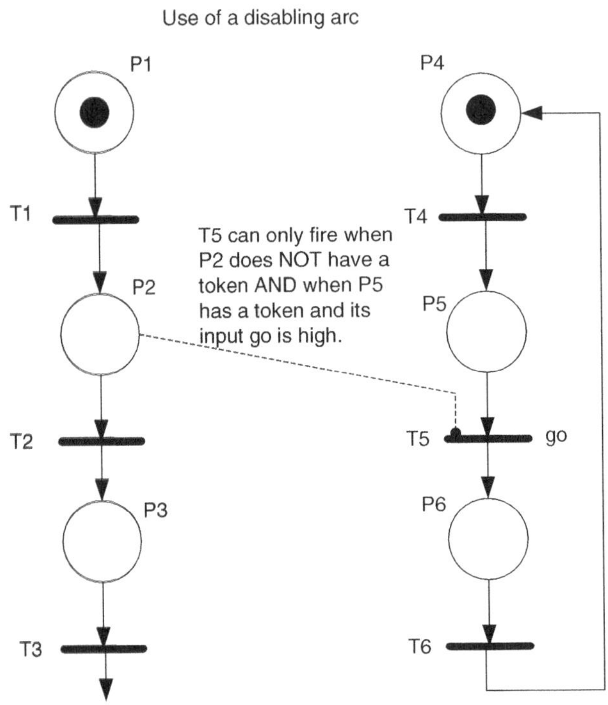

Figure 10.15 Disabling arc to avoid progression at a certain point in the Petri net.

Now consider the disabling arc in the example of Figure 10.15. In this example, the Petri net comprising P1 to P3 can stop the process in the other Petri net P4 to P6 if a token is in P2. This would be represented by the equation

$$T5 = /P2 \cdot go \cdot P5 \cdot /P6.$$

Here, there must *not* be a token in P2, even if P5 has a token and input signal go = 1.

Now an example follows showing how these two ideas could be used in practice.

10.5 SYNCHRONIZATION OF TWO PETRI NETS USING ENABLING AND DISABLING ARCS

In the example of Figure 10.16, the sequence of flow is forced to follow a set sequence:

1. It is assumed that in this system the token will always arrive at P5 first, perhaps because of external circumstances.
2. The token in P5 cannot move on to P6 until the arrival of a token in P2.
3. The token cannot move on from P2 to P3 because T2 is disabled by the disabling arc from P5.
4. As soon as the input signal *go* = 1, the token in P5 can move to P6.
5. This removes the disablement of T2 and the token in P2 can move on to P3.

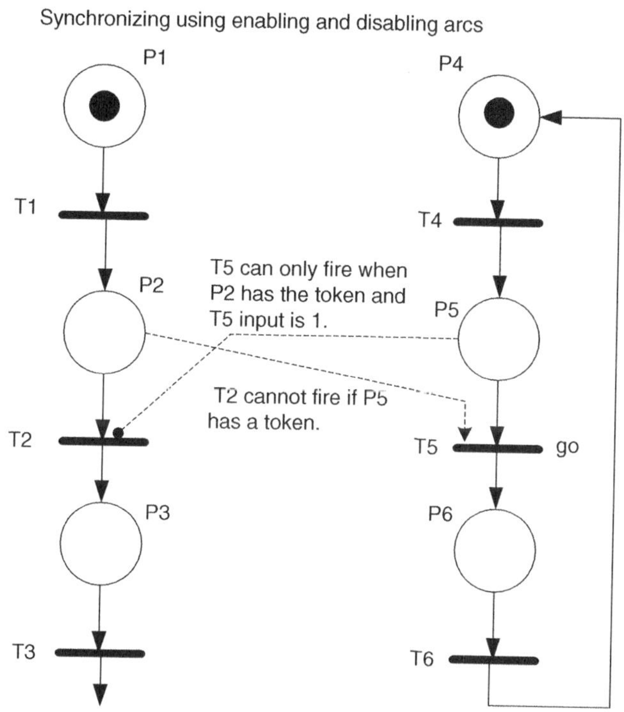

Figure 10.16 Provision of priority to a particular sequencing of two independent Petri nets.

This example illustrates the idea of how the enabling and disabling arcs can be used to check flow.

10.6 CONTROL OF A SHARED RESOURCE

Now consider the more practical example shown in Figure 10.17, which illustrates a system in which two computers, computer A and computer B, share a common resource (e.g. a printer) via a shared data bus. They are separated from the shared resource via tri-state buffers that are controlled by signals EA and EB via a Petri-net controller. Inputs to the Peri-net controller are ra and rb, which are sent by the respective computers. Computer A is to have priority over computer B.

There are a number of ways in which this problem could be resolved, but the most elegant is that shown in Figure 10.18. In this solution, two independent Petri nets have been used: one for processing the ra signal from computer A and the other from computer B.

If computer A accesses its ra signal before computer B accesses its rb signal, then the token in P1 will move to P2 and the disabling arc will disable T3 so that the arrival of a signal on rb will be blocked.

In due course, computer A will lower its ra signal and the token will move back to P1. If rb did arrive during the time that computer A had access to the shared resource, then the token in P3 will not move to P4 because T3 is disabled.

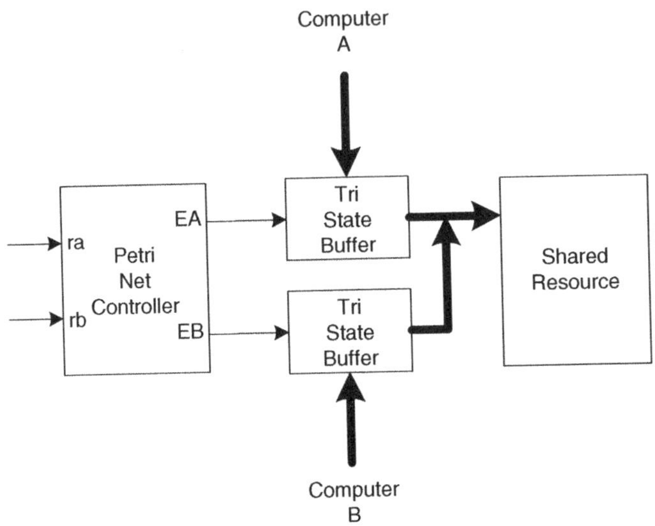

Figure 10.17 Shared resource controller.

Note that if computer A wants to access the shared resource again while computer B has access to it, raising its ra signal will cause the token in P4 to move back to P3 and the token in P1 will move to P2 as well. So, computer A has a priority over computer B.

Of course, if during the time that computer B has access to the shared resource there is no access by computer A, then, when computer B has finished its access, lowering of rb will cause the token to move back to P3.

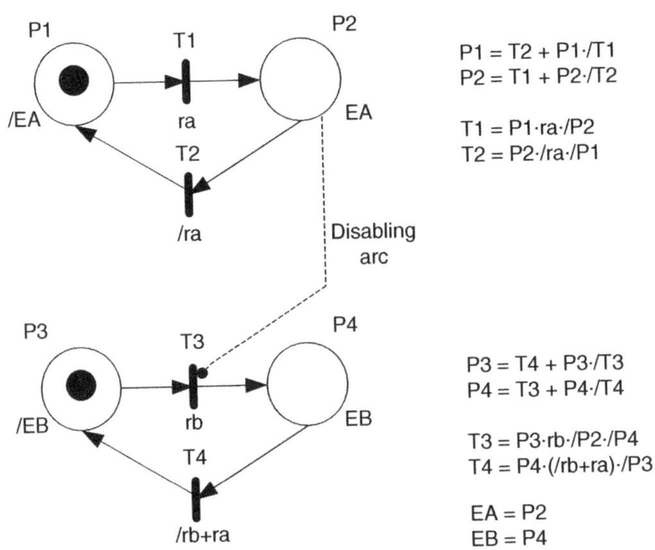

Figure 10.18 A solution to the shared resources problem.

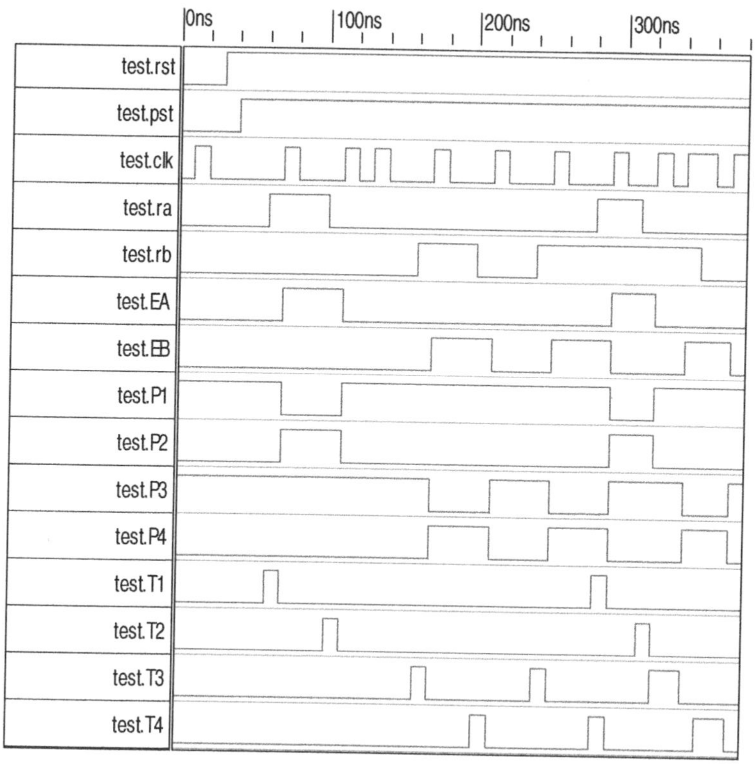

Figure 10.19 Simulation of the shared resource Petri net.

The equations for the Petri-net controller are given in Figure 10.18. In particular, note the equation for T3 with its disabling placeholder /P2 term. T3 can only fire if there is a token in P3 *and* rb is active high *and* there is not a token in P4 *or* P2.

At the start of the simulation (see Figure 10.19), P1 and P3 are active due to the initialization with rst and pst inputs (see Verilog HDL code in shared resource folder of Chapter 10 on the CDROM). Each input ra then rb is asserted in turn to simulate requests for access to the shared resource. At the seventh clock pulse the rb input has become active; then ra is active at the eighth clock pulse (priority request from computer A). This results in computer A gaining access to the shared resource from computer B. Computer A then completes its transaction and, since rb is still active, computer B regains access to the shared resource. In due course, rb returns to its low state and the Petri net returns the token in P4 to P3 to relinquish computer B access to the shared resource.

10.7 A SERIAL RECEIVER OF BINARY DATA

In Section 4.7, an asynchronous binary data receiver was developed using a state diagram implemented with *D*-type flip-flops, together with a shift register, a divide-by-11 counter and a data latch developed using the techniques in Appendix B.

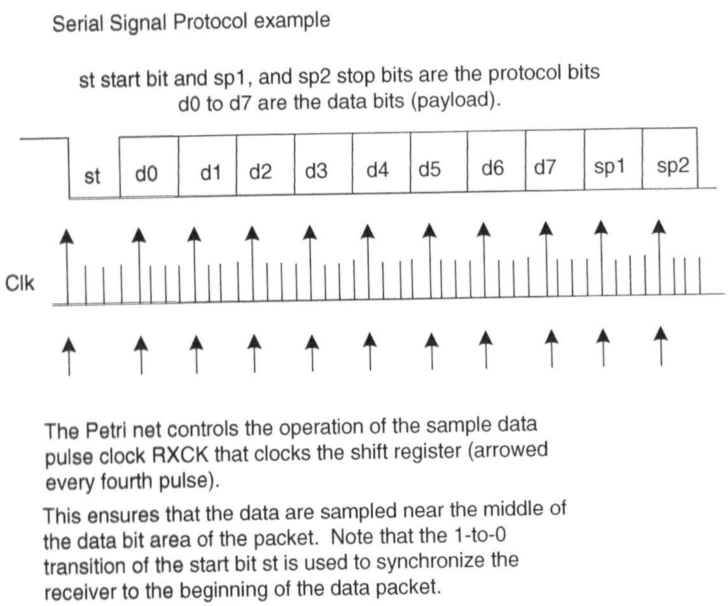

Figure 10.20 Arrangement of the data packet and protocol.

In this section, a similar design is described making use of a Petri-net controller. The design is described in detail, so it can be studied without reference to the one in Chapter 4.

An asynchronous serial receiver is to be developed using a Petri net to allow binary data to be received and converted into parallel data. The Petri net is a good way of implementing the serial receiver, since use can be made of the enabling arc and the design can be implemented using a single interconnected Petri net diagram.

As a reminder of the arrangement used in Chapter 4, the protocol and sample points are illustrated in Figure 10.20. The asynchronous serial protocol is to be one start bit (active low), followed by eight data bits, and two stop bits (11 bits in total). The incoming data need to be shifted into a shift register, and it is important to ensure that this is done when the incoming data have had time to settle. This can be achieved by using a clock that runs faster than the shift register clock so that the point in time that the shift register data is clocked into the shift register is around the middle of the available bit time interval.

In Figure 10.20, the bit time interval is around four clock periods, and at the second clock pulse into the data cell the data at the shift register input are to be clocked into the shift register (indicated by the arrowed clock pulse points). Thus, the shift register clock will be four times slower than the main state machine clock clk. This will be increased in this Petri net version.

Figure 10.21 illustrates a possible Petri-net-based block diagram for the system. In this system, a Petri-net controller is used to control the operation of the system, which consists of an 11-stage shift register with parallel outputs to a data latch. Note that the data into the data latch include only the data bits d0 to d7 (Q0 to Q7), not the protocol bits st, sp1, and sp2. The Divide-by-11 Counter (which could be either an asynchronous binary counter or a synchronous binary

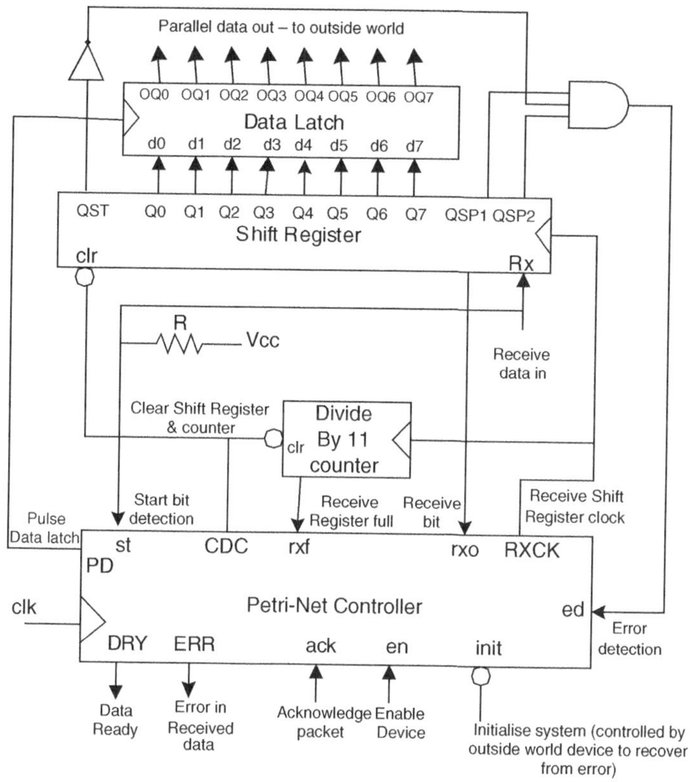

Figure 10.21 Block diagram for the asynchronous serial receiver system.

counter along the lines of those developed in Appendix B) is used to count the number of input data bits received and produces an output rxf indicating to the Petri net that the shift register is full. The shift register is clocked with the RXCK signal derived from the clk signal within the Petri-net controller.

Should an error occur indicated by the ed signal, the error (ERR) output will be asserted high and the system will wait for a reinitialization from some external device ready for the next attempt at receiving a serial data packet. The control here would be via the external device using the serial receive system. The overall system is very similar to the one developed in Chapter 4.

The system is started by the start bit going low, as seen by the serial data in line. Figure 10.22 shows the Petri net diagram developed for the system. This consists of two Petri net diagrams connected by enabling arcs. The first one, comprising P1 to P5, is used to generate the shift register clock rxck. The second, and main, Petri net diagram controls the operation of the system. Both Petri nets are driven from the same clock clk.

Note the use of three enabling arcs. The first one, from PM2, is used to disable the first Petri net via its T1 firing transition until the main Petri net receives a start st data bit. The second enabling arc, from P5, is used to prevent the main Petri net from moving on to PM3 until the first Petri net has generated a shift-register clock pulse RXCK. Note, also, that an enabling arc is used to

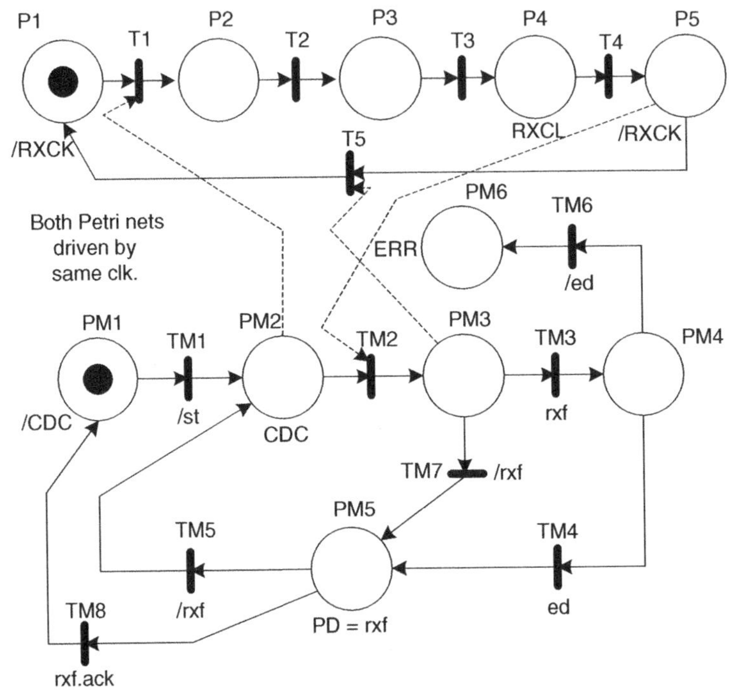

Figure 10.22 Petri net diagram for the asynchronous serial system.

prevent T5 from firing until the main Petri net moves to PM3; otherwise, there is a potential race condition between T5 and TM2.

Thus, the first Petri net can generate shift-register clock pulses at the correct time in the data packet. Note that there are five clock pulses between each RXCK in this realization, rather than the four as suggested in Figure 10.20. Thus, the system clock needs to be five times the required baud rate.

In the main Petri net, the placeholder PM3 and its two transitions TM3 and TM7 test for the shift-register full signal rxf. If low (shift register not full), then the main Petri net loops back to PM2.

Note that while rxf = 0, PM5 will not generate the PD signal (Mealy output). Also, TM5 can fire on rxf = 0. In due course a full data packet of 11 bits will be received. At this point, the main Petri net will move on to PM4 to check the ed signal. This signal should be high if st, sp1 and sp2 are received correctly. This being the case, the main Petri net will move on to PM5, where it will issue a PD signal (since rxf = 1 now) to latch the received data into the data latch ready to be collected by the outside world.

The main Petri net will wait for an ack signal (since rxf = 1 now) from the outside world (indicating that the data have been read) before returning the token to the PM1 placeholder and resetting the shift register and 11-bit counter.

In this Petri net, use has been made of enabling arcs to synchronize the two Petri nets, and a Mealy output for signal PD allows a common loop to be used under different conditions.

10.7.1 Equations for the First Petri Net

$$P1 = T5 + P1 \cdot /T1 \quad T1 = P1 \cdot PM2 \cdot /P2$$
$$P2 = T1 + P2 \cdot /T2 \quad T2 = P2 \cdot /P3$$
$$P3 = T2 + P3 \cdot /T3 \quad T3 = P3 \cdot /P4$$
$$P4 = T3 + P4 \cdot /T4 \quad T4 = P4 \cdot /P5$$
$$P5 = T4 + P5 \cdot /T5 \quad T5 = P5 \cdot PM3/P1.$$

10.7.2 Output

$$RXCK = P4.$$

10.7.3 Equations for the Main Petri Net

$$PM1 = TM8 + PM1 \cdot /TM1 \qquad TM1 = PM1 \cdot /st \cdot /PM2$$
$$PM2 = TM5 + TM1 + PM2 \cdot /TM2 \qquad TM2 = PM2 \cdot P5 \cdot /PM3$$
$$PM3 = TM2 + PM3 \cdot /TM3 \cdot /TM7 \qquad TM3 = PM3 \cdot rxf \cdot /PM4$$
$$PM4 = TM3 + PM4 \cdot /TM4 \cdot /TM6 \qquad TM4 = PM4 \cdot ed \cdot /PM5$$
$$PM5 = TM4 + TM7 + PM5 \cdot /TM5 \cdot /TM8 \qquad TM5 = PM5 \cdot /rxf \cdot /PM2$$
$$PM6 = TM6 + PM6 \qquad TM6 = PM4 \cdot /ed \cdot /PM6$$
$$TM7 = PM3 \cdot /rxf \cdot /PM5$$
$$TM8 = PM5 \cdot rxf \cdot ack \cdot /PM1.$$

10.7.4 Outputs

$$CDC = /PM1 \quad \text{active low}$$
$$PD = PM5 \cdot rxf \quad \text{Mealy active high}$$
$$ERR = PM6 \quad \text{active high.}$$

The simulation of the Petri net for the receiver is illustrated in Figure 10.23. In this simulation, a test-bench module has been developed so that all paths through the Petri net can be checked. This has required manipulation of the rxf, ack, and ed signals that would normally be controlled by the external controller. A study of the waveforms in Figure 10.23 shows the test paths.

Essentially, the simulation shows how the enabling arcs control the sequence of both the shift clock generation produced by P1 to P5, and the main Petri net PM1 to PM6.

Further study of the waveforms reveals the sequence between RXCK pulses, as shown in Figure 10.24. This indicates that, during the serial data receiving phase, a shift register pulse occurs every seven FSM clock pulses. Therefore, for a baud rate of 1×10^6 bits per second, an FSM clock of 7 MHz would be required.

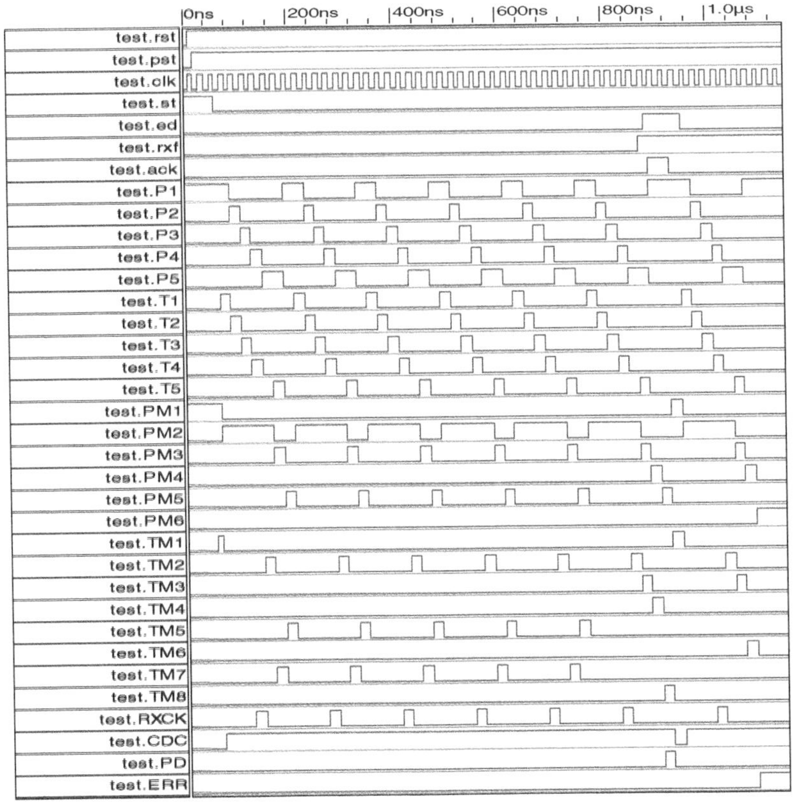

Figure 10.23 Simulation of the Petri net.

The action of the enabling arcs can be clearly seen in Figure 10.23. The simulation ends with an error signal forcing the Petri net into PM6.

The complete Verilog HDL listing can be found on the CDROM in the Chapter 10 folder.

To develop the entire system, the shift register, divide-by-11 counter, the logic AND gate, and data latch also need to be defined and connected together.

10.7.5 The Shift Register

This is an 11-bit device. See Figure B.12a and b in Appendix B for details.

10.7.6 Equations for the Shift Register

For a general shift register of m stages (number of D-type flip-flops)

$$Q_0 \cdot d = \text{din} \quad \text{the data in}$$

$$Q_n \cdot d = Q_{n-1}$$

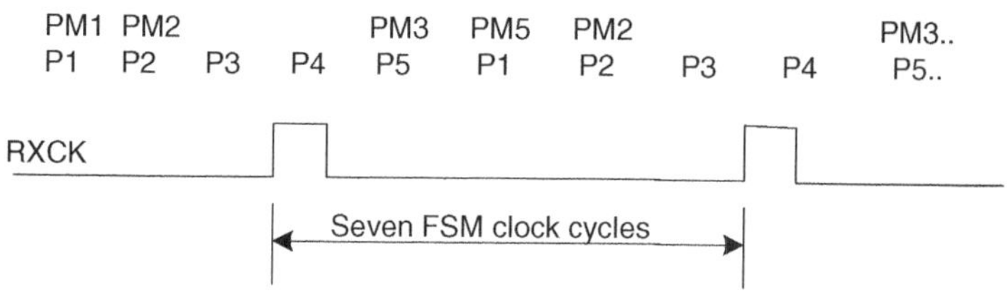

There are seven FSM clock pulses for every rxclk

Therefore the baud rate = FSM clock frequency / 7

In PM2 T1 is enabled and the P1 to P4 cycle can commence.
At P5, TM2 is enabled and the main Petri net can move PM3, then PM5 (rxf = 0) then back to PM2.

The RXCK is produced in P4.

This sequence can continue until rxf goes high (indicating the whole data packet has been received) and the loop is broken. The Petri net will then cycle to PM4 and if ed = 1 (no error) the data will be loaded into the data latch (pd =1) ready for the user to access.

See Figure 10.23 for details

Figure 10.24 Details of Petri net sequence during data receive phase.

for all remaining flip flops where $n = 1$to $n = m - 1$, where m is the number of flip-flops in the shift register.

From this, the equations for the 11-stage shift register are

$$Q_0 \cdot d = \mathrm{rx}$$
$$Q_n \cdot d = Q_{n-1} \quad \text{for } n = 1 \text{to } m - 1 \quad \text{with} \quad m = 11.$$

There is no need to gate the shift-register clock rxck, since it is controlled by the Petri-net controller.

10.7.7 The Divide-by-11 Counter

This can be either a common asynchronous binary counter (ripple through) or a synchronous type. See Appendix B, Section B.9.2 and Figure B.13a and b, for details.

10.7.8 The Data Latch

This is a standard design parallel data latch with eight D-type flip flops each having a data input and data output and all clocked by the pulse data latch signal PD.

Parity detection logic could be added and would follow along the same lines as that used in Chapter 4.

10.8 SUMMARY

The use of Petri nets can provide a means by which parallel control can be realized in hardware. This chapter has explored this area and shown how such systems could be developed and implemented using an HDL. The use of enabling/disabling arcs can help to synchronize parallel Petri net activities.

REFERENCE

1. Fernandes JM, Adamski M, Proeca AJ. VHDL generation from hierarchical Petri net specifications of parallel controllers. IEE Proc Comput Digital Technol 1997; 144(2): 127–135.

AppendixA: Logic Gates and Boolean Algebra Used in the Book

This appendix provides a brief set of notes on Boolean algebra laws and their use. It is assumed that the reader already has knowledge of these laws. The appendix is provided as a reference only for the Boolean algebra used in this book.

A.1 BASIC GATE SYMBOLS USED IN THE BOOK WITH BOOLEAN EQUATIONS

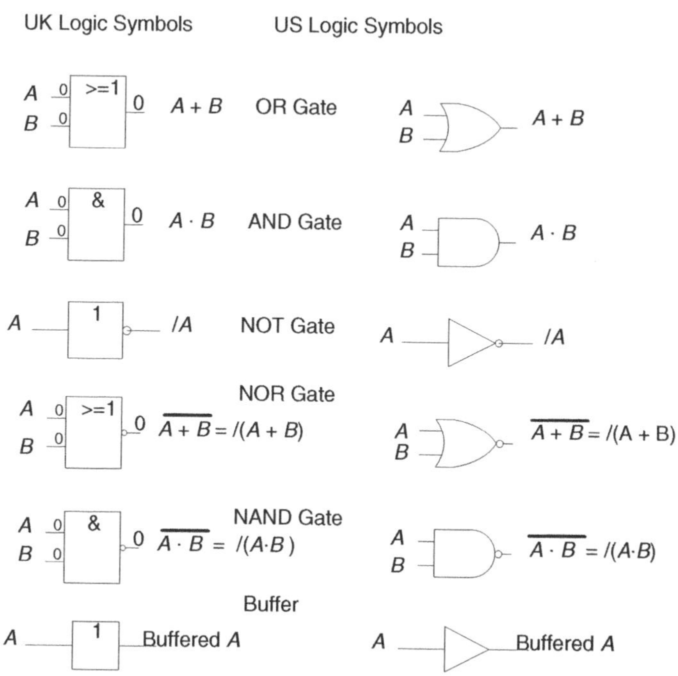

Figure A.1 Basic logic gates.

FSM based Digital Design using Verilog HDL Peter Minns and Ian Elliott

A.2 THE EXCLUSIVE OR AND EXCLUSIVE NOR

The exclusive OR and exclusive NOR (Figure A.2) are well used in logic systems.

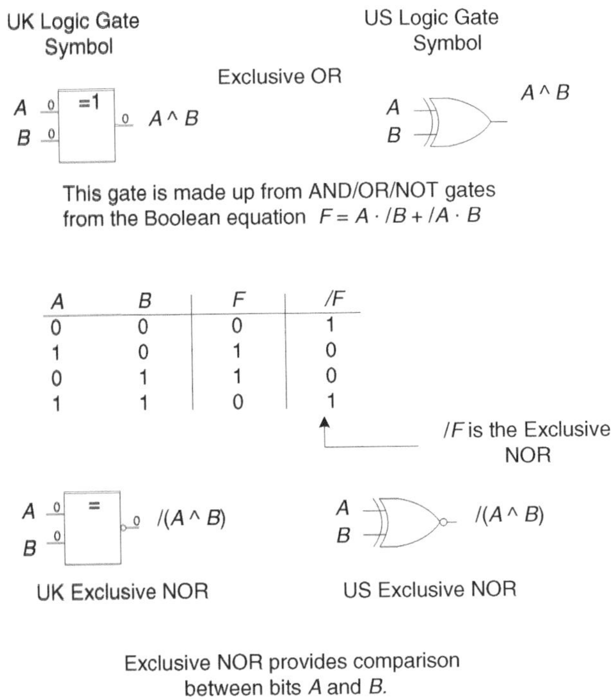

A	*B*	*F*	*/F*
0	0	0	1
1	0	1	0
0	1	1	0
1	1	0	1

Figure A.2 Exclusive OR and exclusive NOR symbols.

A.3 LAWS OF BOOLEAN ALGEBRA

These are presented in terms of the Boolean logic equation and gate circuit.

A.3.1 Basic OR Rules

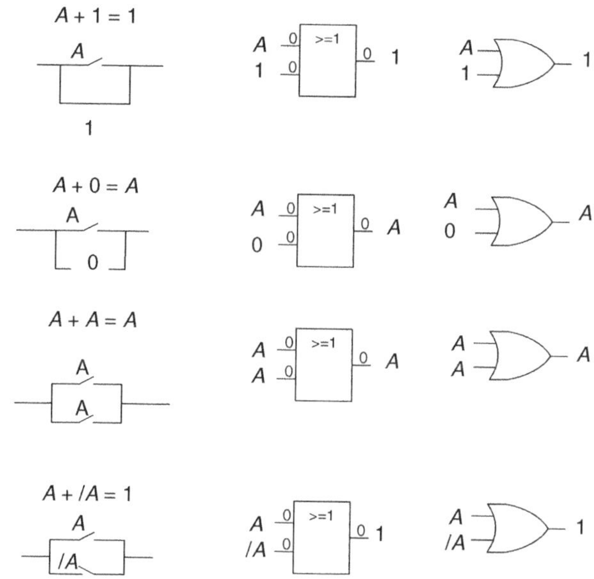

Figure A.3 Boolean algebra basic OR rules.

A.3.2 Basic AND Rules

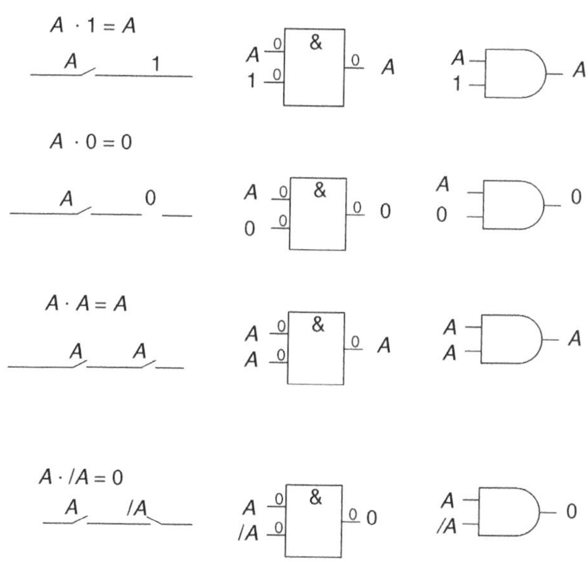

Figure A.4 Boolean algebra basic AND rules.

A.3.3 Associative and Commutative Laws

Associative Laws:

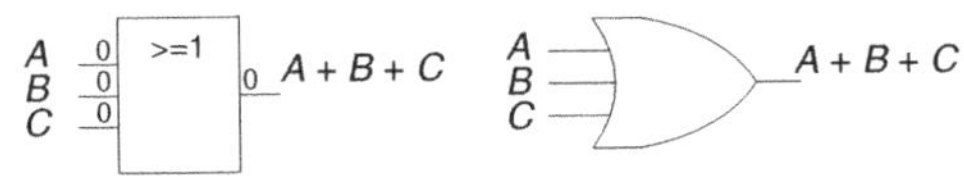

$$(A + B + C) = A + (B + C) = A + B + C$$

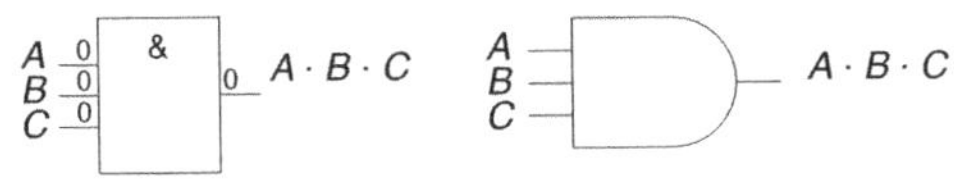

$$(A \cdot B \cdot C) = A \cdot (B \cdot C) = A \cdot B \cdot C$$

Commutative Laws:

$$A + B = B + A$$

$$A \cdot B = B \cdot A$$

These two rules indicate that the order of the literal variables (*A*, *B*, and *C*) can be regrouped (associative rule) and changed (commutative rule).

Figure A.5 Associative and commutative laws.

A.3.4 Distributive Laws

Distributive Laws:

$$F = A \cdot (B + C) = A \cdot B + A \cdot C$$

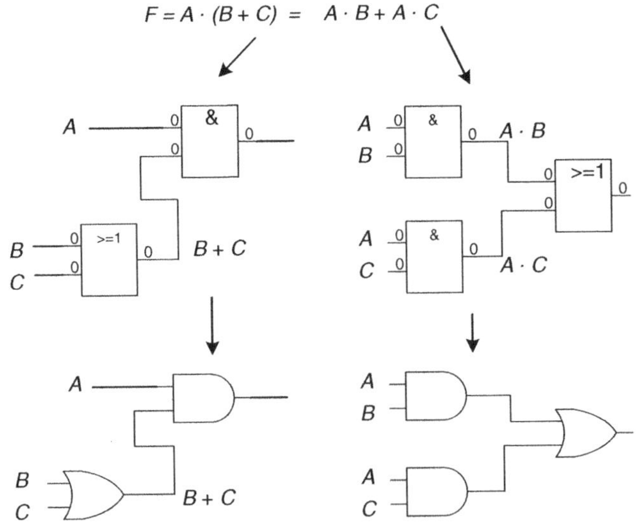

Figure A.6 Distributive laws.

A.3.5 Auxiliary Law for Static 1 Hazard Removal

The auxiliary law (Figure A.7) is particularly significant and much used in this book.

Auxiliary Law: Special Case of Consensus Theorem

A + (B·C) = (A + B) · (A + C) and (A + B)·(A + C) = A + (B·C)

The right hand side shows a circuit reduction.

However, a more interesting example

A + (/A·B) = (A + /A)·(A + B) = 1·(A + B) = A + B

This illustrates a reduction and also the elimination of a static 1 hazard since if B = 1

(A + /A)·(A + 1) = (A + /A).

Try x + (/x·z) = ?

And /p + (q·p) = ?

Figure A.7 Auxiliary law.

A.3.5.1 Proof of Auiliary Rule

The answer to the first question in Figure A.7 is as follows

$$\begin{aligned} Y &= x + /x \cdot z \\ &= (x + /x) \cdot (x + z) \\ &= x \cdot x + x \cdot z + /x \cdot x + /x \cdot z \\ &= x + x \cdot z + 0 + /x \cdot z \\ &= x \cdot (1 + z) + /x \cdot z \\ &= x + /x \cdot z. \end{aligned}$$

The answer to the second question in Figure A.7, $/p + (q \cdot p) = ?$, is, of course, $/p + q$. Note that the auxiliary (Aux) rule is just a special case of the consensus theorem.

A.3.6 Consensus Theorem

Consider the equation

$$Y = x + /x \cdot w \cdot z.$$

In the equation, if $w = z = 1$, then

$$Y = x + /x \cdot 1 \cdot 1 = x + /x,$$

which is 1.

However, it is possible under some conditions of gate delay that a logic 0 glitch can occur (this is known as a static 1 hazard). To avoid this, a cover term can be added, made up from the literals w and z, to make Y stay at logic 1:

$$Y = x + /x \cdot w \cdot z + w \cdot z.$$

Thus, when w and z are both 1:

$$Y = x + /x + 1,$$

thus covering the potential terms $x + /x$ and preventing the glitch.

Now consider the equation

$$Y = x + /xw.$$

Now add a cover term thus:

$$\begin{aligned} Y &= x + /xw + w \\ &= x + w(/x + 1) \\ &= x + w. \end{aligned}$$

So applying the consensus theorem to $Y = x + /xw$ results in $Y = x + w$ as obtained using the auxiliary rule.

In effect the Aux rule is just a special case of the consensus theorem; that is, it is an *auxiliary to the consensus theorem.*

Another example (using the Aux rule):

$$\begin{aligned} Y &= /x + x \cdot z \\ &= /x + z, \end{aligned}$$

since

$$\begin{aligned} /x + x \cdot z &= (/x + x) \cdot (/x + z) \\ &= 1 \cdot (/x + z) \\ &= /x + z. \end{aligned}$$

In the following two examples the term to be removed is crossed out thus $\cancel{/R}$.

$$
\begin{aligned}
P &= q \cdot r + q \cdot /r \cdot s \\
&= q.(r + /r \cdot s) \\
&= q \cdot (r + \cancel{/r} \cdot s) \\
&= q \cdot (r + s) \\
&= q \cdot r + q \cdot s \\
Y &= s \cdot t \cdot /x + s \cdot t \cdot x \cdot z \\
&= s\ .\ t\ .\ (/x + x \cdot z) \\
&= s \cdot t \cdot (/x + z).
\end{aligned}
$$

Therefore:

$$Y = s \cdot t \cdot /x + s \cdot t \cdot z.$$

Note that the reduction here is in the number of inputs. This might be relevant in some cases, but when implementing with PLD devices it may not be so relevant owing to the large number of inputs available in the AND gate array of the PLD.

Remember, however, that the application of the Aux rule eliminates a static 1 hazard and, hence, a potential glitch: in the case of *P*, the term $r + /r$; in the case of Y, $/x + x$.

A.3.7 The Effect of Signal Delay in Logic Gates

In Chapters 3, 4 and 9, it was shown that signal delay can affect the behaviour of a circuit. The basic effect of signal delay is clearly illustrated in Figure A.8, where the points on the two input signals *A* and */A* where an output change can occur are seen for each gate type.

In the case of AND (or NAND) gates, it is the point at which the two overlapping signals are both at logic 1.

In the case of OR (or NOR) gates, it is the point at which the two overlapping signals are both at logic 0.

The effect is to produce an unwanted glitch at the gate output. These so-called 'glitches' can result in maloperation in an FSM. They can occur in a logic system where two signals (they do not need to be the same signals inverted, as shown in Figure A.8) change state at the same time. This could be a change of two state variables; for example, when unit distance coding is not used in a state diagram.

This kind of behaviour can manifest itself within a logic block where two signal changes occur, due to the delays through the logic gates. Further work on this type of behaviour can be found in most textbooks on advanced digital design. These are not discussed any further, as they are beyond the scope of this book.

A.3.8 De Morgan's Theorem

De Morgan's theorem (Figure A.9) is used significantly, particularly in Chapter 9.

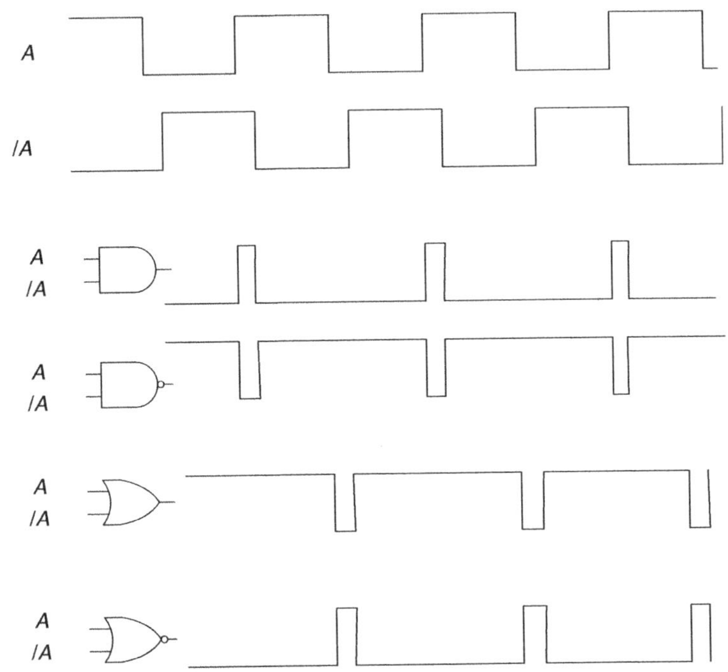

Figure A.8 Effect of delayed signals on gate outputs.

De Morgan's Theorem: very often used.

$$/(A \cdot B \cdot C) = /A + /B + /C$$

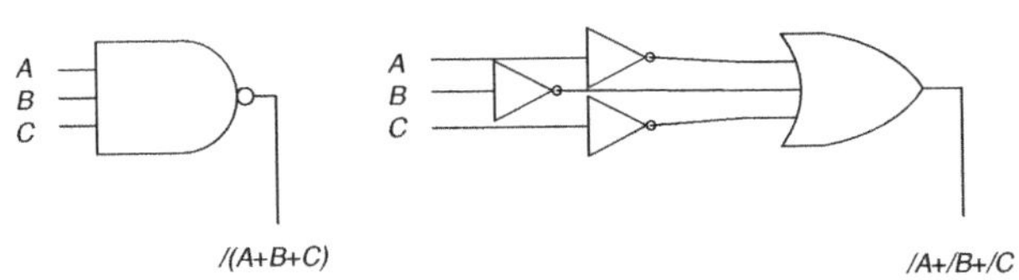

$$/(A + B + C) = /A \cdot /B \cdot /C$$

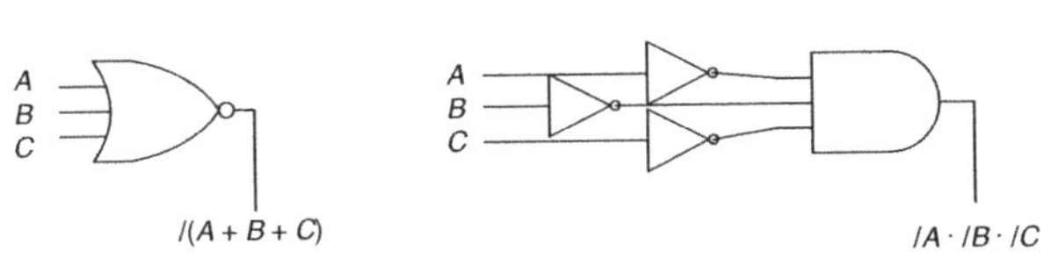

Figure A.9 De Morgan's theorem.

De-Morgan's rules are often used to convert from NAND, $/(a \cdot b)$, to NOR, $/a + /b$, and from NOR, $/(a + b)$, to NAND, $/a \cdot /b$.

A.4 EXAMPLES OF APPLYING THE LAWS OF BOOLEAN ALGEBRA

A.4.1 Example: Converting AND-OR to NAND

$$Z = x \cdot y + /x \cdot /y.$$

Use the De Morgan rule: $/(A + B) = /A \cdot /B$, where A and B can be any product term. So, in the following example, A is replaced by $x \cdot y$,and B is replaced by $/x \cdot /y$:

$$\begin{aligned} Z &= //Z \\ &= //(x \cdot y + /x \cdot /y) \\ &= /(/(x \cdot y) \cdot /(/x \cdot /y)). \end{aligned}$$

A.4.2 Example: Converting AND-OR to NOR

Use the De Morgan rule: $/(A \cdot B) = /A + /B$, where A and B can be any product term. So, in the following example, A is replaced by $x \cdot y$ and B is replaced by $/x \cdot /y$:

$$\begin{aligned} Z &= x \cdot y + /x \cdot /y \\ &= //Z \\ &= //(x \cdot y + /x \cdot /y) \\ &= //(//(x \cdot y) + //(/x \cdot /y)) \\ &= //(/(/x + /y) + /(x + y)). \end{aligned}$$

A.4.3 Logical Adjacency Rule

This rule is used in the Karnaugh map method of Boolean algebra minimization and works on the idea that $b + /b = 1$; so, in the equation

$$\begin{aligned} F &= b \cdot c + b \cdot /c \\ &= b \cdot (c + /c) \\ &= b \cdot 1 \\ &= b, \end{aligned}$$

thus eliminating the literal c (in this case).

The following example makes use of the Logical adjacency rule:

$$\begin{aligned} X &= a \cdot /b \cdot c + a \cdot b \cdot c \\ &= a \cdot c(/b + b) \\ &= a \cdot c \cdot 1 \\ &= a \cdot c. \end{aligned}$$

In the next example, the rule is applied twice:

$$\begin{aligned} Q = /a \cdot /b \cdot /c + /a \cdot /b \cdot c + a \cdot b \cdot /c + a \cdot /b \cdot /c \\ = /a \cdot /b \cdot (/c + c) + a \cdot /c \cdot (b + /b) \\ = /a \cdot /b + a \cdot /c. \end{aligned}$$

The logical adjacency rule is used in most of the work on synchronous FSM design, and also in establishing correct operation in asynchronous (event) FSM design in Chapter 9.

Both the Aux rule and the logical adjacency rules are often used in the reduction of the flip-flop D, and T equations. That is, a typical equation from a state diagram could be

$$\begin{aligned} A \cdot d &= /A \cdot B \cdot \mathrm{st} + A \cdot B + A \cdot /B \cdot \mathrm{sp} \\ &= B \cdot \mathrm{st} + A \cdot B + A \cdot \mathrm{sp}. \end{aligned}$$

This is obtained from

$$\begin{aligned} A \cdot d &= B(/A\mathrm{st} + A) + A(B + /B\mathrm{sp}) \\ &= B(\mathrm{st} + A) + A(B + \mathrm{sp}) \\ &= B\mathrm{st} + AB + AB + A\mathrm{sp}. \end{aligned}$$

Now:

$$AB + AB = AB.$$

Therefore:

$$A \cdot d = B \cdot \mathrm{st} + AB + A \cdot \mathrm{sp}.$$

A.5 SUMMARY

This appendix has looked briefly at the basic laws of Boolean algebra and discussed some of the Boolean techniques used in this book. It provides a reference source for those readers who have not used Boolean algebra for some time. Further information can be found in most books on digital logic.

Appendix B: Counting and Shifting Circuit Techniques

This appendix contains a number of techniques to help in the development of synchronous binary counters and shift registers. These are used in some of the designs covered in chapters throughout the book.

B.1 BASIC UP AND DOWN SYNCHRONOUS BINARY COUNTER DEVELOPMENT

The development of synchronous pure binary up/down counters can be mechanized to produce a general n-stage pure binary counter. This can then be implemented directly using PLDs/complex PLDs (CPLDs)/FPGA devices. To illustrate how this is achieved, a four-stage down-counter is described below.

Table B.1 shows a down-counter with Q0 the least significant bit. This counter is to be designed as a synchronous counter so all flip-flops will be clocked by the same clock edge. Also, the flip-flops will be T flip-flops. Most CPLDs and FPGAs can support the T flip-flop, either directly or by using D-type flip-flops with an exclusive OR input.

The equation for the T input of each flip flop can be obtained by inspection of Table B.1 and entering a product term for every 0-to-1 and 1-to-0 transition required by each flip flop. For example, from Table B.1 the equation for flip flop $q0 \cdot t$ will be

$$q0 \cdot t = s15 + s14 + s13 + s12 + s11 + s10 + s9 + s8 + s7 + s6 + s5 + s4 + s3 + s2 + s1 + s0 = 1.$$

Each state where the T flip-flop is to change state (0 to 1 or 1 to 0) is entered into the equation.

This can then be written in terms of the Q0Q1Q2Q3 outputs, or simply entered into a Karaugh map as illustrated in Figure B.1. The state map of Figure B.1 can then be used to help to minimize the flip-flop equations.

Since all cells will be filled with ones for the $q0 \cdot t$ equation (every cell whose term appears in the $q0 \cdot t$ equation), then the T input for flip-flop Q0 will be logic 1.

The equation for flip flop $q1 \cdot t$ will be

$$q1 \cdot t = s14 + s12 + s10 + s8 + s6 + s4 + s2 + s0 = /Q0$$

FSM based Digital Design using Verilog HDL Peter Minns and Ian Elliott

Table B.1 A down-counter.

Q0	Q1	Q2	Q3	State
1	1	1	1	s15
0	1	1	1	s14
1	0	1	1	s13
0	0	1	1	s12
1	1	0	1	s11
0	1	0	1	s10
1	0	0	1	s9
0	0	0	1	s8
1	1	1	0	s7
0	1	1	0	s6
1	0	1	0	s5
0	0	1	0	s4
1	1	0	0	s3
0	1	0	0	s2
1	0	0	0	s1
0	0	0	0	s0

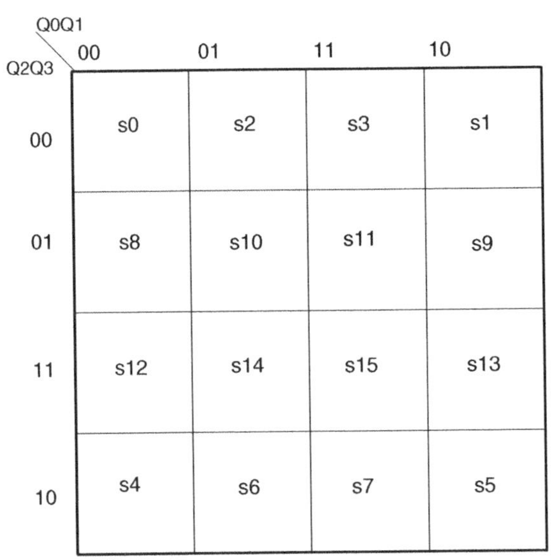

Karnaugh state map showing all states

Figure B.1 State map for the counter.

from the state map. An inspection of the state map of Figure B.1 shows that $q1 \cdot t$ must minimise to /q0, since cells s14, s12, s10, s8, s6, s4, s2, and s0 all contain a 1. Following on in this manner, $q2 \cdot t$ and $q3 \cdot t$ can be obtained thus:

$$\begin{aligned} q2 \cdot t &= s12 + s8 + s4 + s0 \\ &= /Q0 \cdot /Q1 \\ q3 \cdot t &= s8 + s0 \\ &= /Q0 \cdot /Q1 \cdot /Q2. \end{aligned}$$

The patterns of equations follow in a general manner and can be expressed in the form

$$qx \cdot t = /Q(x-1) \cdot /Q(x-2) \cdot /Q(x-3) \cdot \ldots \cdot /Q(x-x). \quad \text{(B.1)}$$

Equation (B.1) describes the *p* terms for a down-counter implemented with *T* flip-flops. These equations can be directly entered into a Verilog HDL file for each flip-flop.

An up-counter can be realized by replacing all the /*q* terms in Equation (B.1) with *q* terms as shown in Equations (B.2) and (B.3):

$$qx \cdot t = Q(x-1) \cdot Q(x-2) \cdot Q(x-3) \cdot \ldots \cdot Q(x-x). \quad \text{(B.2)}$$

Or, in general:

$$qn \cdot t = \prod_{p=1}^{p=n} Q(n-p) \quad \text{(B.3a)}$$

with

$$q0 \cdot t = 1. \quad \text{(B.3b)}$$

For each flip-flop where Π is the product (i.e. AND) of each output term. Note that TFF Q0 has its *T* input at logic 1. This is not covered in Equation (B.3a).

These equations can be obtained directly from a Karnaugh state map similar to that shown in Figure B.1, but counting in the opposite direction.

B.2 EXAMPLE FOR A 4-BIT SYNCHRONOUS UP-COUNTER USING T-TYPE FLIP-FLOPS

The following example, illustrated in Figure B.2, is a design for a 4-bit up-counting synchronous counter using the techniques described above.

The equations for each T flip flop are

$$\begin{aligned} q0 \cdot t &= 1 \\ q1 \cdot t &= Q0 \\ q2 \cdot t &= Q0 \cdot Q1 \\ q3 \cdot t &= Q0 \cdot Q1 \cdot Q2. \end{aligned}$$

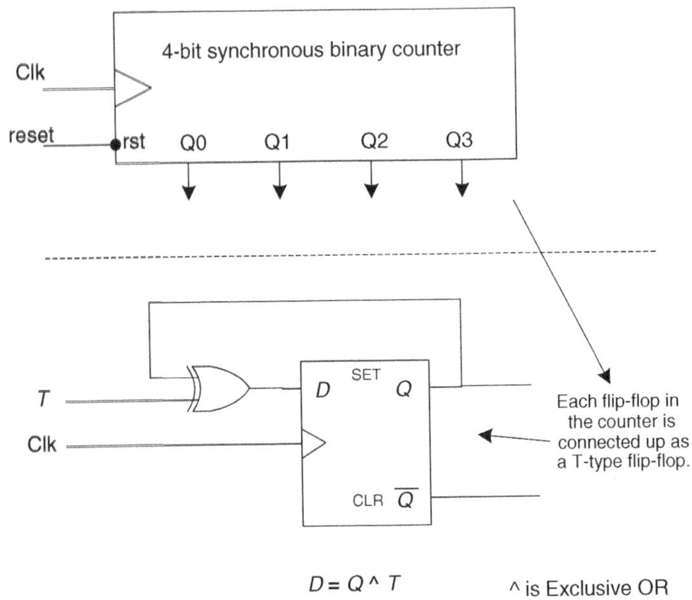

Figure B.2 Block diagram of the 4-bit synchronous binary counter.

This counter can be defined in Verilog HDL as illustrated below in the Verilog source file of Listing B.1.

```
// Four bit counter design.
// Define the TFF.
module T_FF (q,t,clk,rst);
  output q;
  input t,clk,rst;
  reg q; //q output must be registered - remember?
  always @ (posedge clk or negedge rst)
    if (rst == 0)
      q <=1'b0;
    else
      q <=t^q; // TFF is made up with EX-OR gate.
endmodule

// Now define the counter.
module counter(Q0,Q1,Q2,Q3,clk,rst);

input clk, rst; //clk and rst are inputs.
output Q0,Q1,Q2,Q3; // all q/s outputs.
```

```
wire t0,t1,t2,t3; //all t inputs are interconnecting wires.

// need to define instances of each TFF defined earlier.
T_FF ff0(Q0,t0,clk,rst);
T_FF ff1(Q1,t1,clk,rst);
T_FF ff2(Q2,t2,clk,rst);
T_FF ff3(Q3,t3,clk,rst);
// now define the logic connected to each t input.
// we use an assign for this.
assign

t0=1'b1, // this is just following the technique
t1=Q0, // for binary counter design.
t2=Q0&Q1, // will generate AND gates..
t3=Q0&Q1&Q2;
endmodule // end of the module counter.

// Test Bench design to test the circuit under simulation.
module test;
reg clk, rst; // has two inputs which must be registers.
//wire no wires in this part of the design
// since counter is not connected to anything.
counter count(Q0,Q1,Q2,Q3,clk,rst);
initial
  begin
    $dumpfile("counter4.vcd"); // file waveforms..
    $dumpvars; //dump all values to the file.
      rst=0; // initialise circuit with rst cleared.
      clk=0; //set clk to normally low.
      #10 rst=1; // after 10 time units raise rst to remove reset.
      repeat(17)
      #10 clk = ~clk; //change clk 17 times every 10 time units.
      #20 $finish; //Finish the simulation after 20 time units.
  end // end of test block.
endmodule // end of test module.
```

Listing B.1 The Verilog HDL file for the counter, with test bench.

The complete Verilog HDL source file with test-bench module for the counter is shown in listing B.1. This contains the *T*-type flip-flop definition (defined using the behavioural method).

This is followed by the counter definition, which makes use of four instances of the *T* flip-flops and also uses an assign block to define the logic connections between the flip-flop outputs and the *T* inputs of each flip-flop. Note: old-style input and output is used outside of the module header.

Following on from this is the test-bench module. This contains an instance of the 4-bit counter followed by the stimulus to test the counter. Note that there are two **$** commands to save the timing diagram of Figure B.3 so it can be saved to a Word document (for printout) The command **$dumpfile**("counter.vcd"); names the file to be created with the information. The command **$dumpvars**; simply dumps all variables to the file.

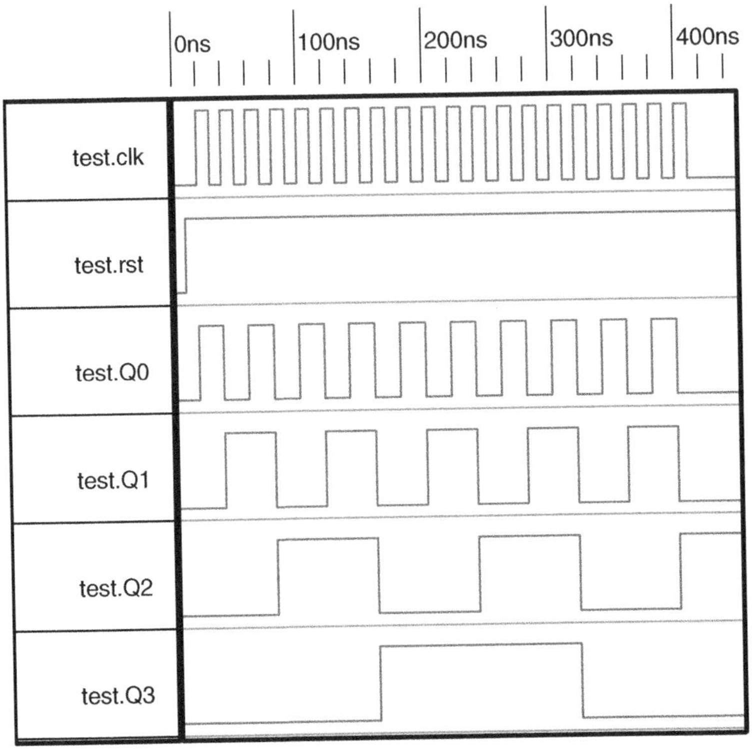

Figure B.3 Simulated 4-bit binary counter.

The file is saved as a 'metafile' and is illustrated in Figure B.3. The waveforms of Figure B.3 clearly show the binary counter sequence.

B.3 PARALLEL-LOADING COUNTERS: USING T FLIP-FLOPS

For a parallel loading counter implemented with cheaper PLDs, a synchronous parallel input may be required if there is not an asynchronous preset and clear input to the flip-flops. This can be done by using additional product terms in the q*x* · *t* equations.

A general bit slice form with the additional inputs is shown in Equation (B.4) for a TFFx:

$$qx.\mathrm{t} = ptermx \cdot /\mathrm{load} + \mathrm{p}x \cdot /\mathrm{Q}x \cdot \mathrm{load} + /\mathrm{p}x \cdot \mathrm{Q}x \cdot \mathrm{load}. \tag{B.4}$$

The load input is used to load the parallel data synchronously into the flip-flop. In this case, the load input is active high.

In Equation (B.4), the product term `ptermx` · /load is the normal product term needed for the counter and is true while the load input is not active. The term p*x* · /Q*x* · load is the parallel input term to set the flip-flop, and the term /p*x* · Q*x* · load is the term to clear the flip-flop.

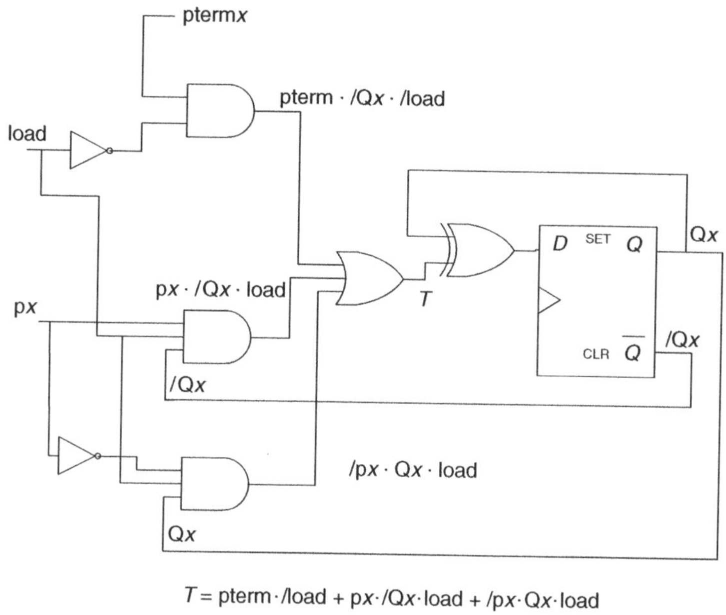

Figure B.4 General structure of a single-flip flop for counting and parallel loading.

Figure B.4 shows a general structure of a single flip-flop. All other flip-flops follow the same general structure. It is assumed here that the active state for the load input is high. Therefore, during counting mode, load would be low (logic 0).

Equations (B.1), (B.2) and (B.4) may be used to produce parallel-loading up/down-counters for many applications, including the address counters for FSMs that control memory.

Thus, it is possible to create not only sequential control of the access of memory, but also random control by way of the parallel inputs.

B.4 USING D FLIP-FLOPS TO BUILD PARALLEL-LOADING COUNTERS WITH CHEAP PROGRAMMABLE LOGIC DEVICES

The *D* flip-flop can be used in place of the *T* flip-flop to implement parallel-loading synchronous counters that do not have preset or clear inputs. There are lots of cheaper PLDs that use only *D* flip-flops and do not have asynchronous preset and clear, so the idea seems attractive.

Consider the circuit of Figure B.5. The bit slice equation for this general model is

$$\mathrm{qx} \cdot d = \mathrm{px} \cdot /l + \mathtt{pterm} \cdot l, \tag{B.5}$$

where l is the parallel loading input and $/l$ the inverted parallel loading input. This defines the general form for the equations for each flip-flop in the counter chain.

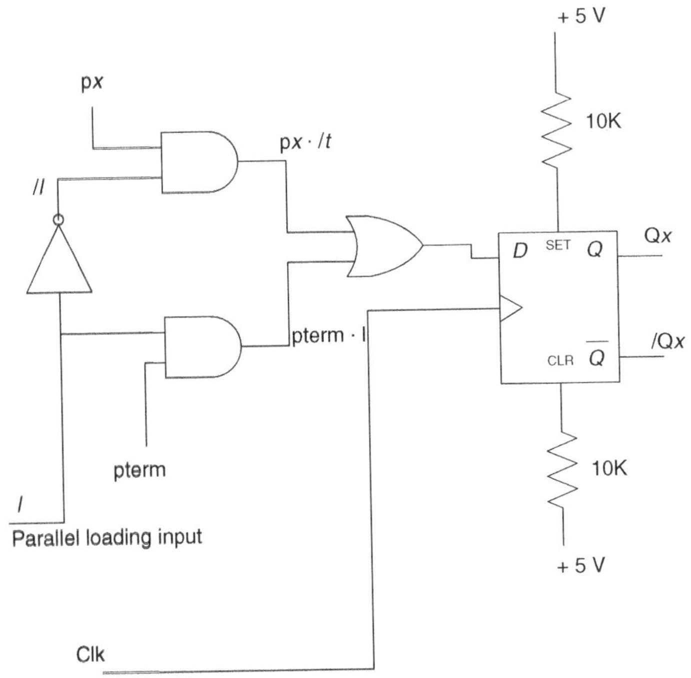

Figure B.5 General bit slice model for of a parallel-loading synchronous counter.

The individual product term `pterm` here will depend upon the sequence table. There is no simple way to do this; therefore, the method is not as easy to implement as that using *T* flip-flops.

As an example, consider a simple three-stage synchronous binary up-counter.

B.5 SIMPLE BINARY UP-COUNTER: WITH PARALLEL INPUTS

To illustrate the form in which a physical circuit will take a simple three-stage parallel-loading pure binary counter is illustrated in Figure B.6.

Looking at Figure B.6, the state sequence illustrates the binary sequence. The state map is used to help simplify the `pterms` (shown here in their simplified form) and, finally, the full equations for the *D* inputs of each flip-flop.

Note that, compared with the method for designing synchronous parallel-loading up/down-counters using *T* flip-flops, this arrangement requires the development of each flip-flop `pterm`. In general, there is no systematic way to do this other than to work out the logic for each flip-flop.

However, one advantage of using *D* flip-flops is that the count sequence is not restricted to pure binary count sequences (i.e. one could develop unit distance code sequences, for example).

Of course, the counter could be developed from the Verilog HDL behavioural description direct, and this would be the more usual way of doing it. The above method, however, gives an insight into the Boolean equations involved in such counters.

Q0	Q1	Q2	State
0	0	0	s0
1	0	0	s1
0	1	0	s2
1	1	0	s3
0	0	1	s4
1	0	1	s5
0	1	1	s6
1	1	1	s7

State sequence

Q2 \ Q0 Q1	00	01	11	10
0	s0	s2	s3	s1
1	s4	s6	s7	s5

State map

pterms

q0 · d = /Q0

q1 · d = Q0 · /Q1 + /Q0 · Q1

q2 · d = Q2 · /Q1 + Q2 · /Q0 + /Q2 · Q1 · Q0

q0 · d = p0 · /l + (/Q0) · l

q1 · d + p1 · /l + (Q0 · /Q1 + /Q0 · Q1) · l

q2 · d + p2 · /l + (Q2 · /Q1 + Q2 · /Q0 + /Q2 · Q1 · Q0) · l

Full equations with parallel loading inputs

Figure B.6 Illustrating the form of the equations for the three-stage pure binary synchronous counter with parallel inputs.

B.6 CLOCK CIRCUIT TO DRIVE THE COUNTER (AND FINITE-STATE MACHINES)

There are many circuit arrangements for crystal oscillators, but the one shown in Figure B.7 is a common one that is often used. It is included for completeness.

The circuit in Figure B.7 provides overtone suppression via the two capacitors C1 and C2 with values to keep the capacitive reactance small, as indicated in Figure B.7.

B.7 COUNTER DESIGN USING DON'T CARE STATES

In some designs, use can be made of states that do not appear in the count sequence. This can lead to a reduction in the number of gates used in the logic of the counter.

Consider the twisted ring counter, so called because it has each flip-flop connected in the form of a ring, but with a twist in the connection between the last flip-flop and the first. Figure B.8 illustrates the state sequence and a design method using a state map to highlight the don't care states.

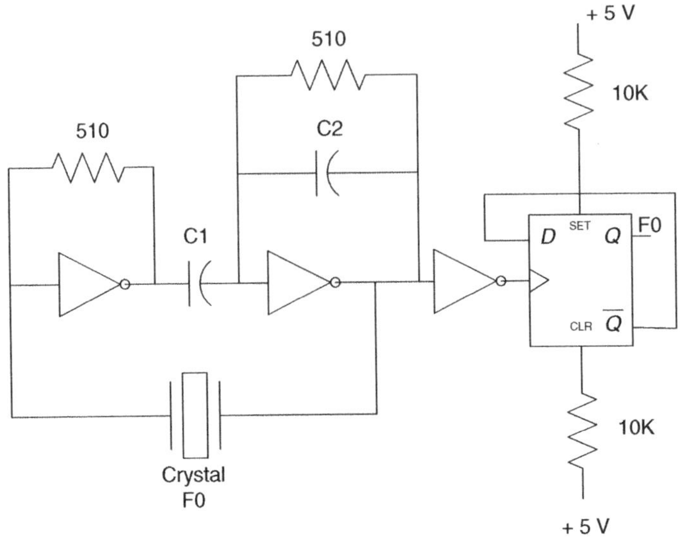

Xc1 @ F0 should tend towards 0

Xc2 @ F0 proportional to 510 ohms

Figure B.7 Typical crystal oscillator circuit.

Q0	Q1	Q2	Q3	State
0	0	0	0	s0
1	0	0	0	s1
1	1	0	0	s3
1	1	1	0	s7
1	1	1	1	s15
0	1	1	1	s14
0	0	1	1	s12
0	0	0	1	s8

State sequence

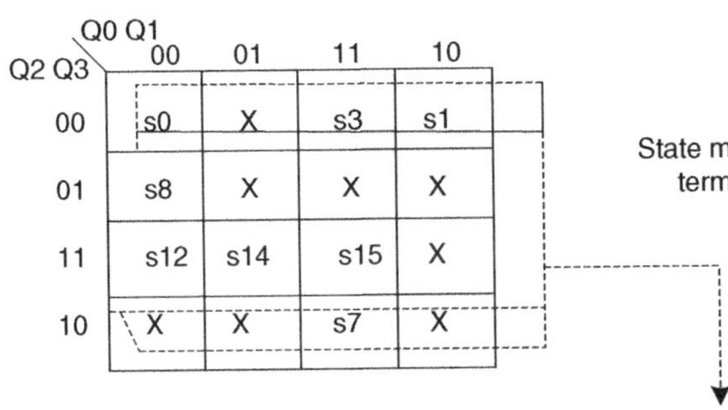

State map with don't care terms (X) included.

Q0 · *d* = s0 + s1 + s3 + s7 + (don't care terms) = /Q3

Q1 · *d* = s1 + s3 + s7 + s15 + (don't care terms) = Q0

Q2 · *d* + s3 + s7 + s15 + s14 + (don't care terms) = Q1

Q3 · *d* + s7 + s15 + s14 + s12 + (don't care terms) = Q2

Figure B.8 Twisted ring counter design making use of don't care terms.

The state sequence table in Figure B.8 shows the required sequence for the counter. From this it is apparent that states s2, s4, s5, s6, s9, s10, s11 and s13 are not part of the sequence, so these are made don't care terms (marked as X) in the state map.

From the state sequence table, and state map of Figure B.8, the equations for each flip-flop *D* input ($Qx \cdot d$) can be obtained, looking for 0-to-1 and 1-to-1 transitions in each column of the sequence table. The don't care terms are then added to the end of each equation. Finally, the state map is used to obtain the minimized equations.

For example, in equation $Q0 \cdot d$, states s0, s1, s3 and s7 are combined with don't care terms s2, s4, s5 and s6 to obtain /Q3 (as highlighted by the dotted lines in Figure B.8). The other equations are dealt with in a similar manner.

B.8 SHIFT REGISTERS

A special form of synchronous counter is the shift register. Quite often, a parallel-loading shift register is required (see examples in Chapter 4). The bit slice form for each stage of the parallel-loading shift register is obtained from Equations (B.6a) and (B.6b):

$$Q0 \cdot d = \text{din} \cdot \text{ld} + \text{p0} \cdot /\text{ld} \quad \text{(B.6a)}$$

$$Qx \cdot d = Q(x-1) \cdot \text{ld} + \text{p}x \cdot /\text{ld}, \quad \text{(B.6b)}$$

where in this case the active state for the load input ld is low and din is data input.

Note that if serial input is to be zero, make din = 0. The shift register design is using *D* flip-flops.

These equations could be used to create a four bit parallel loading counter thus:

$$Q0 \cdot d = \text{din} \cdot \text{ld} + \text{p0} \cdot /\text{ld} \quad \text{(B.7)}$$

$$Q1 \cdot d = Q0 \cdot \text{ld} + \text{p1} \cdot /\text{ld} \quad \text{(B.8)}$$

$$Q2 \cdot d = Q1 \cdot \text{ld} + \text{p2} \cdot /\text{ld} \quad \text{(B.9)}$$

$$Q3 \cdot d = Q2 \cdot \text{ld} + \text{p3} \cdot /\text{ld} \quad \text{(B.10)}$$

$$\text{Sft\_clk} = \text{clk} \cdot \text{ld} \quad \text{(B.11)}$$

In Equation (B.7), the first term is the serial data input. In Equations (B.8)–(B.10), the first term denotes that the output of each flip-flop will connect into the input of the next (i.e. a standard shift-register connection). In addition, Equation (B.11) defines the shift clock. This is disabled during parallel loads.

Figure B.9 shows the four-state shift register developed from the Equations (B.7)–(B.11). Note that, in practice, the equations would be converted into Verilog HDL code direct for synthesization. The equations converted into Verilog HDL are:

```
Q0d = din&ld | po&~ld;
Q1d = Q0&ld | p1&~ld;
Q2d = Q1&ld | p2&~ld;
Q3d = Q2&ld | p3&~ld;
Sft_clk = clk&ld;
```

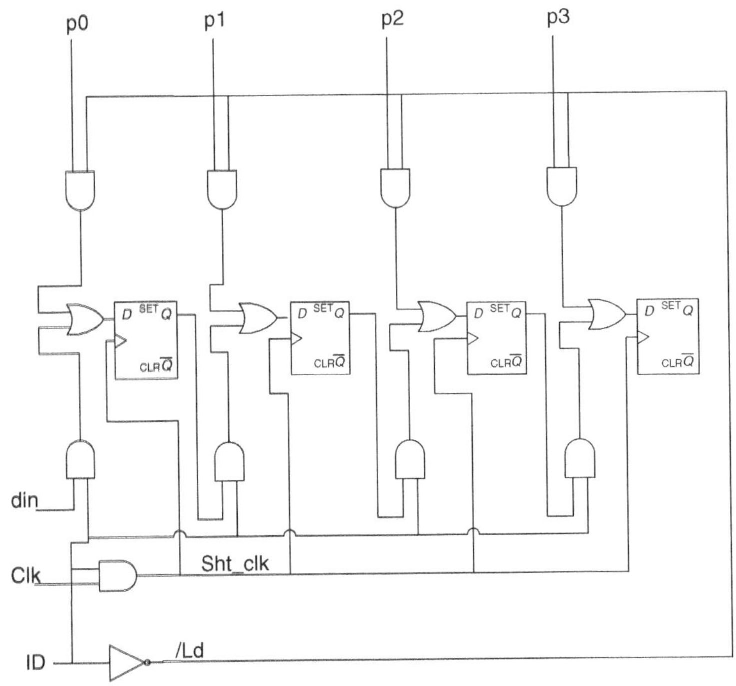

Figure B.9 Four-stage shift register developed from Equations (B.7)–(B.11).

The above shift register, once converted into Verilog HDL code, can then be simulated for correct operation. Figure B.10 shows such a simulation. The Verilog coding is available in the Appendix B folder on the CDROM.

In Chapter 4, the asynchronous serial receiver system made use of a shift register to store the incoming binary data and present them to a data latch. In addition, a divide-by-11 counter was used to keep track of the number of binary bits received and alert the FSM when a complete packet was received (receive shift-register full).

The details and Verilog code for the two modules are now described.

B.9 ASYNCHRONOUS RECEIVER DETAILS OF CHAPTER 4

Figure B.11 (which is Figure 4.21 repeated here for convenience) illustrates the different module blocks needed to make up the complete receiver. Each module in this diagram and its Verilog modules will be described below.

The associated test-bench modules and complete code for the asynchronous receiver are available on the CDROM disk that is supplied with this book. The FSM is described in detail in Section 4.7, with the state diagram Figure 4.22.

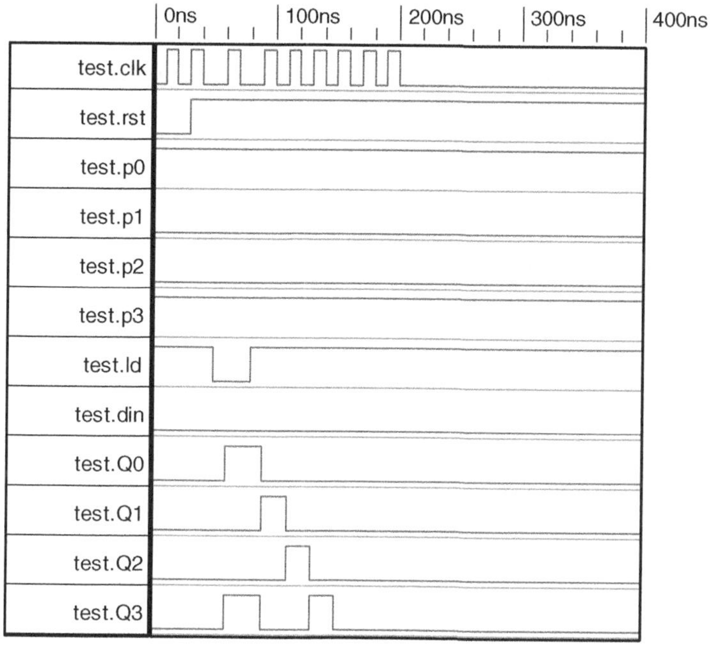

Figure B.10 Simulation of a four-stage shift register with din = 0.

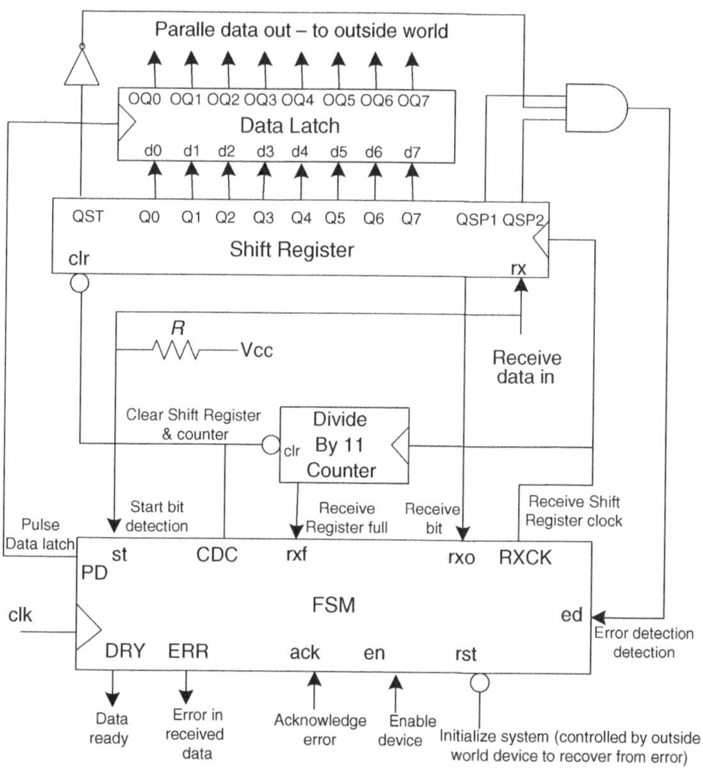

Figure B.11 Asynchronous receiver block diagram from Chapter 4.

B.9.1 The 11-Bit Shift Registers for the Asynchronous Receiver Module

This is an 11-bit shift register with a start bit, eight data bits (d0 to d7), and two stop bits (sp1 and sp2).

The incoming data (din) connect to the sp2 flip-flop and are shifted into the sp1 flip-flop. The last flip-flop in the shift register is the start-bit flip-flop, since this is the first data bit into the shift register. This is illustrated in Figure B.12a.

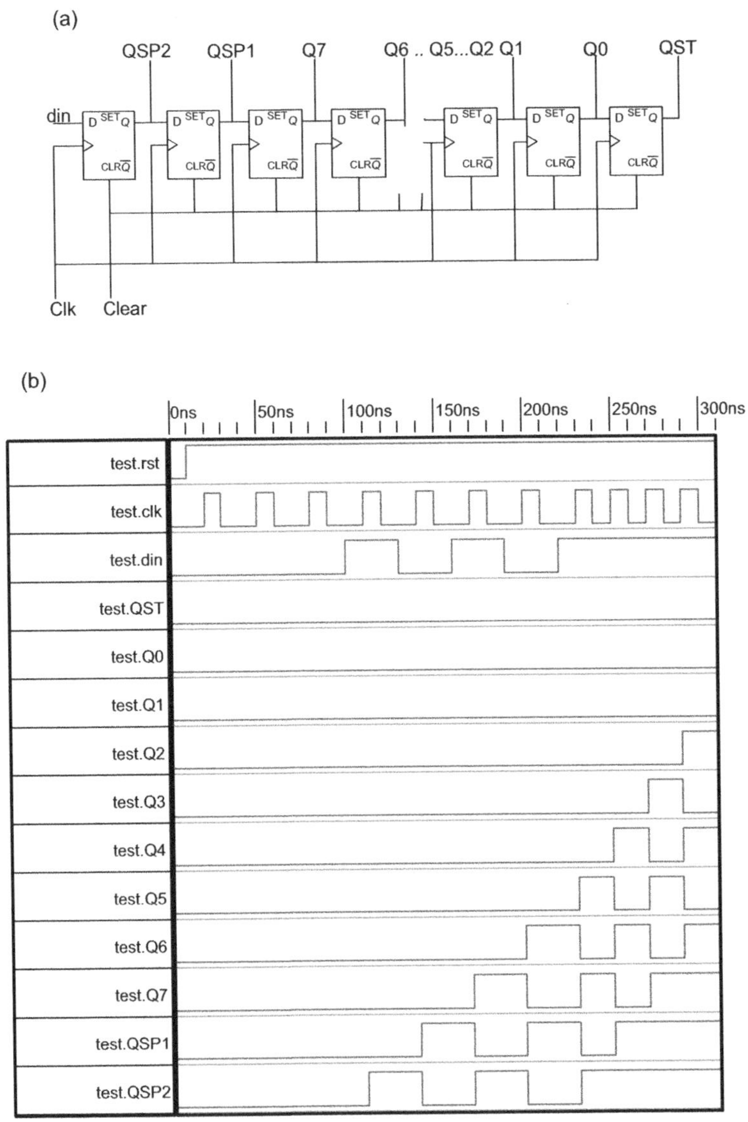

Figure B.12 (a) The shift-registers circuit. (b) Simulation of the shift-register module.

The Verilog HDL code for the shift register is shown in Listing B.2.

```
// Define DFF
module D_FF(q,d,clk,rst);
  output q;
  input d,clk,rst;
  reg q;
  always @ (posedge clk or negedge rst)
    if (rst==0)
      q<=1'b0;
    else
      q<=d;
endmodule
```

Listing B.2 Verilog module for the shift register.

Listing B.3 gives the module used to build the shift register.

```
//-------------------------------------------
// define shift register
// The shift register clock is rxclk which
// is controlled by the fsm.
// The protocol bits (st, sp1, and sp2) are
// shifted into their own FF's.
//-------------------------------------------

module shifter(rst,clk,din,QST,Q0,Q1,Q2,Q3,Q4,Q5,Q6,Q7,QSP1,QSP2);
  input clk,rst,din;
  output QST,Q0,Q1,Q2,Q3,Q4,Q5,Q6,Q7,QSP1,QSP2;
  wire dst,d0, d1, d2, d3, d4, d5, d6, d7, dsp1, dsp2 ;

  D_FF_qstd(QST,dst,clk,rst);
  D_FF q0d(Q0,d0,clk,rst);
  D_FF q1d(Q1,d1,clk,rst);
  D_FF q2d(Q2,d2,clk,rst);
  D_FF q3d(Q3,d3,clk,rst);
  D_FF q4d(Q4,d4,clk,rst);
  D_FF q5d(Q5,d5,clk,rst);
  D_FF q6d(Q6,d6,clk,rst);
  D_FF q7d(Q7,d7,clk,rst);
  D_FF qsp1d(QSP1,dsp1,clk,rst);
  D_FF qsp2d(QSP2,dsp2,clk,rst);
  assign
  // note the way that the flip flops have been connected up.
  dst= Q0,
  d0 = Q1,
  d1 = Q2,
  d2 = Q3,
  d3 = Q4,
  d4 = Q5,
  d5 = Q6,
```

```
    d6 = Q7,
    d7 = QSP1,
    dsp1 = QSP2,
    dsp2 = din;
endmodule
```

Listing B.3 Test-bench module for the shift register.

A simulation of the shift register, illustrated in Figure B.12b, indicates that it is working correctly.

A study of the din waveform and the output from the shift register at around the 300 ns point shows that the shift register has received the incoming data, together with the protocol bits.

B.9.2 Divide-by-11 Counter

The counter uses a synchronous pure binary up-counting sequence that counts up to 11 (1101 binary) and then stops. Its output is the RXF signal. This goes high when the eleventh clock pulse is received.

Figure B.13a illustrates the divide-by-11 counter. This is made up of four *T*-type flip-flops (shown here as *D* types with exclusive OR gate feedback in the circuit diagram). The

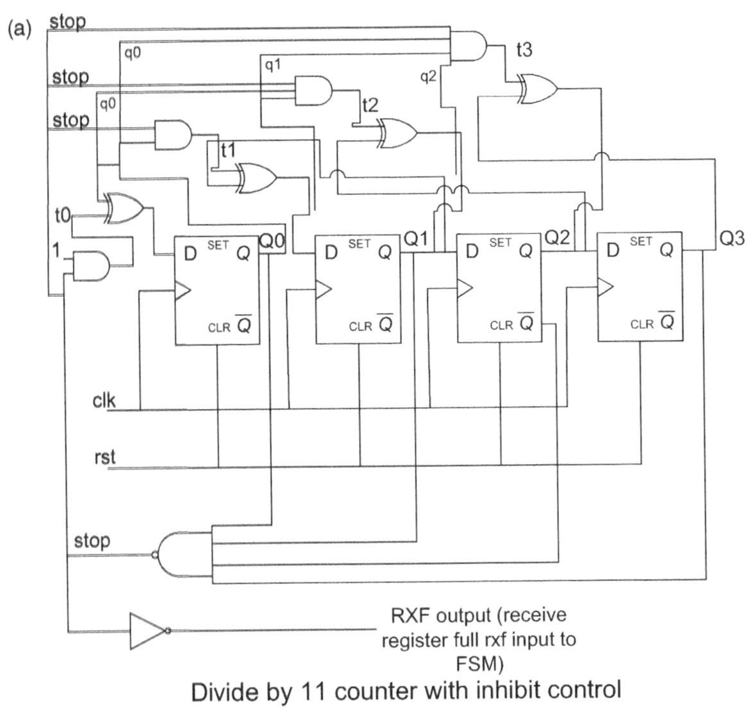

Figure B.13 (a) Schematic circuit diagram of the divide-by-11 counter with inhibit control. (b) The divide-by-11 counter simulation.

(b)

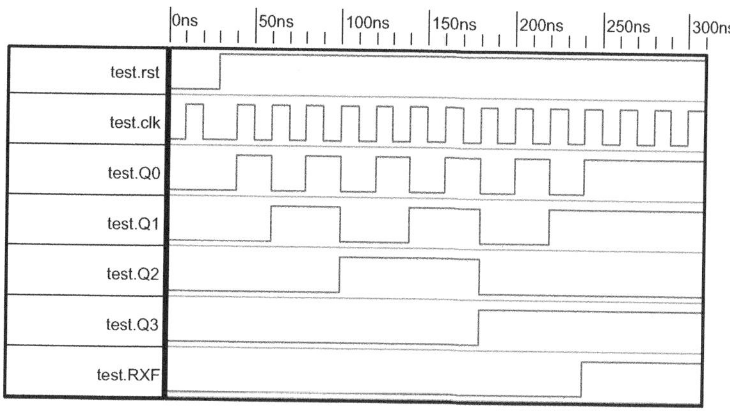

Divide by 11 counter showing details of counter outputs

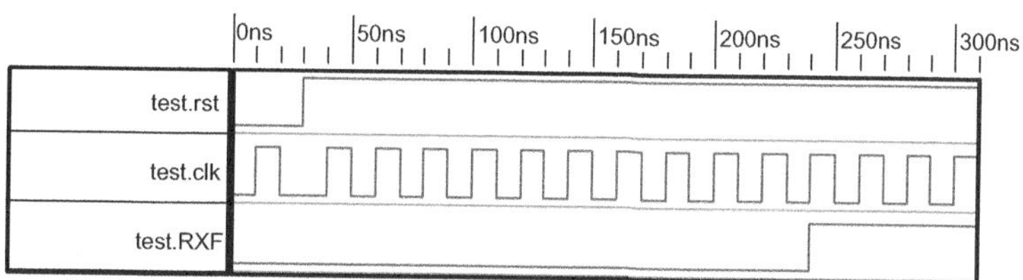

Divide by 11 counter showing only the terminal inputs and outputs to the asynchronous receiver

Figure B.13 (*Continued*)

four-input NAND gate provides a stop control to inhibit the counter when the count value reaches 11 (Q3Q2Q1Q0 = 1011). The reset input rst is used to reset the counter back to zero.

The Verilog code for this module is illustrated in Listing B.4 (all variables in lower case).

```
// define TFF
// Needed for the divide by 11 asynchronous counter.
module T_FF (q,t,clk,rst);
  output q;
  input t,clk,rst;
  reg q;
  always @ (posedge clk or negedge rst)
    if (rst == 0)
      q<=1'b0;
    else
      q<=t^q;
endmodule

// Now define the counter.
module divideby11(Q0,Q1,Q2,Q3,clk,rst,RXF);
```

```
input clk, rst; //clk and rst are inputs.
output RXF,Q0,Q1,Q2,Q3; // all q/s outputs.
wire t0,t1,t2,t3,stop; //all t inputs are interconnecting wires.

// need to define instances of each TFF defined earlier.
T_FF ff0(Q0,t0,clk,rst);
T_FF ff1(Q1,t1,clk,rst);
T_FF ff2(Q2,t2,clk,rst);
T_FF ff3(Q3,t3,clk,rst);

// now define the logic connected to each t input.
// use an assign for this.
assign
t0=1'b1&stop, // this is just following the technique
t1=Q0&stop, // for binary counter design.
t2=Q0&Q1&stop, // will generate AND gates..
t3=Q0&Q1&Q2&stop,
stop = ~(Q0&Q1&~Q2&Q3), // to detect 11the clock pulse.
RXF = ~stop;
endmodule // end of the module counter.
```

Listing B.4 Verilog module for the divide-by-11 counter.

Note that the simulation stops at the eleventh clock pulse due to the NAND gate. This is used to raise the RXF signal via an inverter operation. The RXF (receive register full flag) is used to inform the FSM that the receiver shift register is full. It is cleared by the FSM after transferring the shift register data bits to the octal data latch.

The simulation of this module is illustrated in Figure B.13b.

B.9.3 Complete Simulation of the Asynchronous Receiver Module of Chapter 4

The complete asynchronous receiver with FSM defined in Section 4.7 can now be simulated. The complete Verilog code is contained on the CDROM.

The simulation of the asynchronous receiver is shown in Figure B.14. Here, the only signals visible are those of the complete block, although the secondary state variables are also displayed to show the FSM state sequence. The simulation starts by asserting en high, then the FSM section (signals not seen here) controls the operation of the shift register, divide-by-11 counter, and output data latch.

The data are presented to the user when signal DRY goes high and acknowledged by the user bringing signal ack high. The FSM, in response, lowers DRY (and PD), and the user (optionally) lowers ack to acknowledge the end of the transaction. Prior to loading received data into the data latch its contents are unknown (or the last received).

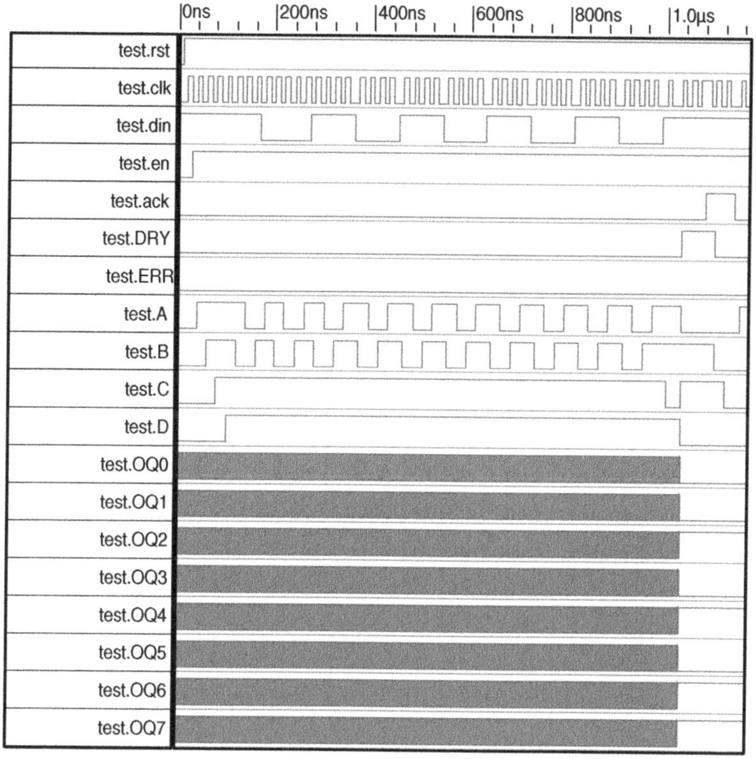

Figure B.14 The complete asynchronous receiver simulation.

B.10 SUMMARY

This appendix has introduced simple ways to develop synchronous up and down pure binary counters, with or without parallel-loading inputs that can be used in a PLD or FPGA device. It has also described how parallel-loading shift registers can be developed and used.

These techniques may be used to develop Verilog HDL modules for use in some of the designs covered in this book. Bit slice equations have been developed to allow counters and shift register circuits to be constructed directly in equation form in Verilog HDL.

Finally, some of these ideas have been used in the development of an asynchronous serial receiver, complete with their Verilog modules.

AppendixC: Tutorial on the Use of Verilog HDL to Simulate a Finite-State Machine Design

C.1 INTRODUCTION

This appendix quickly describes an FSM model in Verilog code and then simulates it using SynaptiCAD's VeriLogger Extreme simulator. The code for the model, VeriLogger Extreme, and the code for most of the examples in the book are contained on the CDROM provided with the book.

A more detailed account of the Verilog HDL is provided in Chapters 6–8, where the language is developed at a slower and more defined pace.

C.2 THE SINGLE PULSE WITH MEMORY SYNCHRONOUS FINITE-STATE MACHINE DESIGN: USING VERILOG HDL TO SIMULATE

The design of a single-pulse generator with memory is outlined and then a Verilog HDL file is created. This Verilog file will use the most basic of the Verilog methods so as to keep it simple.

C.2.1 Specification

Whenever input *s* is asserted high, a single pulse is to be generated at the output *P*. Signal *s* must be returned low and then reasserted high again before another pulse can be generated. In addition, a memory output *L* is to go high to indicate that a pulse has been generated; going low again when the *s* input is returned to logic 0.

C.2.2 Block Diagram

Figure C.1 illustrates the block diagram of the system.

FSM based Digital Design using Verilog HDL Peter Minns and Ian Elliott

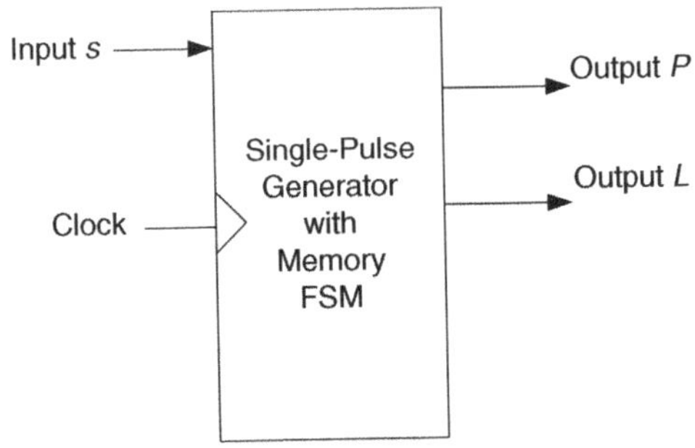

Figure C.1 Block diagram of the system.

C.2.3 State Diagram

A state diagram is implemented as illustrated in Figure C.2.

C.2.4 Equations from the State Diagram

The equations can be derived directly from the state diagram of Figure C.2, in this case using D-type flip-flops:

$$\begin{aligned} A\cdot d &= \mathrm{s0}\cdot s + \mathrm{s1} \\ &= /A/B\cdot s + A/B \\ &= /B\cdot s + A/B \\ B\cdot d &= \mathrm{s1} + \mathrm{s2} + \mathrm{s3}\cdot s \\ &= A/B + AB + /AB\cdot s \\ &= A + B\cdot s. \end{aligned}$$

The output equations are

$$P = \mathrm{s1} = A/B \quad \text{and} \quad L = \mathrm{s2} + \mathrm{s3} = B.$$

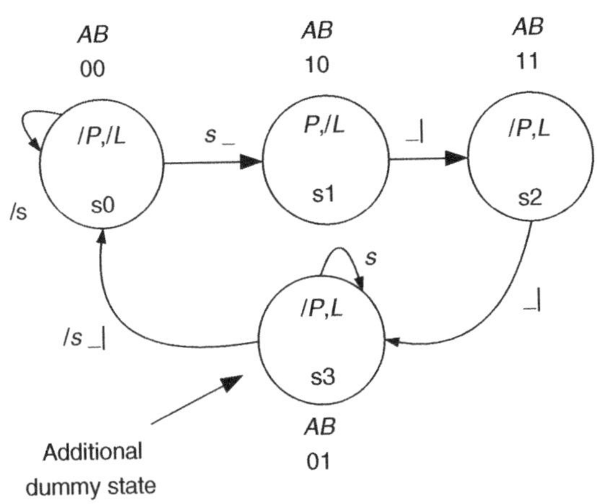

Figure C.2 State diagram of the system.

C.2.5 Translation into a Verilog Description

These equations can be translated into their Verilog form as shown below:

```
ad = ~B&s | A&~B,
bd = A | B&s,
P = A&~B,
L = B;
```

Here, the AND operator (·) is replaced with (&), the OR operator (+) replaced with (|), and the NOT operator (/) replaced with (~). Also, each equation ends with a comma (,) except for the last equation which is ended with a semicolon (;). Finally, the whole equation set must be placed into a continuous assignment thus:

```
assign
ad = ~B&s | A&~B,
bd = A | B&s,
P = A&~B,
L = B;
```

Now the Verilog HDL file will be created. To create a design using the equations just derived, a Verilog HDL file using the data-flow mode of design will be created; that is, the Verilog HDL file is developed using the predefined logic equations.

Alternative ways would be to develop the Verilog HDL file using the logic gates required to build the design or to use a behavioral structure. These alternative methods are described in Chapters 6–8.

In Verilog, a design is built up using one or more modules. A module can have one or more inputs and one or more outputs that define its terminal properties. These can, in turn, be connected together by wires.

In this example, the Verilog description is made up of three modules:

- the module that describes the behavior of the D type flip flop used in the design;
- the module that describes the FSM;
- the module that describes the tests to be carried out on the design (usually referred to as a 'test fixture', or 'test bench', or 'test module').

The first module consists of a behavioural description of the *D*-type flip-flop used in the design. Despite what has been said about the behavioural method, the D-type flip-flop is a standard circuit element that will behave as expected. This *D* flip-flop is created as a module called D_FF. It is illustrated in Listing C.1.

```
module D_FF(output q, input d,clk,rst);
reg q;
always @ (posedge clk or negedge rst)
  if (rst == 0)
    q<=1'b0;
```

```
6    else
7      q<=d;
8  endmodule
```

Listing C.1 The module to define the D-type flip-flop used in the design.

The behavioural description of the *D*-type flip-flop is a description of its terminal behavior. The key words **`module`** and **`endmodule`** define the beginning and end of the module. D_FF is its name, and the signals between the parentheses are the terminal signals of the flip-flop. In this case, *q* is the output and *d*, clk, and rst are inputs. The keywords **`output`** and **`input`** are needed to define the signal types. The line numbers are provided for reference purposes only; they are not entered when creating this Verilog code.

The flip-flop output needs to remember its last (present) state, so is further declared as a register using the keyword **`reg`** in line 2. Note that each of the lines 1, 2, 5, and 7 ends with a semicolon.

In line 3, an **`always`** keyword is used to define the conditions under which lines 4 to 7 will occur. Verilog is defining hardware, and each part of the hardware description needs to be able to execute in parallel. Thus, the **`always`** keyword with `@` is used to define the conditions under which the assignments on lines 5 and 7 will occur. In this case, the conditions are either when there is a logic 0 to logic 1 transition on the clk signal (referred to as **`posedge`**), *or* there is a logic 1 to logic 0 transition on the rst signal (referred to as **`negedge`**).

In line 4, the conditions upon which of the two assignments on lines 5 and 7 will occur are specified using the **`if`** and **`else`** keywords. Here, if the input signal rst is logic 0, then the assignment on line 5 will occur, i.e. `q <=1'b0;`, otherwise the assignment in line 7 (under the **`else`**) will occur, i.e. `q <=d;`. Note the use of <= rather than =. This is preferred in a sequential block (see Chapter 6 for details on why this is the case).

The assignment in line 5, i.e. `q<=1'b0;`, assigns the logic value 0 to the output *q*. The `1'b0` is the way that Verilog defines a single binary bit to logic 0. Logic 1 would be `1'b1`. Hence, the syntax is `<number of bits>'b<binary value, 0 or 1>`.

The assignment in line 7, i.e. `q <=d;`, simply makes the *q* output equal to the input signal value of *d*, this being the required behaviour for the *D*-type flip-flop.

Finally, line 8 defines the end of the module.

Thus, it is seen that the module D_FF defines the terminal behavior of a *D*-type flip-flop. If the rst (reset) is taken low (**`negedge`** `rst`), then the **`if`** `(rst == 0)` will be true and the assignment of line 5 will occur `q<=1'b0`; to reset the flip-flop. Thereafter, **`if`** `rst` is taken to logic 1 (reset removed from the rst input of the flip-flop), every time the clock input clk receives a logic 0 to 1 transition (**`posedge`** `clk`) the *q* output will take on the logic value of the *d* input.

The behavioural methods to define logic circuits and systems are fully described in Chapters 6 and 7.

In the second module (shown in Listing C.2), called the FSM module, two instances of the D_FF are created from the D_FF module, one called FFA in line 3 and the other FFB in line 4. These are connected to the circuit of the single-pulse FSM; see signals inside the parentheses of FFA and FFB.

```
1 module fsm(input S,clk, output P,L,A,B);

2 wire ad, bd;
3 D_FF FFA(A,ad,clk,rst);
4 D_FF FFB(B,bd,clk,rst);

5 assign
6 ad=~B&s | A&~B | A&~s,
7 bd=A | B&s,
8 P=A&~B,
9 L=B;
10 endmodule
```

Listing C.2 The FSM module.

Note that the terminal signals for the FSM are declared between the parentheses in line 1; they are also defined as inputs and outputs. Note also that the flip-flop outputs *A* and *B* are defined here as well. Each instance of the *D* flip-flop needs to be connected to external gates defined in the assignment block so they need to be defined for each flip flop instance.

In line 2, the signals ad and bd (the *d* inputs to each flip-flop) are defined as wires, since they are internal to the FSM module. Lines 3 and 4 define instances of the two flip-flops, using the D_FF name, followed by an instance name (FFA for flip-flop *A* and FFB for flip-flop *B*).

Note here that the *A* output is placed first in the parameter list since it is a *q* output from the flip-flop, the data *d* input is the ad, the clock input clk, and finally the reset input rst. Flip-flop *B* follows in the same manner.

So, by defining the signals used by the D_FF as *q*, *d*, clk, and rst in the behavioral description of the *D* flip-flop, these signals can then be connected up to the signals *A*, ad, clk and rst of the FFA, and *B*, bd, clk and rst of the FFB used in the FSM. The order of these signals is important.

The logic equations follow in lines 6–9; note that this continuous assignment begins with a Verilog keyword **`assign`**, which is needed so that the Verilog compiler can distinguish the following logic equations.

Each line ends with a comma, except the last line 9, which should end with a semicolon, thus defining the end of the continuous assignment. The assignments make use of the = (blocking assignment), since the equations can take place in any order (see Chapter 6 for explanation).

In line 10 the Verilog keyword **`endmodule`** is used to terminate the module that describes the FSM.

Note that in earlier versions of Verilog the modules were defined as shown in Listing C.3. Here, the inputs and outputs are defined after the module header in lines 2 and 3, rather than on the module header in line 1. This is a minor difference, and some older versions of Verilog use this arrangement and not the one shown in Listing C.2.

```
1 module fsm(s,clk, P,L,A,B); // signals here are not specified as
  inputs or outputs.
2 input s,clk; // inputs defined here, not in the header.
3 output P,L,A,B; // outputs defined here, not in the header.
4 wire ad, bd;
```

```
D_FF FFA(A,ad,clk,rst);
D_FF FFB(B,bd,clk,rst);

assign
ad = ~B&s | A&~B | A&~s,
bd = A | B&s,
P = A&~B,
L = B;
endmodule
```

Listing C.3 Alternative 'older' way to define a module.

The two modules defined so far, DFF and FSM, are all that are required to define the FSM. However, in order to test the FSM to ensure that it is correct and performs in the way intended in the specification, a third module is required. This is the test-bench module.

C.3 TEST-BENCH MODULE AND ITS PURPOSE

Figure C.3 illustrates the arrangement of the test-bench module in relation to the FSM module, from which it can be seen that the test-bench module provides test signals (outputs that are registered) to the FSM module. These outputs from the test-bench module are used to test the FSM by applying the signals *s*, rst, and clk in such a way as to test the operation of the state diagram, and hence the FSM.

The test-bench sequence is created by observing the requirements of the state diagram and applying the signals *s*, rst, and clk so that it can test for all conditions. It is the test-bench module that will define the sequence of signals that will be applied to the FSM in order to verify that the state diagram structure is followed correctly. Test-bench modules can be defined in a much more concise form than the one shown here, and you will learn about these in Chapters 6 and 7.

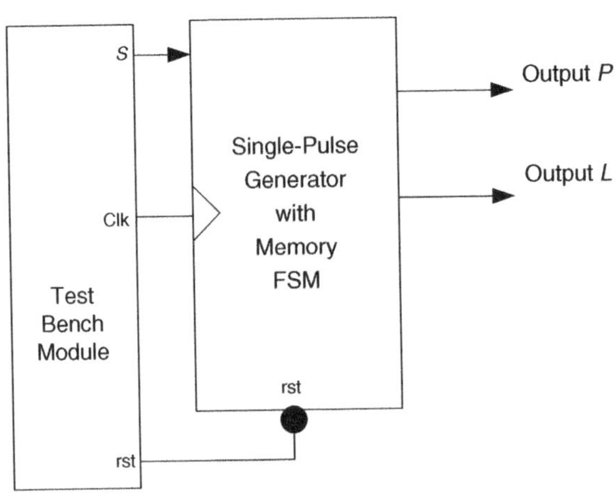

Figure C.3 Connection of test-bench module to the FSM for testing.

The test-bench module is shown in Listing C.4. This is a simplistic way to define the test-bench module and is the easiest to understand. Chapters 6–8 contain other ways to define these.

```
//The Test Bench module.
module test;
reg s,clk,rst;
fsm single_pulse(s,clk,rst,P,L,A,B);

initial
   begin
     $dumpfile("single_pulse.vcd");
     $dumpvars;
// initialise the inputs.
     s=0;
     rst=0;
     clk = 0;   // clk normally low.
//should stay in s0 since reset still on.
     #10 clk=~clk;
     #10 clk=~clk;
//release reset, should stay in s0.
     #10 rst=1;
     #10 clk=~clk;
     #10 clk=~clk;
// set s to 1 to move to s1.
     #10 s=1;
     #10 clk=~clk;
     #10 clk=~clk;
// move to s2 on next clock pulse.
     #10 clk=~clk;
     #10 clk=~clk;
//  and on to s3 on next clock pulse.
     #10 clk=~clk;
     #10 clk=~clk;
// should stay in s3 on next clock pulse
// since s still 1.
     #10 clk=~clk;
     #10 clk=~clk;
// let s=0 to allow fsm to return to s0
// on the next clock pulse.
     s=0;
     #10 clk=~clk;
     #10 clk=~clk;
// go around the loop again using
// repeat loop with 4 clk pulses.
```

```
28        #10 s=1;
29        repeat(8)
30        #10 clk=~clk;
     // back to s3
31        #10 s=0;
32        #10 clk=~clk;
33        #10 clk=~clk;
     // back to s0.
     // finish the simulation.
34      #10 $finish;
35        end
36   endmodule
```

Listing C.4 The test-bench module.

The module of Listing C.4 starts at line 1 and is simply called `test`. It does not have, nor indeed does it need, any input parameters apart from those in line 2, i.e. the signals to connect to the FSM. These are *s*, clk, and rst and are defined as registers using the **`reg`** keyword. This is because the signal values to be defined within the test module need to be remembered (stored in a register type) during the test sequence.

The FSM module is instantiated in line 3, and given the name `single_pulse`.

The signal assignment between the parentheses must follow the same order as that in the FSM module definition.

What follows in lines 4–33 is the sequence of signal values to be applied to the inputs of the FSM to test that the state sequence (and outputs) are correct. The sequence is obtained by looking at the state diagram and applying signal values that allow the FSM to be completely tested. The comment lines (beginning with //) indicate the test being carried out.

The state diagram (Figure C.2) and Listing C.3 should be studied to see how the test sequence has been obtained from the state diagram.

The keyword **`initial`** in line 4, followed by **`begin`** in line 5, defines the start of an initialization block that ends with the keyword **end** in line 35. There are more elaborate ways to do this, and these are discussed in Chapters 6 and 7.

In the **`initial`** block, the logic level of the outputs *s*, rst, and clk are defined in lines 8, 9, and 10, all set to logic 0. In the case of the clock signal clk, this defines the clk to be initially at logic 0, so that any clock pulses will be 0 to 1 transitions.

In line 11, the clock signal clk is toggled to logic 1, then in line 12 it is toggled to logic 0 again. This is how a clock pulse is produced. `~clk` simply inverts the logic level of the clk signal.

The purpose of the test in lines 11 and 12 is to ensure that with the reset rst = 0 the FSM will remain in state s0 (see state diagram in Figure C.2).

The rst signal is raised to logic 1 in line 13, but notice that the assignment is

```
#10 rst = 1;
```

The significance of the #10 is that it will delay the assignment 10 time-units before it will allow rst to become logic 1. The actual delay value can be specified, but for now assume it to be 10 ns into the simulation. So what has happened here is that the signals *s*, rst, and clk were assigned the value 0 at time 0 ns, then after 10 ns the signal rst was assigned the value 1. In this way, a sequence of test signals can be applied sequentially to the FSM under test by changing signal levels after a certain time interval. The clock pulse in lines 11 and 12 can now be seen to create a clock pulse of 10 ns duration.

In lines 14 and 15, another clock pulse is produced (clk going $0 \rightarrow 1 \rightarrow 0$); however, since *s* is still at logic 0, the FSM should remain in state s0.

In line 16, *s* is made equal to logic 1 and the FSM can now clock through from s0 to s3 on the clock pulses produced in lines 17–24.

At line 23 and 24, the FSM should remain in s3, since *s* is still at logic 1.

At line 25, $s = 0$ and the FSM can return to s0 on the next clock pulses in lines 26 and 27.

At line 28, *s* is again raised to logic 1, and in lines 29 and 30 a **`repeat`** block is used to cause the FSM to step through states s0 to s3 to produce four clock pulses. This arrangement allows a number of operations (in this case clock pulses) to be produced in a loop. In line 31 the input *s* is cleared to 0. The next two clock assignments in lines 32 and 33 cause the FSM to move back to its initial s0 state. Finally, the simulation finishes at line 34 with the keyword **`$finish`** (this could be replaced with **`$stop`**).

The whole Verilog file is compiled and simulated. If there are any errors in the design (these could be syntax errors, i.e. spelling mistakes or errors in the design), then these need to be eliminated and the process of compiling and simulation repeated.

Note that pressing the compile button (see Figure C.4) brings up a file-modified window. Click 'yes' and then click the simulate button to resimulate after errors have been corrected.

This FSM has been simulated using the SynaptiCAD's VeriLogger Extreme simulator. Figure C.4 is a screenshot of the VeriLogger Extreme with the source Verilog code displayed in the left-hand window and the waveforms of the simulation displayed in the right-hand window.

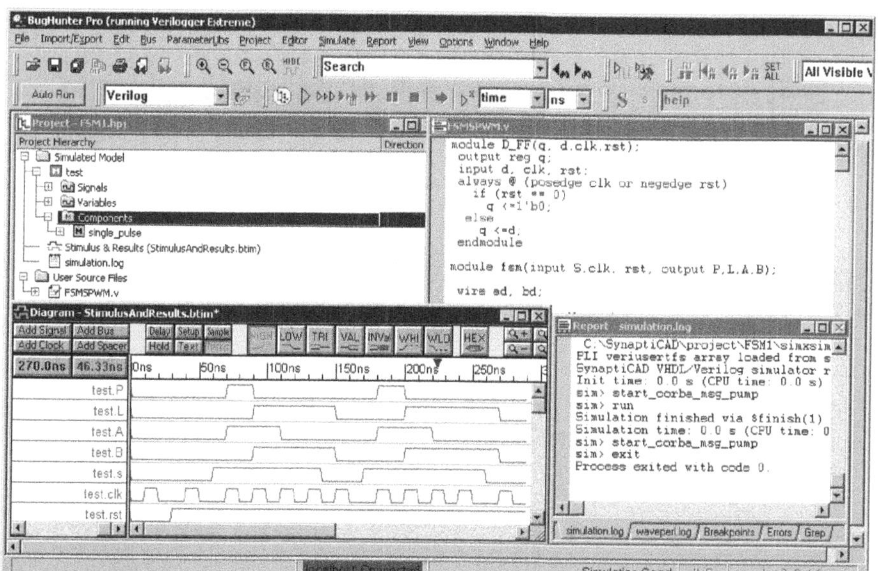

Figure C.4 Screenshot of VeriLogger Extreme running under the BugHunter graphical debugger.

C.4 USING SYNAPTICAD'S VERILOGGER EXTREME SIMULATOR

Install SynaptiCAD's VeriLogger Extreme Simulator located on the CDROM provided with the book or on SynaptiCAD's website at www.syncad.com. This installation is the evaluation version of the program, which is capable of simulating small Verilog projects and displaying the results. You may contact SynaptiCAD directly to purchase a full version (or student version) that can simulate larger models and save the results files.

Run VeriLogger Extreme:

- Choose **Start** > **SynaptiCAD** > **Simulation Debug** > **VeriLogger Extreme** + **BugHunter** menu to launch the simulator with the graphical debugger.
- Notice that the **Help** > **BugHunter VeriLogger Manual** menu launches a help program with the full simulator instructions.
- Also notice that **Help** > **Tutorials** > **Basic Verilog Simulation** is a tutorial on how to use the graphical interface and test-bench generation features of VeriLogger Extreme.

Create a project to store the list of files to be simulated:

- Choose the **Project** > **New Project** menu to open the *New Project Wizard* dialog (Figure C.5).
- In the **Project Name** box, type in **FSM1**, then click the **Finish** button. This will create a project file named FSM1.hpj in the directory specified.

Copy or Create-&-Add the source file to the project:

- Either copy the source file, by right clicking on the **User Source Files Folder** and choosing **Copy HDL Files to Source File Folder** from the context menu which will open a file dialog

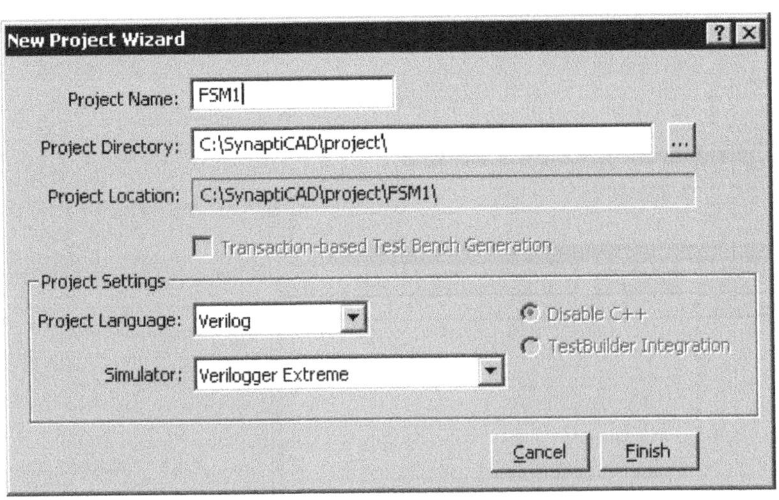

Figure C.5 Project wizard screen.

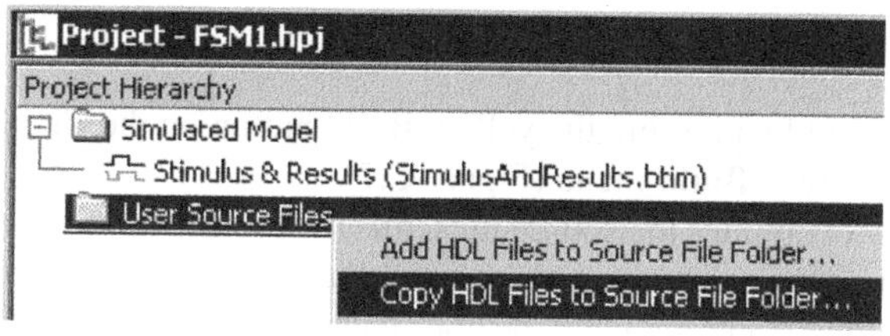

Figure C.6 Showing how to copy source file of Listings C.1–C.4 to project.

(Figure C.6) and then use the browse button in the dialog to find the **FSMSPWM.v** on the CDROM.

- Or create a file called **FSMSPWM.v** by choosing the **Editor** > **New HDL File** menu option to open an editor window. Type in your source code printed in this appendix and save the file. Then add the file to the project by right clicking on the **User Source Files Folder** and choosing **Add HDL Files to Source File Folder** from the context menu.

Figure C.7 Tool bar to build and simulate the code.

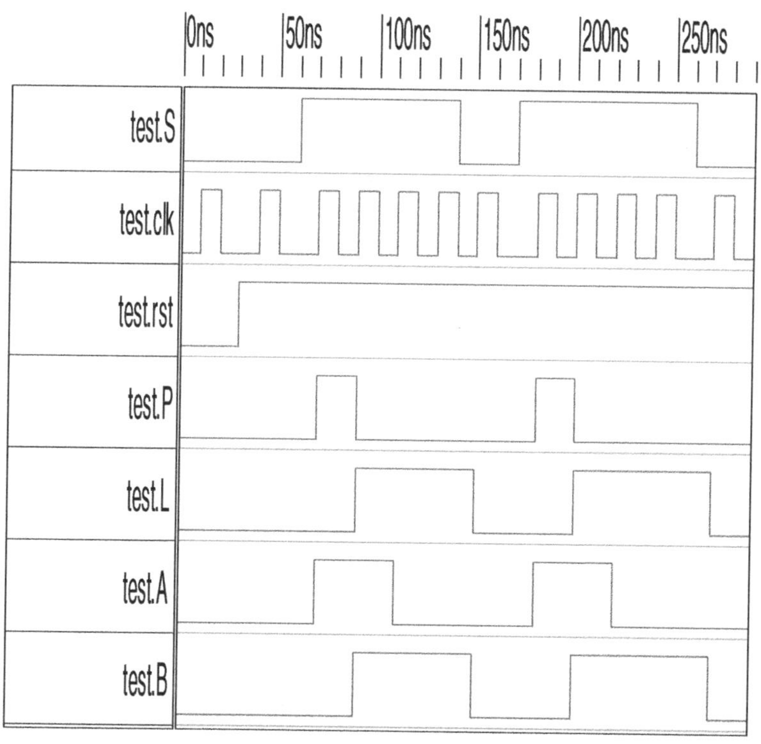

Figure C.8 Verilog simulator output waveforms.

Build and simulate the code:

- First, build the project by pressing the yellow Build button on the simulation button bar or selecting the **Simulate** > **Build** menu (Figure C.7).
- Building the project compiles the source files, fills the project window with the hierarchical structure of the design, and sets watches on all the signals and variables in the top-level component. A build will automatically be done each time the simulation is run, but having a separate build button enables you to create the project tree without having to wait for a simulation to run. After the build you are also able to set the top-level component for the project and/or select additional signals to watch using the project tree context menus. Watch signals are those listed in the *Stimulus and Results* diagram.
- Check the **Report** window to find any syntax errors found by the build.
- Next, start the simulator by pressing one of the green buttons on the Build and Simulate button bar. *Section 2.1 Build and Simulate* in the on-line help explains the differences between the types of single stepping and running.
- The simulated signals should appear in the Waveform window (Figure C.8).
- To produce waveforms with vertical transitions, first select the **Options** >**Drawing Preferences** menu to open a dialog, and then in the Edge Display section and check the **Straight** radio button.

C.5 SUMMARY

This tutorial looks at only one of many ways in which to develop a Verilog description of an FSM design. Other ways are discussed in Chapters 6–8. This tutorial also shows how you can use SynaptiCAD's VeriLogger Extreme simulator to verify your FSM design.

The design method is very easy to apply, with all the design information contained within the state diagram. This information can then be 'extracted' via the equations in order to synthesize a given design. The simulation of the circuit can then be used to confirm the design. The latest version of the software used in this tutorial can be downloaded free from SynaptiCAD at http://www.syncad.com. The version of VeriLogger Extreme may be updated from time to time, and a more recent copy of the demo version can be down loaded from the above website.

AppendixD: Implementing State Machines using Verilog Behavioural Mode

D.1 INTRODUCTION

In Chapters 1–5, state machines have been implemented using the equations obtained from the state diagram. This approach ensures that the logic for the state machine is under complete control of the designer.

However, if the state machine is implemented using behavioural mode, the Verilog compiler will optimize the design.

There is a very close relationship between the state diagram and the behavioural Verilog description that allows a direct translation from the state diagram to the Verilog code.

D.2 THE SINGLE-PULSE/MULTIPLE-PULSE GENERATOR WITH MEMORY FINITE-STATE MACHINE REVISITED

In this system there are two inputs: s to start the system and x to choose either single-pulse or multiple-pulse mode. In single-pulse mode, the L ouput is used to indicate to the user that a single pulse has been generated. In multiple-pulse mode, L is suppressed. Figure D.1 illustrates the state diagram for this system.

Rather than derive the equations directly from the state diagram, a Verilog description can be obtained directly from the state diagram of Figure D.1. This is illustrated in Listing D.1.

```
// Behavioural State Machine.
module pulsar(s,clk,rst,P,L,ab);
  input s,clk,rst;
  output [1:0] ab,P,L;
  reg[1:0] state, P, L;

  parameter s0=2'b00, s1=2'b01, s2=2'b11, s3=2'b10;

  // now define state sequence for FSM (from state diagram).
  always @ (posedge clk or negedge rst)
  if (~rst)
    state <= s0;
```

FSM based Digital Design using Verilog HDL Peter Minns and Ian Elliot

```
8  else
9    case(state)
10     s0: if (s) state <=s1; else state <= s0;
11     s1: state <= s2;
12     s2: if (~x) state <= s3; else state <= s1;
13     s3: if (~s) state <=s0; else state <= s3;
14   endcase

   // now define the outputs @ each state.

15 always @ (state)
16 case(state)
17   s0: begin P=1' b0; L=1' b0; end
18   s1: begin P=1' b1;
19             L= (state==s1) & ~x; //mealy output.
20       end
21 s2: begin L= (state==s2) & ~x; P = 1' b0; end
22 s3: begin P = 1' b0; L = (state==s3) & ~x; end
23 endcase
24 assign ab = state;
25 endmodule
```

Listing D.1 Verilog description of the state diagram.

A comparison between the state diagram of Figure D.1 and the state machine behaviour can be made.

In Listing D.1, lines 5–14 define the state diagram sequence in terms of states and input signals, and lines 15–24 define the outputs in terms of each state. Both of these sets of lines are, of course, happening at the same time (i.e. in parallel).

After declaring the module and its signals in lines 1 and 2, the inputs and outputs are defined.

Outputs are defined in line 2. Note the 2-bit vector `[1:0] ab` that will be used to show the state of the FSM (Figure D.2). Line 3 defines the state vector `[1:0] state` used to keep track of the

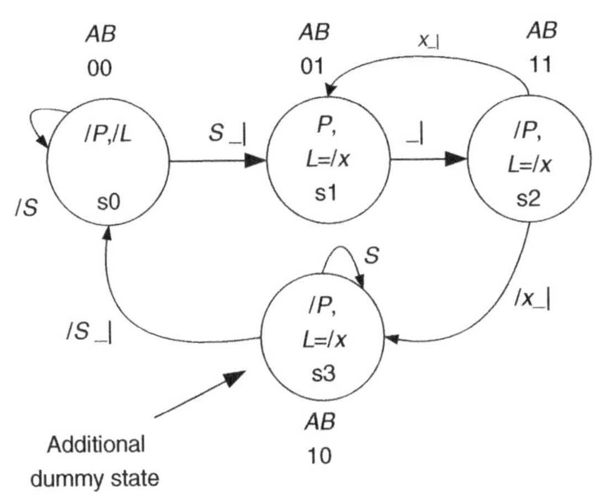

Figure D.1 State diagram for the single pulse with memory FSM.

current state of the FSM, as well as the outputs P and L as register type. Line 4 defines the secondary state variables for each state of the state diagram (Figure D.1). Note, these follow the binary values defined for each state in Figure D.1.

In line 5, an **always** statement defines the conditions under which the **case** statement (lines 9–14) will be used. In line 6, the reset state is defined if rst transition is 1 to 0; otherwise the **else** will allow the **case** statement block (lines 10–14) to occur.

The **case** statement defines the four possible states that the FSM can reside in, and the conditions under which the state transitions occur. The **case** uses the present state to select one of the four possible next states. Initially, after a reset, state will be s0 `(2' b00)` at line 10 and the **if** statement evaluates input s.

If input s is logic 1, then the next state will be s1; otherwise it will be s0.

On the next **posedge** of the clock signal the **case** statement will be activated again. If the current state is s1, then at line 11 the next state will be s2.

Then, on the next **posedge** of clk, line 12 will ensure that the next state will be either s3 if $x = 0$, or s1 if it is logic 1. In line 13, the test of input s will decide whether the FSM moves to s0 ($s = 0$) or remains in s3.

Thus, the state sequence is determined by the **case** statement and use of the **if** statement to evaluate input conditions, or just define the next state if no input condition is needed.

The output conditions are defined in a separate **case** block using outputs for each state as defined in the state diagram. Note that, in line 19, the Mealy output for L in state s1 is defined as

```
L= (state==s1) & ~x;
```

This ensures that L will be determined by the value of input $x(x = 0)$ and conditional on being in state s1, i.e. $L = s1 \cdot /x$ as defined in the state diagram.

Thus, in this way, ouput L will only be logic 1 if the input $x = 0$ and the FSM is in s1; otherwise it is suppressed. Output P will be assigned a logic 1 value.

Other **case** conditions are defined in a similar manner and assigned the output values defined in the state diagram of Figure D.1.

Comparing the behavioural statements from lines 5–14, and lines 15–25 with the state diagram of Figure D.1, it is possible to see a strong relationship between the Verilog description and the state diagram.

Using this approach allows an FSM design to be created without the need to define any hardware logic. It is in fact a 'terminal behaviour and sequence' for the FSM.

Listing D.2 defines the test-bench module for the system. This follows along the lines of other test-bench modules used elsewhere in the book.

```
// now define the test bench fixture..
module test;
reg s,clk,rst;
wire[1:0] ab;
wire p, l;
pulsar uut(s,clk,rst,p,l,ab);
initial
   begin
     rst=0;
     clk=0;
     s=0;
```

```
x=0;
#10 clk=~clk;
#10 clk=~clk;
//stays in s0.
#10 rst=1;
#10 s=1;
#10 clk=~clk;
num;10 clk=~clk;
// moves to s1.
#10 clk=~clk;
#10 clk=~clk;
// moves to s2.
#10 clk=~clk;
#10 clk=~clk;
// stays in s2.
#10 s=0;
#10 clk=~clk;
#10 clk=~clk;
// moves back to s0.
// now make x=1 to produce multiple pulses.
#10 x=1;
#10 s=1;
#10 clk=~clk;
#10 clk=~clk;
// move to s1
#10 clk=~clk;
#10 clk=~clk;
// move to s2
#10 clk=~clk;
#10 clk=~clk;
// move to s1 again
#10 clk=~clk;
#10 clk=~clk;
// and back to s1
#10 clk=~clk;
#10 clk=~clk;
// s2 again
#10 clk=~clk;
#10 clk=~clk;
#10 x=0; //prepare to stop multiple pulse mode.
#10 clk=~clk;
#10 clk=~clk;
// to s3.
#10 clk=~clk;
#10 clk=~clk;
#10 s=0;
#10 clk=~clk;
#10 clk=~clk;
// back to s0.
#10 $stop;
```

```
    end
endmodule
```

Listing D.2 The test-bench module.

Figure D.2 illustrates the simulation waveforms for the design. Note the vector ab defining the state of the FSM at each clock transition. Compare this with the state diagram of Figure D.1.

You should follow the waveform sequence to determine the paths taken through the state diagram sequence. Both modes are tested and seen to work:

with $x = 0$, the FSM is seen to produce a single output pulse;

with $x = 1$, the FSM produces a series of output pulses as it moves between s1 and s2, finally returning to state s0 when $x = 0$.

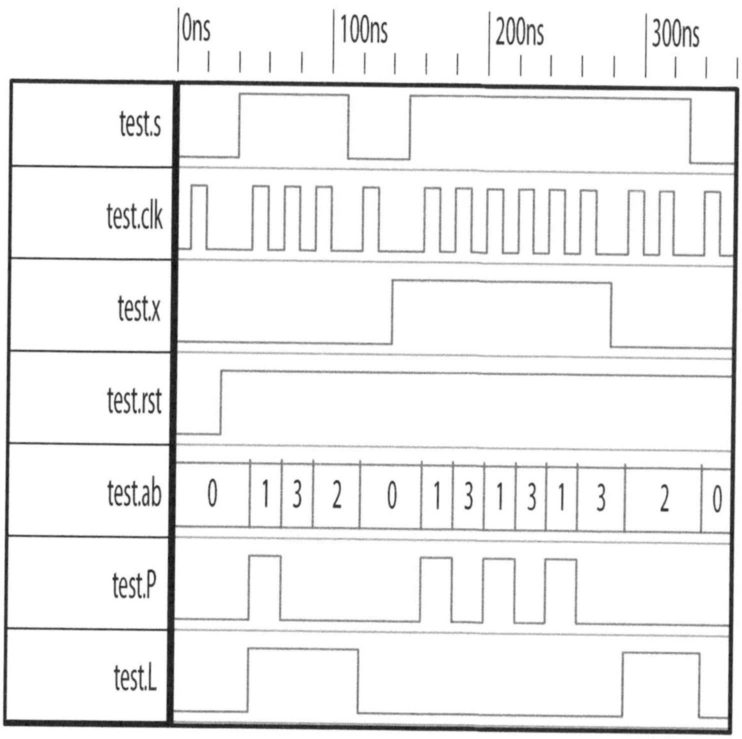

Figure D.2 Simulation of the FSM using Listings D.1 and D.2.

D.3 THE MEMORY TESTER FINITE-STATE MACHINE IN SECTION 5.6

This example can be coded directly in Verilog as a behavioural description using the state diagram of Figure 5.15. The listing is illustrated in Listing D.3.

```
module
memory_tester_state_machine(clk,rst,st,fab,full,RC,P,CS,RD,WR,OK,
ERROR,abc);
   input clk,rst,st,fab,full;
   output[3:0] abc,RC,P,CS,RD,WR,OK,ERROR;

   reg[3:0] state, RC,P,CS,RD,WR,OK,ERROR;
// assign secondary state variable values (as used in Figure 5.15)
   parameter s0=4'b0000, s1=4'b1000, s2=4'b1010, s3=4'b0010,
             s4=4'b0110, s5=4'b1110, s6=4'b1100, s7=4'b1101,
             s8=4'b1001, s9=4'b1011, s10=4'b0100;

// the state machine sequence..

always @(posedge clk or negedge rst)
   if (~rst)
     state= s0;
   else
   case(state)
     s0: if(st) state <=s1; else state <= s0;
     s1: state <= s2;
     s2: state <= s3;
     s3: state <= s4;
     s4: state <= s5;
     s5: state <= s6;
     s6: if(fab) state <= s7; else state <= s10;
     s7: state <= s8;
     s8: if(full) state <= s9; else state <= s1;
     s9: state <= s9;
     s10: state <= s10;
   endcase

// the outputs for each state.

always @ (state)
   begin
   { RC,P,CS,RD,WR,OK,ERROR} = 7'b0011100;
   case(state)
     s0: begin assign RC=1'b0;
            P=0;
            CS=1;
            RD=1;
            WR=1;
            OK=0;
            ERROR=0;
       end
     s1: begin assign RC=1'b1;
            CS=1'b0;
            P=0; end
     s2: begin WR=1'b0; end
     s3: begin C S=1'b0;
```

```
            WR=1'b1; end
       s4: begin end
       s5: begin RD=1'b0;end
       s6: begin end
       s7: begin RD=1'b1; end
       s8: begin CS=1'b1;
            P= 1'b1; end
       s9: begin OK=1'b1;
            P=0; end
       s10: begin ERROR=1'b1; end
endcase

end
assign abc = state;
endmodule
module test;
   reg st,clk,rst,fab,full;
   wire[ 3:0] abc;

   memory_tester_state_machine
uut(clk,rst,st,fab,full,RC,P,CS,RD,WR,OK,ERROR,abc);

   initial
     begin
       rst=0;
       clk=0;
       st=0;
       fab=0;
       full=0;
       #10 rst=1;
       #10 st=1;
       #10 repeat(14)
          #10 clk=~clk;
       #10 rst=0;
       #10 rst=1;
       #10 repeat(10)
       #10 clk=~clk;
       #10 fab=1;
       #10 repeat(16)
       #10 clk=~clk;
       #10 clk=~clk;
       //#10 rst=0;
       //#10 rst=1;
       #10 full=1;
       #10 repeat(20)
       #10 clk=~clk;
       #10 $finish;
     end
endmodule
```

Listing D.3 Behavioural description of memory tester from state diagram of Figure 5.15.

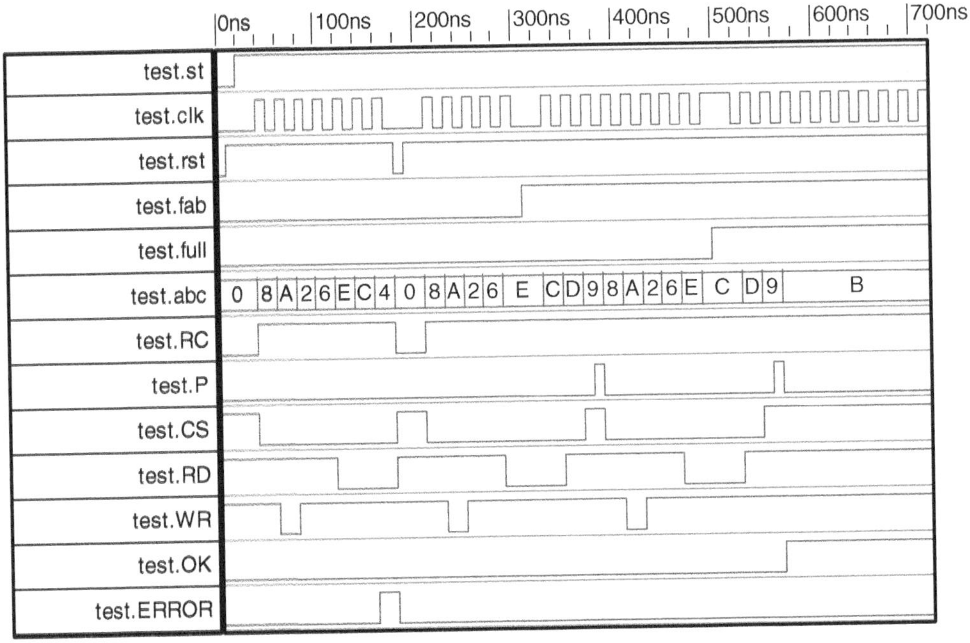

Figure D.3 Simulation of the memory tester FSM of Listing D.3.

The simulation is shown in Figure D.3.

The abc vector illustrates the FSM states in terms of the secondary state variables allocated in the state diagram of Figure 5.15. These were assigned to provide a unit distance code, and the numbers shown in the simulation in Figure D.3 are the hexadecimal values.

The FSM is seen to move around the state diagram from state s0 (0) through s1 (8), s2 (A), s3 (2), s4 (6), s5 (E), s6 (C), then s10 (4) the error state. The FSM is then reset back to state s0 and recycles around the state diagram to s8 (9) and on to s9 (B) the OK state.

The write cycle (8, A, 2) is followed by the read cycle (E, C), and again in (8, A, 2, 6, E, C, D).

D.4 SUMMARY

The use of a behavioural description has the advantage of using the information in the state diagram directly. It can be seen that there is a one-to-one correspondence between the state diagram structure and the Verilog structure, and this shows how the state diagram method can allow a realization of the design directly (via the behavioural method) or via the Boolean equations.

Both methods, owing to their formal descriptions, are suitable for computer implementation. This has been exploited in a number of programs developed at Northumbria University by postgraduate students.

Index

Printed and bound by CPI Group (UK) Ltd, Croydon, CR0 4YY
16/09/2026
14963389-0004